T0317093

CALIFORNIA NATURAL HISTORY GUIDES

FIELD GUIDE TO THE SPIDERS OF CALIFORNIA AND THE PACIFIC COAST STATES

California Natural History Guides

Phyllis M. Faber and Bruce M. Pavlik, General Editors

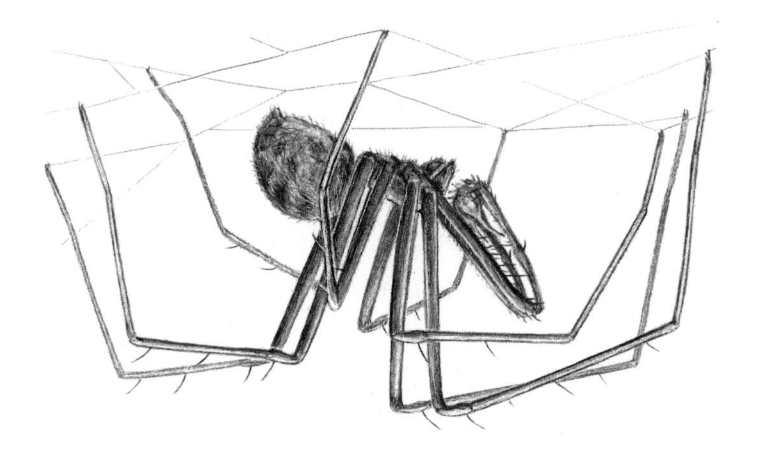

Field Guide to the SPIDERS of California and the Pacific Coast States

Text by
R. J. Adams

Illustrations by
Tim D. Manolis

UNIVERSITY OF CALIFORNIA PRESS
Berkeley Los Angeles London

University of California Press, one of the most distinguished university presses in the United States, enriches lives around the world by advancing scholarship in the humanities, social sciences, and natural sciences. Its activities are supported by the UC Press Foundation and by philanthropic contributions from individuals and institutions. For more information, visit www.ucpress.edu.

California Natural History Guide Series No. 108

University of California Press
Berkeley and Los Angeles, California

University of California Press, Ltd.
London, England

© 2014 by the Regents of the University of California

Library of Congress Cataloging-in-Publication Data

Adams, R. J. (Richard John), 1970-
Field guide to the spiders of California and the Pacific Coast states / text by R.J. Adams ; illustrations by Tim D. Manolis.
p. cm. — (California natural history guides ; no. 108)
Includes bibliographical references and index.
ISBN 978-0-520-27660-4 (cloth : alk. paper)—ISBN 978-0-520-27661-1 (pbk. : alk. paper)
1. Spiders—California—Identification. 2. Spiders—Pacific States—Identification. I. Manolis, Tim, 1951- II. Title. III. Title: Spiders of California and the Pacific Coast states. IV. Series: California natural history guides ; 108.
QL458.41.U6A33 2014
595.440979—dc23 2013012465

Manufactured in China

23 22 21 20
10 9 8 7 6 5 4 3

The paper used in this publication meets the minimum requirements of ANSI/NISO Z39.48-1992 (R 1997) (*Permanence of Paper*). ♾

Cover illustrations, clockwise from upper left: *Aphonopelma steindachneri* male, *Microlinyphia dana* female, *Theridion* female, *Diaea livens* female. Illustrations by Tim D. Manolis.

The publisher gratefully acknowledges the generous contributions to this book provided by

the Gordon and Betty Moore Fund
in Environmental Studies
and
the General Endowment Fund of the
University of California Press Foundation.

CONTENTS

ACKNOWLEDGMENTS

This book would never have been possible without the help and support of a great many people. Our spouses, Monika Davis and Annette Manolis, deserve special appreciation for their continuous support, both at home and in the field. We would also like to thank Dr. Sarah Crews for her numerous reviews and constructive comments throughout the writing of this book, and Wendell Icenogle for so freely sharing his time, specimens, and expertise. Vital in the production of this book were the specimens, knowledge, and reprints housed at the California Academy of Sciences, and for their encouragement and assistance, the entire arachnology department deserves our deep appreciation, especially Dr. Charles Griswold, Darrell Ubick, Anthea Carmichael, Vic Smith, and Dr. Joel Ledford. We must thank Rod Crawford for freely sharing his unrivaled expertise on the spiders of Washington. For specimen identification we would like to thank Dr. G. B. Edwards, and for access to specimens, we extend our appreciation to Cheryl Barr (Essig Museum, University of California at Berkeley) and Rosser Garrison (California Department of Food and Agriculture). For help in the field, we would like to thank Kathy Biggs, Pete Carmichael, Cary Kerst, Jonathon Martin, Kenneth Martin, Bruce Webb, and Stan Wright. Kevin Pfeiffer assisted with the translation of German arachnological publications. Numerous other people, all experts in their fields, generously provided feedback, access to publications, and insights. Without their help this book could never have been completed. These include Dr. Marshal Hedin, Rick Vetter, Dr. Daniel Mott, Dr. Greta Binford, Dr. Jason E. Bond, Jozef Slowik, Dr. David Bixler, Dr. Herbert W. Levi, Thomas Prentice, Dr. Robb Bennett, Dr. David Shorthouse, and Chris Hamilton. The skills and discipline needed to write this book were developed and guided during my graduate school years in Dr. Dale Clayton's laboratory with additional support from Dr. Sarah Bush. For their unwavering support and belief in the importance of this project, the author would also like to thank his parents, Rick Adams and Clancy McCassey, his stepdaughter Meghan Ueland who fought her tarantula phobia to help with this book, Kevin Kaos, Andy Pacejka, Joanne Robbins, Ken Schneider, Valin Mathis, Geoff and Debbie Verhoeven, and Michelle Torres-Grant. Any errors in the text are the author's alone.

INTRODUCTION

WITH OVER 40,000 DESCRIBED SPECIES, spiders have adapted to nearly every terrestrial environment across the globe. Considering the wide variety of habitats in California, Oregon, and Washington, from montane meadows and desert dunes to redwood forests and massive urban centers, it is not surprising that over half of the world's spider families live within this region. Any place you look, you will almost certainly find spiders, and when you take time to notice their many lifestyles and forms, it will be difficult not to appreciate the incredible diversity of this ancient arachnid order.

The purpose of this guide is to help the interested naturalist and the curious observer alike to identify many of the region's spiders to the family level and, when possible, to the level of genus and species. Identifying spiders can be extremely difficult, often requiring a detailed examination of the spider's reproductive structures under a microscope. While color and pattern can provide useful clues for identifying many spiders, color patterns can also vary greatly, even within a single species. For that reason, this book focuses on all of the families, covers most of the genera, and includes many of the conspicuous, unique, and most common species found in the contiguous US Pacific coast states. You should keep in mind that the distribution and habitat requirements of most invertebrates, unlike birds, mammals, and other vertebrates, are poorly known. In some cases, the only published information on a species is its original description. It is not unheard of to find spiders many miles outside their described ranges, especially among species that disperse by ballooning.

Because this book is meant to be useful in the field, most of the features discussed are those that can be seen with a good-quality 10× hand lens. For example, fig. 1 shows some easily seen differences in cheliceral (fang) alignment. In some cases though, smaller structures, such as the spider's jaw elements or its claw arrangement, are mentioned if these characteristics are likely to improve your success in identification. As small, affordable digital cameras are increasingly built with more powerful macrophotographic lenses, important features such as a spider's eye arrangement and clypeus height can often be seen with relative ease, even on freely moving individuals.

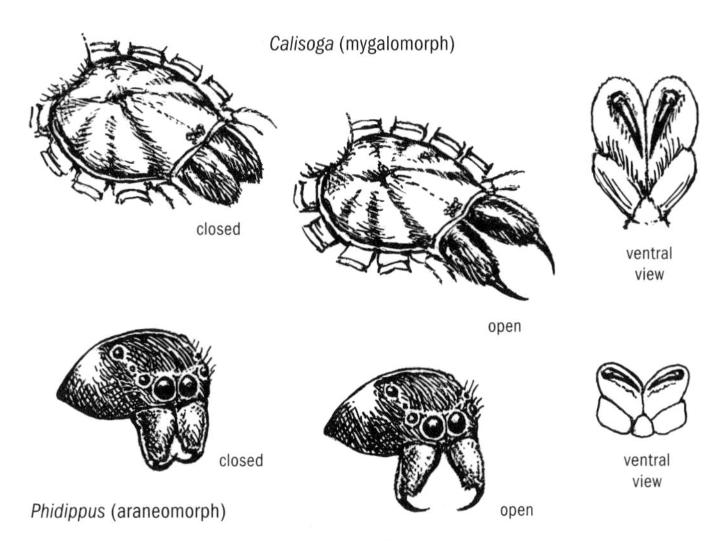

Figure 1. Comparative jaw morphology of mygalomorph and araneomorph spiders.

Understanding Spider Systematics

Like all living things, spiders are classified using Linnaean taxonomy. In the Linnaean system, organisms are ideally clustered together based on shared features that reflect their common ancestry. The categories within this system start very broadly and become increasingly specific. To illustrate this point, the taxonomy of the Western Lynx Spider *(Oxyopes scalaris)* is shown below:

KINGDOM Animalia (all animals)
PHYLUM Arthropoda (invertebrates with a chitinous exoskeleton and jointed appendages, including spiders, insects, centipedes, and crustaceans among others)
CLASS Arachnida (arthropods with two body segments and eight legs, including spiders, scorpions, and ticks)
ORDER Araneae (spiders)
SUBORDER Opisthothelae (spiders with nonsegmented ab-

domens and with spinnerets near the posterior abdominal margin, including all New World spiders)
INFRAORDER Araneomorphae ("true" spiders)
FAMILY Oxyopidae (lynx spiders)
GENUS *Oxyopes* (one of three North American lynx spider genera)
SPECIES *Oxyopes scalaris* (Western Lynx Spider)

One of the most important divisions in spider systematics separates all North American spiders into two infraorders: Araneomorphae, as in the list above, and Mygalomorphae. Mygalomorphae includes the tarantulas and trapdoor spiders. They have several features held over from early in the evolutionary lineage of spiders, including four book lungs and parallel chelicera. Araneomorphae contains what are occasionally referred to as the "true" spiders and includes all of the most common "household" spiders. Except for members of the family Hypochilidae (p. 49), araneomorphs have only two book lungs and have chelicerae that move at an angle toward one another. They are further divided into entelegyne and haplogyne families. A female entelegyne spider, with a few exceptions, has an intricate, sclerotized reproductive organ called the epigynum, while the adult male has architecturally complex palps for transferring sperm. The structure of these organs is extremely important in identifying nearly all spider species. The haplogyne spiders, much like the mygalomorphs, have comparatively simple reproductive organs. They lack the entelegyne's ornate epigynum, and the male's palps are little more than bulbs tipped by long, hollow emboli.

Morphology

Like all arachnids, spiders have bodies that are divided into two main parts, the cephalothorax and abdomen, as shown in fig. 2A. When you measure a spider or when its size is given in the text, its length is the combined length of these two body parts and does not include the legs.

The cephalothorax holds the mouthparts, eyes, and legs. A spider's jaws consist of their chelicerae and a pair of hinged

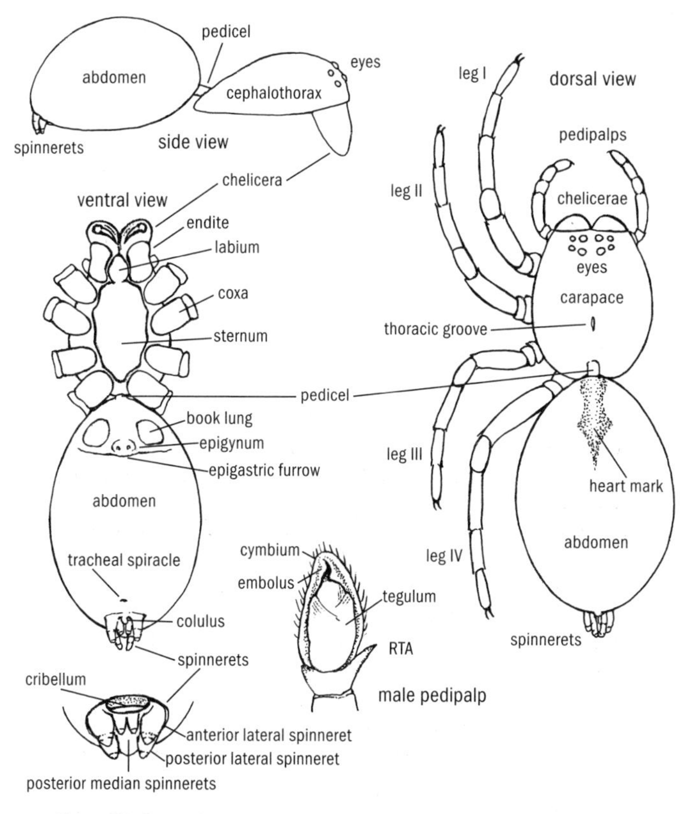

Figure 2A. General spider anatomy.

fangs. When at rest, the fang lies within a cheliceral furrow, which is often bordered by numerous teeth and smaller denticles. The arrangement of a spider's cheliceral dentition is often used to differentiate closely related genera and species. Also part of the spider's mouthparts are the pedipalps, often referred to simply as the palps. Resembling a fifth pair of legs, the palps are used for tactile and chemosensory purposes as well as food manipulation. On an adult male, the tips of the palps house the spider's reproductive organs. The palps are also

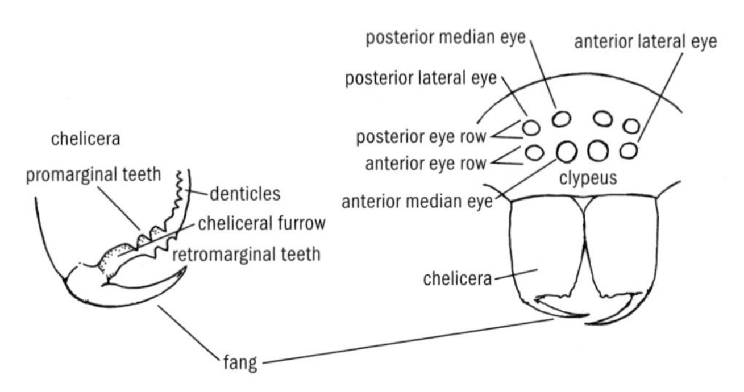

Figure 2B. Spider head anatomy.

coated with chemoreceptive hairs that are used to find and follow female spiders' pheromone trails. On females, these hairs are not as dense and are mainly used to "taste" their prey before eating it. At the base of each palp is an expanded plate called an endite that borders the mouth. Along an endite's anterior margin are occasionally found swellings or rows of minute teeth, forming a structure called the serrula. Posterior to the mouth is the labium. Rows of setae on the labium and endites act as filters, removing particulate material from the spider's predigested, liquefied meal.

The area between the top of a spider's chelicerae and its eyes is the clypeus, and its shape can be important when you are separating some groups of spiders (fig. 2B). The size and arrangement of a spider's eyes are incredibly important diagnostic features. While the majority of spiders have eight eyes, many have only six. A few spiders, typically species that dwell in caves or deep leaf litter, have four or fewer eyes. Spider eyes are generally divided into four pairs based on their shape and location. In front are the anterior median eyes, which are bordered by the anterior lateral eyes. Behind these are the posterior median and posterior lateral eyes.

The carapace, which covers the upper surface of the cephalothorax, often carries an invagination called the fovea or thoracic groove (see fig. 2A). The thoracic groove provides an internal attachment point for the stomach muscles, and its shape can be useful for distinguishing some similar genera and families. On the underside of the cephalothorax is the sternum, which

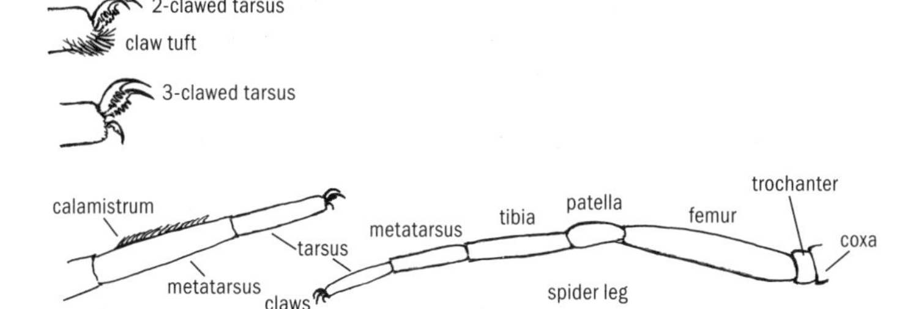

Figure 2C. Spider leg anatomy.

is surrounded by the spider's legs. In some genera, there are small, sclerotized points called precoxal triangles between the sternum and the legs. Each leg is composed of eight segments; the most basal segment, where the leg attaches to the body, is the coxa (fig. 2C). Next is the trochanter, which distally attaches to the femur, which is often the thickest leg segment. Past the femur are the patella and tibia. The number and arrangement of spines on the tibia are often very informative features for identifying spiders. The last leg segments are the metatarsus and then the tarsus, which is tipped by the spider's claws. While the claws are extremely small, the number of them on each leg is an important characteristic, both ecologically and taxonomically. Many of the two-clawed spiders have conspicuous tufts of setae between their claws and are cursorial (running) or arboreal (tree-based) hunters. The three-clawed spiders, on the other hand, generally use webs to capture their prey and use the tiny central claw to help grip the strands of silk. Spider legs (and leg segments) are often referred to by their location on the spider's body. Going down one side of a spider, the leg closest to the front is known as leg I while at the rear is leg IV (see fig. 2A). Using this method, if a description refers to "femur II," it is referencing the femur on the second leg from the front.

Connecting the cephalothorax to the abdomen is the pedicel. In some species, a dark stripe runs down the midline of the abdomen's dorsal surface. This stripe is the heart mark, or cardiac mark (see fig. 2A). In other species, there may be shiny plates called scuta partially or entirely covering the abdominal surface. At the posterior end of the abdomen are the spinnerets.

While most mygalomorphs have only two pairs of spinnerets, the majority of araneomorphs have three pairs. The anterior lateral spinnerets are generally the largest, followed by the posterior lateral spinnerets. The posterior median spinnerets are the smallest and are often hidden by the other pairs. Known as cribellate spiders, some species have a plate (sometimes divided in two) called the cribellum just anterior to the spinnerets. The surface of the cribellum is covered with hundreds of tiny spigots that produce an unusual grayish-blue silk that regularly has a teased, or carded, appearance. On a cribellate spider there is almost always a distinctive row of short, curved setae, called the calamistrum, along metatarsus IV (see fig. 2C), which is used to comb silk from the cribellum. A spider without a cribellum (an ecribellate spider) often has in its place a small, fleshy protuberance or setal cluster known as the colulus.

Anterior to the spinnerets in the araneomorphs is the tracheal spiracle (see fig. 2A). Part of the spider's respiratory system, the trachea enters the body cavity where its thin walls allow gas exchange to occur between the air and the spider's hemolymph. Respiratory functions are also carried out by the spider's book lungs. Among the mygalomorphs and members of the araneomorph family Hypochilidae, spiders have two pairs of book lungs, while in most of the remaining families, each spider has only one. The entrance to each book lung is a narrow slit on the underside of the abdomen, often marked by a bare patch of surrounding cuticle. Internally, book lungs are made up of many thin, hollow sheaths, similar to the pages of a book. Between these sheaths the spider's hemolymph flows and gas exchange occurs. On an araneomorph, the book lungs are located on either side of the epigastric furrow.

Life History

Female spiders bundle their eggs together into egg sacs. An egg sac can vary from a few silken threads to a multilayered packet made of several kinds of silk along with protective layers of dirt and debris. In many cases the egg sac's appearance can be tied to a particular family, genus, or even species. When an egg hatches, the spiderling is in an immobile postembryonic state

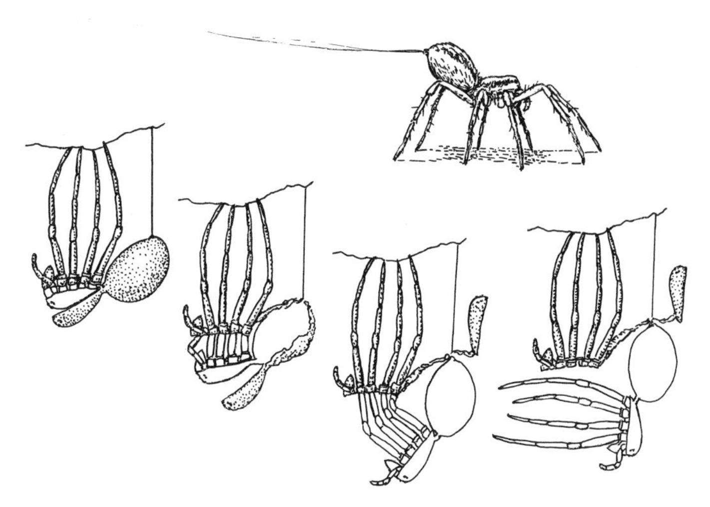

Figure 3. *Upper:* Small wolf spider (Lycosidae) ballooning. *Lower sequence:* Spider molting.

and is only able to leave the egg sac after undergoing one or two additional molts. With some exceptions, the spiderlings generally disperse soon after exiting the sac. Many araneomorph spiderlings balloon away from their natal grounds by climbing to the tops of twigs and releasing several long strands of silk into the wind (fig. 3). Carried away by the breeze, most spiderlings are deposited close to their takeoff points; however, others can travel hundreds of kilometers and some have been collected thousands of meters above the ground. Ballooning is the origin of the shiny gossamer threads commonly seen floating through the air and blanketing large areas in summer and fall. Adults of some small species will also balloon away from nonproductive hunting grounds. Other spiders, including most mygalomorphs, are unable to balloon. Instead, they disperse by crawling away from their natal sites.

As a spider grows, it must molt its rigid exoskeleton. The number of molts required to reach adulthood varies from as few as five for small spiders to more than 10 for the largest. When molting, most spiders suspend themselves upside down (see fig. 3). The old carapace then opens along its lateral seams and falls back like a dropping trapdoor. Next, the sides of the abdominal integument begin to tear, freeing up the constricted abdomen within. Last, the spider pulls its legs and palps free from its old skin. Mygalomorphs follow the same basic routine, but unlike araneomorphs, they put down silk sheets and molt while lying on their backs. The molt before adulthood is referred to as penultimate and is marked on a male spider by a noticeable swelling of the palpal tarsi and on a female entelegyne spider by a darkening of the area around the epigastric furrow. While nearly all araneomorphs stop molting after reaching maturity, female mygalomorphs will continue to molt throughout their comparatively long lives.

After reaching adulthood, a male spider weaves a small sheet of silk known as a sperm web, on which he deposits a drop of semen. He then dips the tips of his palps into the semen and draws it into the emboli, which are syringe-like structures used to transfer sperm to the female. His palps now "charged," the male leaves the safety of his web or burrow, pursuing females by following the pheromones wafting off their webs or, in the case of wandering spiders, their draglines. In some cases, when a male finds a penultimate female, he will guard her from other males in an attempt to be the first to mate. Upon encountering a receptive female, male spiders immediately begin courtship behavior. This represents one of the most complex areas in spider biology, ranging from simple web plucking and light touches with the fore legs to exuberant dance-like displays with flashes of color, palpal drumming, and sound production. Courtship rituals both aid in mate selection and help identify the normally smaller males as something other than prey items. During mating, the male transfers sperm either into the female's gonopore (in mygalomorph and haplogyne species) or into her epigynum (in entelegyne species), where the sperm is stored in the spermathecae until she is ready to lay her eggs. The sperm is released and the eggs are fertilized only when they are being deposited into the egg sacs. Rarely does the male stay with the female after mating, and while an adult male may attempt to mate with multiple females, he has reached the last portion of his life and inevitably will die soon after.

The relationship between female spiders and their young varies considerably across taxa. Some species abandon their egg sacs soon after construction, while others guard them for extended periods. Although many females die before their eggs hatch, some do live to see their young emerge. Female mygalomorphs (and a few araneomorphs) can be particularly long-lived and may parent multiple generations over their lifetimes. While many araneomorphs appear to have annual life cycles, surviving for one year and overwintering either in the egg sac or as spiderlings, others may take several years to mature, or there may even be different generations that mature at different rates. Life cycles can also be affected by the environment. In some cases, the same species can demonstrate different life history patterns depending on whether the population is in a temperate coastal climate or in a more seasonally affected montane one.

Webs and Silk

While other organisms can make silk, spiders are unique in that silk is produced throughout their lives from abdominal glands. The surprising physical properties of spider silk are well documented. In some cases it can stretch up to three times its resting length, and ounce for ounce, it can be stronger than steel. What is less appreciated is the wide variety of silks spiders produce. Seven kinds have been recorded, and the different kinds are used for different purposes. These include swathing silk for wrapping prey and egg sacs, dry strands for web scaffolding, and droplets of sticky silk for prey capture, although no one spider produces all types of silk. Cribellate silk is made by the members of numerous araneomorph families and is bluish gray with a fuzzy or tangled look to it. Unlike silk that captures prey with sticky droplets, cribellate silk attaches to an insect's cuticle because of molecular forces. Members of several orb-weaving families often include thick bands of UV-reflective silk through the hubs of their webs. Known as a stabilimentum, this reflective area normally either is a vertical bar or has an X shape and may serve a number of purposes, including warning birds away from the web, attracting insects, or aiding in thermoregulation.

Spider webs vary dramatically between families, and in some cases, genera (fig. 4). Most araneomorph webs fall into one of five broad categories. A tube web consists of a sock-

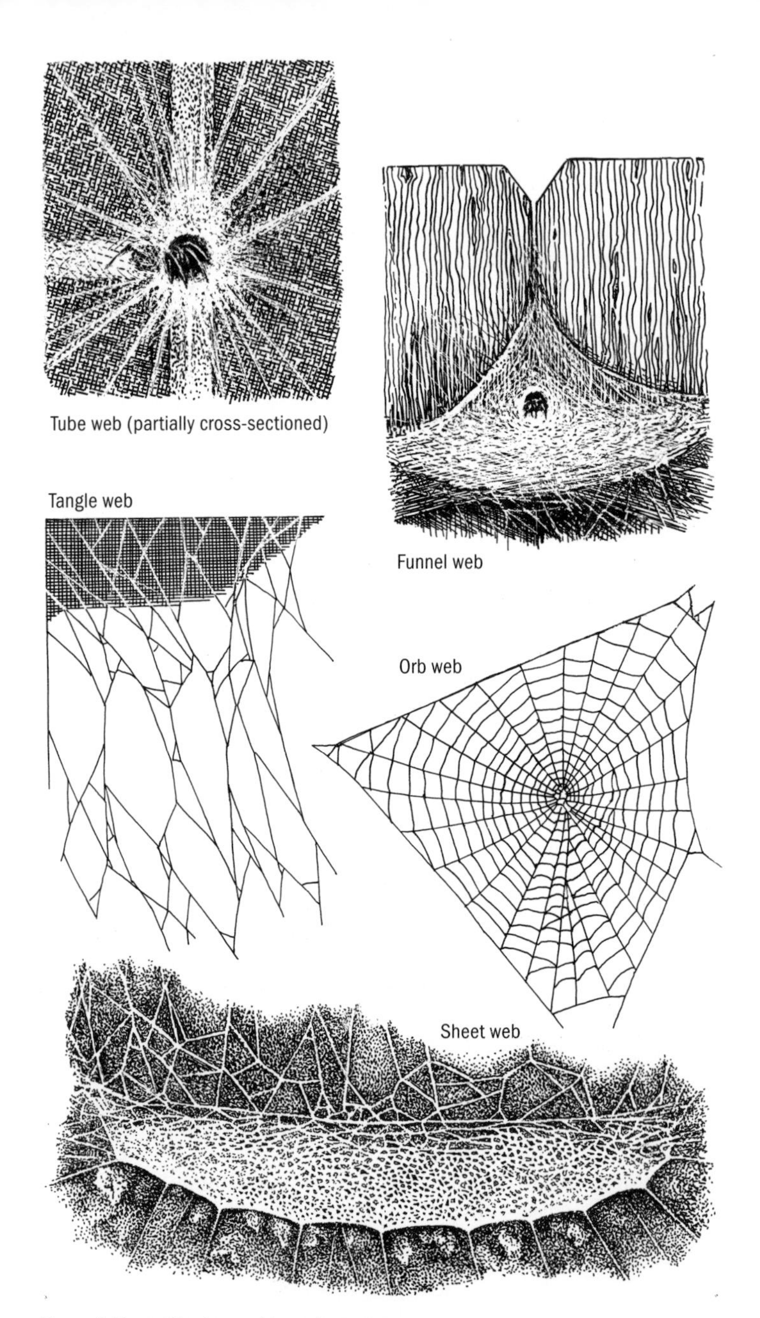

Figure 4. Types of webs used to capture prey.

like retreat with a radiating collar of signal threads around its entrance. Tube webs are often hidden in cracks or along the seams of buildings and are made by several families, including Filistatidae (p. 52) and Segestriidae (p. 54). A funnel web consists of a tubular retreat with a platform spreading out from its entrance; funnel webs are made by members of the family Agelenidae (p. 232). Depending on the species of spider, a sheet web consists of one to several flat or curved sheets suspended above the substrate. These webs are made by the families Pimoidae (p. 131) and Linyphiidae (p. 133). With their iconic circular design, orb webs are among the most familiar of spider webs. A number of families build these, including Araneidae (p. 110) and Tetragnathidae (p. 126). Despite initial appearances, tangle webs are well-organized structures with different kinds of silks playing different roles in prey capture and protection. They are built by members of the family Theridiidae (p. 93). Some spiders incorporate multiple elements into their webs, including members of the Araneidae genus *Metepeira*. They surround their cuplike retreats and orb webs with expansive tangles. The Marbled Cellar Spider (*Holocnemus pluchei,* Pholcidae, p. 73) builds a tangle around a filmy sheet web. With practice, you will often easily recognize the typical webs of each family, and they are described in the family accounts in Part 2 of this book. Because webs are regularly damaged, many spiders repair them on a daily basis. Orb web weavers (Araneidae, p. 110) regularly devour much of their old webbing and recycle its chemical components into new webs.

In addition to catching prey, webs serve a number of other purposes. They act as extensions of the spider's sensory field: just as a female spider at one side of a large orb web can feel the vibrations of an entangled fly, she can also feel the rhythmic plucking of a male well before her potential mate is visible. Laced with pheromones, spider silk can inform others about the age, sex, health, and even diet of the spider that made it. Male spiders normally find mates by following the pheromonal "scents" on the females' silk. In some spiders, the male ties the female down with silk as part of the premating courtship ritual; known as a bridal veil, the silk ties do not constrain her, but pheromones on the silk may play a role in keeping her sedate during the mating process. Pheromones also provide some information between species, as some spiders will respond defensively when encountering the fresh dragline of a potential predator. Ballooning spiders use silk to help them disperse. Even wandering spiders that don't build webs weave retreats in protected nooks for resting, molting, and brooding their young.

Spiders as Predators

Although crab spiders and jumping spiders have both been noted drinking nectar in addition to ingesting animal prey, and several spider families are known to gain nutritional benefits by eating pollen grains caught in their webs, nearly all of the world's spiders are believed to be almost entirely carnivorous. Not surprisingly for such a large and diverse group, spiders have evolved numerous hunting strategies with prey selection ranging from generally broad to highly specialized. At first glance most spider webs appear to be indiscriminate filters catching anything small enough to get tangled in the weave. However, when their ecological parameters are combined with their genetically determined web architecture, many spiders disproportionately capture only a small subset of the total variety of possible prey items. In most cases, a spider's diet thus consists of only a few kinds of prey at any given time, with other items making up a minor percentage of its overall intake. Exhibiting one of the most extreme examples of prey specialization are members of the genus *Mastophora* (Araneidae, p. 110), who spin gluey bolases laced with pheromones that attract male moths of certain species. Many spiders specialize in eating ants. To accomplish this, some construct webs around ant mounds, while others go so far as to infiltrate ant colonies. The desert-dwelling *Septentrinna steckleri* (Corinnidae, p. 146) lives within ant nests, its cuticular hydrocarbons mimicking those of the ants among whom it dwells.

Among cursorial and arboreal hunters, many use vibrations to determine a potential prey item's size and location before attacking. Others, including jumping spiders (Salticidae, p. 167) use their vision to hunt. Several families of trapdoor spiders and many crab spiders (Thomisidae, p. 180) are ambush predators. They stay in one spot, sometimes for most of their lives in the case of trapdoor spiders, waiting for potential prey to come near. Spitting spiders (Scytodidae, p. 64) subdue their prey by spraying it with a sticky mix of glue and venom. While most prey items are insects, many other small animals, including other spiders, earthworms, isopods, and rarely small vertebrates, can be part of spiders' dietary regimes.

Once an insect is captured in a spider's web, the spider will normally respond in one of several ways. If the prey is soft bodied and harmless, such as a moth or a fly, the spider will quickly bite it, immobilizing it before it escapes. If the prey is large and potentially dangerous, such as a wasp or grasshopper, the spider may choose to throw silk over it from a distance, only biting it after it is thoroughly ensnared. In some cases, when the prey is especially aggressive or difficult to subdue, the spider may avoid the captured creature entirely and simply cut it out of the web.

When a spider bites, it injects venom into the prey though tiny holes at the tips of its fangs. Spider venoms can be divided into two broad categories: neurotoxins, which affect the nervous system, and cytotoxins, which destroy the surrounding tissues. Once the prey is immobilized, the spider regurgitates digestive fluids into it and the prey's internal organs begin to dissolve. Additionally, some spiders use their cheliceral teeth to crush their prey's bodies, allowing the digestive fluids to act more efficiently. As a prey's organs liquefy, the pumping action of the spider's stomach sucks in the fluid while hairs on the spider's labium and endites filter out the larger particles.

Spiders as Prey

A great many animals eat spiders, including mammals, reptiles, amphibians, insects, birds, and even other spiders. Pirate spiders (Mimetidae, p. 88) specialize in hunting and feeding on spiders (a behavior known as araneophagy). Other spiders, including the Long-bodied Cellar Spider (*Pholcus phalangioides*, Pholcidae, p. 73), make spiders a major part of their diet. In areas with winters cold enough to dramatically reduce insect populations, overwintering spiders are one of the main food sources for small birds. Against these predators, passive defenses such as camouflage are extremely common, but spiders also employ a wide range of behaviors to dissuade or confuse potential predators. Some species, especially those in the families Corinnidae (p. 146) and Salticidae (p. 167), use color and behavior to mimic stinging ants and wingless wasps. When threatened in their webs, some spiders drop to ground, while others rapidly vibrate, making themselves more difficult to capture. Others might play dead, or a spider might even drop a grabbed leg, giving the predator something to feed on while the spider gets away. Velveteen tarantulas (Nemesiidae, p. 33) have a more assertive defense, rearing up aggressively when disturbed. Threatened tarantulas (Theraphosidae, p. 28) will kick off from their abdomens special barbed setae known as urticating hairs that can get caught in a predator's skin and, if the predator is close enough, will severely irritate its eyes and respiratory system; some small rodents have even suffocated from urticating hairs caught in their throats. Female Green Lynx Spiders (*Peucetia viridans*, Oxyopidae, p. 220) are exceptional in that they can spray a burst of venom from their fangs when threatened.

Spiders are also highly susceptible to parasites, both external and internal. Several common wasps (order Hymenoptera), including spider wasps (family Pompilidae) and thread-waisted wasps (family Sphecidae), paralyze spiders and lay eggs on their living, but immobilized, bodies (fig. 5). The wasp then entombs the spider, providing a fresh source of food for its developing young. Some wasps in the family Ichneumonidae attach an egg directly to a spider's abdomen. The ectoparasitic larva then feeds on its still-active host's bodily fluids. Other ichneumonids and some genera of mantidfly (order Neuroptera, family Mantispidae) insert their eggs into spiders' egg sacs. The young parasitoids then feed on the spiders' eggs, eventually emerging as adults. Among the numerous families of flies (order Diptera) that parasitize spiders and their egg sacs are the small-headed flies (family Acroceridae). Female small-headed flies scatter their eggs on the ground, and after hatching, the maggots crawl in search of hosts. When a suitable spider is found, a maggot moves up its leg and burrows into its abdomen through its book lungs. Relatively unobtrusive for most of its development, the

Figure 5. *Left:* A spider wasp (Pompilidae) stalking a wolf spider (Lycosidae). *Upper right:* A spider with a wasp larva (Ichneumonidae) attached to its abdomen.

larva devours its host from the inside during its last instar. It then pupates in the expired host's retreat.

Studying Spiders

There are three main ways to study spiders: alive in the field, in captivity, or preserved as part of a permanent collection. A great deal is still unknown about many North American spiders, including their food preferences, egg sac designs, growth rates, courtship behaviors, and other natural history aspects. Finding spiders, though, isn't difficult. They are in nearly every habitat, and different genera and families often live in fairly specific ecological niches. Just as members of the genus *Tetragnatha* (Tetragnathidae, p. 126) most commonly build their large, open orb webs over creeks, seeps, and other wet places, the Common House Spider (*Parasteatoda tepidariorum*, Theridiidae, p. 100) is almost always found around homes and other dwellings. By taking note of their behaviors, diets, periods of activity, and habitats, you will almost certainly reveal new information about how spiders live. Spiders can be studied alive in the field in several ways. With advances in close-focus binoculars and the ability of small digital cameras to take good-quality macrophotographs, it is becoming increasingly easy to make basic behavioral observations. Another highly effective technique for looking at small features on living spiders requires a clear, sandwich-size, resealable plastic bag and a 4-inch embroidery hoop. An embroidery hoop consists of two tightly interlocking wooden rings, a solid inner one and an adjustable outer one, and is available at any fabric store (see fig. 6B, p. 23). Once the spider is coaxed into the bag, seal it and place the spider near the center of the outer ring. Then insert the inner ring, which should pull the bag tight. This will immobilize the spider without injuring it. With a 10× hand lens or jeweler's loupe, you can then examine many important anatomical features that would otherwise be impossible to see on a still-living spider.

When studying spiders, taking notes and making field sketches will both significantly improve the quality of your observations and add to their scientific value. While note taking is a very personal activity, many wildlife professionals employ the Grinnell method. Named after Joseph Grinnell, the first director of the University of California at Berkeley's Museum of Vertebrate Zoology, this method uses several steps to promote detailed, accurate record keeping. The first step is creating a field notebook, where immediate notes, including the date, time of day, habitat, elevation, species lists, and any field observations, are made in one section. This information is later transcribed into a more readable field journal where quickly written notes are organized into clearly written daily entries. If you are interested in a particular species, then a separate section, with pages dedicated to that species, should be maintained. Another section is a record (linked to earlier sections) of specimens collected. By keeping such detailed notes, you will greatly accelerate your learning curve, and it is very likely that new biological and distributional information will come to light. Recording observations is also one of the best ways that amateur naturalists can contribute to scientific knowledge. If well maintained, your notes can provide extremely valuable information on the plants and animals of an area for many years after you first write them.

Some spiders are diurnal and others are mainly nocturnal, so an outing at night will reveal an entirely different assemblage than a walk during the day. A nocturnal orb weaver can be seen in the heart of its web, a trapdoor spider waits at the edge of its burrow, and the reflected eye shine of wolf spiders often looks like glittering stars scattered across the ground. When you look for spiders at night, using a powerful headlamp will keep your hands free. At night (and during the dry parts of the day) nearly invisible webs can be revealed with the use of a spray bottle of water on its mist setting or by using a "duster." This can be made by putting a small bit of cornstarch in the bulb of a turkey baster. When the bulb is squeezed, a puff of cornstarch will settle on any nearby webs, revealing their structure in intricate detail. Another trick is to put cornstarch inside an old sock (which in turn should be carried in a sealed plastic bag). When the sock is lightly beaten over a web, a mist of cornstarch falls onto it, highlighting its otherwise invisible architecture.

Because most species are identifiable only as adults, it is often necessary to rear immature spiders to maturity. Many species are easy to keep in captivity, if you follow a few simple rules, and they even make fascinating pets. Most spiders are solitary creatures and each should be kept in its own container. If you attempt to breed a pair, the male can be introduced into the female's enclosure but should be removed after mating, to prevent cannibalism. A wide variety of containers can be used to house captive spiders, from small plastic cups to large terrariums. Remember that a burrowing spider requires soil that is at least twice as deep as the spider is long and that a web-building spider needs a scaffold that is appropriate for the kind of web it constructs. Spiders also need hiding places incorporated into their enclosures, such as small rocks, bark, and shards of pottery. Generally spiders eat insects slightly smaller than themselves once or twice a week. They should also have a constant supply of water; larger spiders do well with a small dish of water refilled daily, while a cotton or sponge plug on a vial of water will prevent smaller spiders from falling in and drowning. As spiders prepare to molt, they often become lethargic and secretive and stop eating. Molting is a risky period, especially for young spiders in captivity, and injuries are relatively common. Not only is the process itself challenging, but immediately after molting, spiders are soft and fairly defenseless. For this reason, all food items should be removed from a spider's enclosure before it begins molting, and none should be put back in for at least a day, to give the spider's newly exposed exoskeleton a chance to harden.

Efficiently capturing spiders can be done in several ways. In weedy or grassy fields, sweep nets are used. For collecting spiders from taller, woody vegetation, a beat sheet is invaluable (fig. 6A). This is a square of white cloth pulled taught on a wooden frame. When the branches of a bush are shaken over it, insects and spiders drop onto the sheet. If a beat sheet isn't available, a large-mouthed muslin net can also be very useful.

Wandering spiders are most easily collected with a pitfall trap (fig. 6B). One is easily constructed from a 2-liter plastic soda bottle by cutting the top off of a bottle just below where it stops curving and becomes straight sided. By inverting the top back into the bottle, you create a funnel. Anything that falls into it then drops down into the main body of the bottle. Bury a pitfall trap with its upper edges flush with the ground. Then place a wooden board, slightly raised with pebbles or screws at the corners, over it, both to protect the trap and to encourage exploration by small invertebrates. If your goal is to capture live spiders, the pitfall must be checked daily. Placing several balls of crumpled paper at the bottom will provide the spiders with places to hide. Otherwise, you will generally find one large spider and the remains of several others at the bottom of your trap. If you are collecting spiders for a permanent or comparative collection, a few inches of preservative fluid should be poured into the bottom of the trap. The best choices are either 70 percent ethyl alcohol (available at most pharmacies) or propylene glycol (a common antifreeze), mixed with some water to reduce its surface tension. Never use ethylene glycol, another common antifreeze, which is both attractive and toxic to mammals. It is important to mark the trap so you can find it again, and always remove the pitfall once you're done collecting at that location.

Spiders are also abundant in leaf litter. The easiest way to find these is by spreading out small handfuls of duff on a white cloth or plastic sheet (cut-up shower liners work very well) and looking for movement as the spiders scurry away. The litter can also be placed in a mesh bag and sifted over the sheet.

Many spiders are collected using an aspirator, also called a pooter (see fig. 6A). In its simplest form, an aspirator consists of several feet of elastic tubing with a few inches of stiff plastic pipe sticking out from one end. The stiff piece extends back an inch or two into the tubing and is capped on the inside with a small piece of screen or mesh. To collect a spider, you suck in at the tubing end and vacuum the spider into the pipe. To avoid sucking mold or spores into your lungs, you should avoid aspirating directly from the ground or from rotting logs. A slightly more complex version consists of two pieces of tubing extending from a vial plugged with a rubber stopper. With this arrangement, the spider is trapped in the vial rather than at the end of the tubing.

Maintaining a spider collection can be very important, both as a means of verifying a specimen's identity and because it contains voucher specimens representing specific dates and localities. Unlike insects, which are often pinned, spiders are preserved in alcohol. Although the alcohol often bleaches out the spiders' coloration, it does preserve their anatomic structures, which are more useful in identifying spiders. The best fluid for preserving spiders is 70 to 80 percent ethanol, or ethyl alcohol, although in a pinch isopropyl, or rubbing, alcohol can be used. Avoid alcohols with perfumes or dyes. While never pleasant, killing spiders is necessary to make a preserved collection. Many people place the living spider directly into alcohol in the field. Another technique that works well is keeping a jar of alcohol in the freezer. A living spider dropped in the cold alcohol is quickly killed. A few seconds later it can be removed and placed in a vial of room-temperature alcohol for permanent storage.

The next step is to write data labels to accompany the specimens. This is extremely important, as a specimen without associated data is scientifically worthless. At a minimum, the information should include the date, written with the month either partially spelled out or in Roman numerals; where the spider was found, including the specific locality, county, state, and if known, its GPS coordinates; and the collector's name. Additional desirable information includes habitat and brief behavioral notes, elevation, method of collection, and time of day collected (if known). A separate identification label should include the scientific name of the spider, the name of the scientist who first described it, the year it was described, who identified it (the determiner), and the year the determination was made. This label should also include a note regarding the numbers and

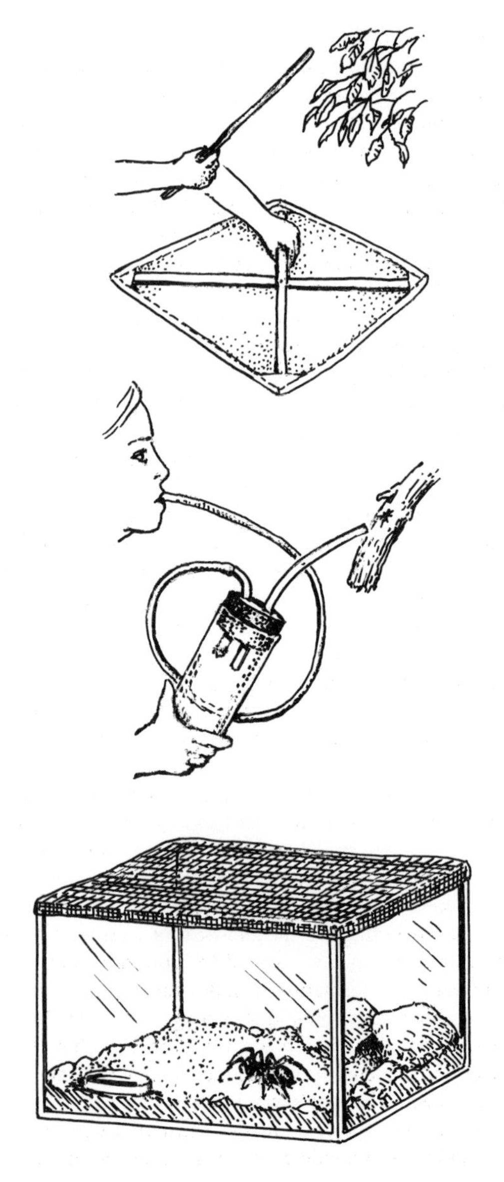

Figure 6A. Techniques for studying spiders. *Upper:* Using a beat sheet to collect spiders from vegetation. *Middle:* Using a pooter/aspirator to collect a small spider from a branch. *Lower:* Housing a spider in a small terrarium.

Figure 6B. Tools for studying spiders. *Upper:* Pitfall trap constructed from a plastic soda bottle. *Middle:* Embroidery hoop and plastic bag used to immobilize a spider. *Lower:* Specimen vial and sample collection labels.

sexes of spiders of the same species in the vial. It is important not to mix up collections, to ensure that all spiders stored in the same vial were collected at the same place and time. The labels should be written on 100 percent cotton paper (easily found in art supply stores) and written in archival or india ink. Pencil is also relatively stable in alcohol and may be used. Never make your labels with ballpoint pens or computer printers, as these quickly fade, dissolve, or leach into the alcohol. Slip the two or more labels into the vial with the spiders as a permanent record (see fig. 6B). The best vial for preserving spiders is glass with a screw cap lined with a Polyseal insert to prevent evaporation. Tightly sealed small jars can also be used for larger specimens as long as the alcohol levels are checked regularly and refilled when necessary. Nearly all entomological research supplies can be purchased through companies such as BioQuip (see Additional Resources at the back of the book for its URL). Exposure to light will quickly fade specimens, so when they are not in use, keep them in a dark cabinet or closed cupboard. Because alcohol is flammable, keep specimens away from heat sources, and have a fully charged fire extinguisher nearby.

Spiders around the Home

The most familiar spiders are those found around homes. These are generally synanthropic species that are adapted to living in houses and other buildings. Nearly all of these spiders are from other parts of the world, having traveled to North America aboard boats, on trains, and in agricultural supplies. Although any spider bite can be serious if it becomes infected or if the person bitten is exceptionally sensitive to the venom, almost none of the region's common household spiders, listed below, are considered consistently dangerous, with the exception of the native Western Black Widow (*Latrodectus hesperus,* Theridiidae, p. 95).

1. Wood Louse Hunter (*Dysdera crocata,* Dysderidae, p. 62, pl. 7). Most likely native to the Mediterranean region. Common in California and Oregon, less so in Washington.

2. Long-bodied Cellar Spider (*Pholcus phalangioides,* Pholcidae, p. 73, pl. 9). Native to central Europe. Common in California, Oregon, and Washington.
3. Marbled Cellar Spider (*Holocnemus pluchei,* Pholcidae, p. 73, pl. 9). Native to the Mediterranean region. Widespread in California and Oregon.
4. Wall spider (*Oecobius navus,* Oecobiidae, p. 86, pl. 12). Cosmopolitan. Place of origin unknown. Found in California, Oregon, and Washington.
5. Western Black Widow (*Latrodectus hesperus,* Theridiidae, p. 95, pl. 14). Native to western North America. Common in California, Oregon, and Washington.
6. Brown Widow (*Latrodectus geometricus,* Theridiidae, p. 95, pl. 14). Cosmopolitan. Place of origin unknown. Extremely common in Southern California.
7. False Black Widow (*Steatoda grossa,* Theridiidae, p. 96, pl. 14). Native to Europe. Found in California, Oregon, and Washington.
8. Common House Spider (*Parasteatoda tepidariorum,* Theridiidae, p. 100, pl. 14). Cosmopolitan. Place of origin unknown. Found in California, Oregon, and Washington.
9. Long-legged sac spiders (*Cheiracanthium* spp., Miturgidae, p. 143, pl. 27). Native to African and Mediterranean regions. Widespread in California; uncommon to very rare farther north.
10. Mouse Spider (*Scotophaeus blackwalli,* Gnaphosidae, p. 161, pl. 31). Native to Europe. Common in coastal California, Oregon, and Washington.
11. Johnson Jumper (*Phidippus johnsoni,* Salticidae, p. 168, pl. 33). Native to North America. Common throughout the Pacific coast states.
12. Wall jumpers (*Menemerus* spp., Salticidae, p. 174, pl. 36). Native to the Old World tropics and Mediterranean region. Widespread in California.
13. Zebra Jumper (*Salticus scenicus,* Salticidae, p. 171, pl. 36). Holarctic, possibly introduced to North America. Common in California, Oregon, and Washington.
14. False Wolf Spider (*Zoropsis spinimana,* Zoropsidae, p. 220, pl. 44). Native to the Mediterranean region. Common and spreading throughout California's San Francisco Bay Area.

15. Funnel web weavers (*Tegenaria* spp., Agelenidae, p. 232, pl. 48). Native to Eurasia. Common throughout California, Oregon, and Washington.
16. Gray House Spider (*Badumna longinqua,* Desidae, p. 243, pl. 50). Native to Australia. Widespread along the coasts of California and Oregon.
17. *Metaltella simoni* (no common name, Amphinectidae, p. 244, pl. 50). Native to South America. Very common in Southern California.

SPIDER FAMILY ACCOUNTS

THERAPHOSIDAE — Tarantulas or Baboon Spiders

Pl. 1

IDENTIFICATION: While the family Theraphosidae contains the largest spiders in North America, it also includes several small desert-dwelling species. All tarantulas have urticating hairs on their abdomens and thick claw tufts on each tarsus. Adult males have two-pronged tibial mating spurs on their first legs. The larger species have thick coats of long hairs over an underlying pubescence, giving them a distinctly furry look, while on smaller species, the long hairs are finer and more diffuse, giving these spiders a more velvety appearance.

SIMILAR FAMILIES: Members of the family Theraphosidae are most likely to be mistaken for velveteen tarantulas (Nemesiidae, p. 33). Superficially similar, nemesiids lack claw tufts and urticating hairs, have noticeably longer posterior lateral spinnerets, and have three tarsal claws. Additionally, these two families have a fairly segregated distribution, overlapping only across a narrow band running through Central and Northern California. Some of the larger wolf spiders (Lycosidae, p. 223) could be mistaken for small tarantulas, but they are generally more overtly patterned, with a distinctive eye arrangement. Wolf spiders also run with their long, thin legs splayed out, while tarantulas generally hold their thick legs closer to their bodies, elevating their cephalothoraxes and abdomens higher off the ground.

PACIFIC COAST FAUNA: One genus with 18 currently described species. The only tarantula genus native to North America, *Aphonopelma* (pl. 1) is in need of a careful and thorough revision. The vast majority of California's tarantula species were described by Chamberlin and Ivie (1939), Chamberlin (1940), and Smith (1994). Unfortunately, many of these descriptions were based on only one or two specimens and used features susceptible to a fair degree of individual variation. This has led to a great deal of confusion, leaving many individuals unidentifiable, even when the original literature is consulted. When you attempt to identify a tarantula, even characteristics such as color and size that may be potentially helpful must be used cautiously. A freshly molted tarantula will be darker and more intensely colored than the same spider a few months later. While size can also be an informative feature, enough individuals must have been measured that the range of variation is known for that species. Because of these challenges, many different names have been given to spiders that almost certainly represent only a few, widespread species. In his revision of the *Aphonopelma* of the northern and western Mojave Desert, Prentice (1997) found that all of the area's previously described large tarantula "species" were actually representatives of the widespread *A. iodius*.

Tarantulas are common denizens of Southern and Central California's lowlands and live in a variety of habitats, from coastal scrub and chaparral to open grasslands and deserts. Along the Coast Ranges, they are found north to the Jasper Ridge region of San Mateo County. They also live in the foothills around the south end of San Francisco Bay and north to Mount Diablo in Contra Costa County. In the western foothills of the Sierra Nevada they have been recorded as far north as the Placer–El Dorado County region east of Sacramento. Tarantulas were formerly more common within the Central Valley itself, but a great deal of the appropriate habitat has been lost to development and agriculture. East of the Sierra Nevada, they can be found in the Owens Valley to the south and in patches of Great Basin Desert to the north, including around Susanville in Lassen County (T. Prentice, pers. comm., 2009).

The Pacific coast's tarantulas can be broken down into several groups, the most speciose of which is the *A. eutylenum* complex (pl. 1) with twelve currently recognized species. These are large spiders, 25 to 66 mm (.98 to 2.6 in.) in length, and are found both along the coast and in the interior of the state. They are generally brownish in color (which varies in relation to molt) and share similar patterns of metatarsal scopulation and a fall breeding season. From mid-September through mid-November freshly molted adult males are often seen wandering in search of receptive females.

The second group, known as the *A. reversum* complex, contains three described species of large tarantulas, including *A. steindachneri* (pl. 1). In this complex, males emerge in mid-July and for the most part have finished mating and died off by early September. When compared with individuals of similar age, sex, and molt, members of the *A. reversum* complex are generally darker than members of the *A. eutylenum* complex (W. Icenogle, pers. comm., 2009). Another feature separating these two complexes is the length of the scopulae on metatarsi IV. On *A. eutylenum* tarantulas, the scopulae cover more than half the length of the fourth metatarsi. On members of the *A. reversum* complex, the scopulae are distinctly shorter, covering only the distal half (or less) of the tarantula's metatarsi IV.

Less conspicuous than their larger cousins are three described species of "dwarf" tarantulas: *A. joshua, A. mojave* (pl. 1), and *A. radinum*. These comparatively small spiders are mostly black with a faint blue or silvery-gray luster and range from 14.5 to 28 mm (.57 to 1.1 in.) in length. *A. joshua* is a summer-breeding species, almost entirely endemic to Joshua Tree National Park. The species narrowly overlaps there with the more widespread *A. mojave*, which lives throughout most of the Mojave Desert, with distinctive eastern and western populations separated by the Death Valley Basin, and has a fall breeding season. The third species, *A. radinum,* was described by Chamberlin and Ivie (1939) based on a single male from Manhattan Beach, Los Angeles County, California. Unfortunately, no additional specimens have ever been identified, and the veracity of the location data associated with this spider is questionable.

NATURAL HISTORY: Despite their charismatic appearance and broad distribution, there has been only limited research on North America's Theraphosidae fauna, especially when compared with the region's other mygalomorph families. Overall, tarantulas are quite sedentary creatures, rarely leaving the immediate vicinity of the burrow except when forced to do so or, in the case of adult males, when looking for mates. During the breeding period, large numbers of adult males are often seen wandering across roads and fields, initiating the commonly used misnomer "tarantula migration." In actuality, these males are seeking out females, whose burrows are often patchily distributed across the landscape. Radio telemetry studies on *A. hentzi*, a large species from the midwestern United States, found that male tarantulas will occasionally travel over a half mile in their search for a receptive female (Janowski-Bell and Horner, 1999). The courtship and mating behaviors of only a few species have been studied in detail. However, they demonstrate enough similarity that a general pattern can be described. Behavioral observations strongly suggest that the webbing along the outside rim of a female tarantula's retreat provides pheromonal information regarding her state of sexual receptiveness and may act as a beacon, attracting potential mates. Once a male finds the burrow of a mature female, he announces his presence by tapping on the ground with his pedipalps and fore legs. The male may also stridulate by raising and lowering his body in a rapid bobbing motion, creating a high-pitched squeaking sound. An interested female will respond with fore leg drumming of her own. After their initial contact, the female rears up toward the male, forcing him to grapple with her and secure her fangs with the tibial spurs on his front legs. In this position, with her body elevated and fangs held apart, mating occurs. Upon completion, the male quickly disentangles himself and retreats from the larger female. Male tarantulas will attempt to mate with multiple females but inevitably die soon after reaching maturity. Females, in contrast, can live for many years, producing numerous broods throughout their lifetimes (life spans of 30-plus years in captivity have been recorded). As they are so long-lived, it's not surprising that they are also slow to develop, often taking between four and seven years to reach maturity. While they are widespread across the southwestern United States, patterns of reproductive isolation are apparent. When two or more species of *Aphonopelma* are found in the same area, often either they are significantly different in size, as *A. iodius* and *A. mojave* are, or they segregate by breeding season, as *A. eutylenum* and *A. steindachneri* do. In places where *A. eutylenum* and *A. steindachneri* occur together, they also separate by microhabitat. In this situation, *A. eutylenum* is most common on the valley floors and lower foothills while *A. steindachneri* lives in the upper foothills (C. Hamilton, pers. comm., 2011).

A tarantula is an ambush predator, hunting in a territory that barely extends beyond the entrance of its burrow. Its vision is extremely poor, but its ability to detect the vibrations of a potential prey item passing nearby is exceptional. Once it cues onto its quarry, the tarantula lunges forward and, scooping with its fore legs, impales the prey with its outstretched fangs. The tarantula will then consume its kill undisturbed in the recesses of its burrow.

In addition to their predatory prowess, tarantulas have several defensive tools, the most obvious of which are their impressively large fangs. Although the region's tarantulas are fairly docile, they will rear up and bite when disturbed. While their venom isn't harmful to humans, their bite can be painful. Just as fangs guard the front of the spider, a patch of urticating hairs on the abdomen protects it from behind. The tarantula creates a mist of these barbed setae by rapidly brushing its hind legs over the back of its abdomen. If these hairs are inhaled by a predator or get into its eyes, they cause an exceptionally painful burning sensation, and on some people, contact with urticating hairs will produce an uncomfortable rash. The predators most closely associated with North America's *Aphonopelma* are the tarantula hawk wasps in the genera *Pepsis* and *Hemipepsis* (Pompilidae). These are exceptionally large blue-black wasps with bright-orange wings. Adult female wasps seek out tarantulas, paralyze them with venomous stings, and then bury them with attached eggs. Once the larval wasp hatches, it feeds on the immobilized but still-living tarantula until it pupates to become an adult wasp.

Tarantulas are rather sedentary, often spending many years without straying more than a few feet from the entrance of the burrow. While all *Aphonopelma* are capable of digging their own lairs, displaced individuals will readily take up residence in abandoned burrows or other acceptable cavities. A tarantula burrow is more or less vertical in its major orientation, ranging from 25 cm to 1 m (10 to 39 in.) in depth, and regularly includes horizontal side chambers filled with the accumulated remains of prey and other debris. It usually terminates in a horizontal den, creating an environment of fairly consistent temperature and humidity. During winter months, many of the region's tarantulas close the entrances to their burrows with thick plugs of soil and silk, only emerging months later in spring. A tarantula will line the outer perimeter and rim of its burrow with a silk collar, and during the daylight hours, it often puts a silk curtain across its entrance. At night, the tarantula opens up the burrow by pressing the curtain against the walls, steadily building up a thick accumulation of silk at the burrow's opening. Unlike the larger tarantulas, *A. mojave* and *A. joshua* regularly build prominent cone-shaped turrets around the entrances to their burrows. The purpose of the turrets is unknown, but they may act as pheromonal beacons in addition to funneling prey toward the burrow entrance.

NEMESIIDAE — Velveteen Tarantulas, Aggressive False Tarantulas

Pl. 2

IDENTIFICATION: Several common names have been used to describe North America's Nemesiidae, but *velveteen tarantula* is arguably the most descriptive. They are large spiders, 16 to 30 mm (.63 to 1.2 in.) in length, with a distinctive smooth "velvety" pubescence covering their bodies. They have exceptionally long posterior lateral spinnerets and three tarsal claws, lack claw tufts, and range in color from silvery gray to brown with purplish iridescence on their palps and fore legs. They are also more aggressive than other Pacific coast mygalomorphs. When disturbed, they readily raise their front legs and expose their large fangs. If further provoked, they will bite, and while not dangerous to humans, the bite can be painful.

SIMILAR FAMILIES: Superficially similar to California's "true" tarantulas (Theraphosidae, p. 28), nemesiids lack long hairs, giving them their more velveteen appearance. Theraphosids also have much shorter spinnerets, only two tarsal claws, and claw tufts on each of their tarsi. Helpfully, Theraphosidae and Nemesiidae only overlap in distribution across a narrow portion of Central and Northern California.

PACIFIC COAST FAUNA: One genus with two generally recognized species (although five species are currently described). The taxonomy of this group is complicated. When Bentzien (1976) examined the original specimens of each of the currently described "species," he determined that four of the five were indistinguishable, meaning that the North American specimens represented at most two, and possibly even a single, morphologically variable species. However, because this work was never published, all five names are still officially considered valid. Raven (1985) argued that all of North America's velveteen tarantulas belong in the genus *Calisoga* (pl. 2), but because two species, *Brachythele longitarsis* and *B. anomala,* were never formally transferred according to the rules of zoological nomenclature, they are still officially included as members of the otherwise European genus *Brachythele*. This is discussed in order to explain incongruities both in the literature and on the Internet. For the purposes of this account, two very similar species, *Calisoga longitarsis* (pl. 2) and *C. theveneti*, are recognized. *Calisoga* is almost entirely endemic to Central and Northern California.

The larger and more common of the two species is *C. longitarsis*. The male has a carapace length of 7 to 14 mm (.28 to .55 in.), and metatarsus I is curved, with two or fewer spines on its ventral surface. The *C. theveneti* male has a carapace length of only 5 to 7 mm (.20 to .28 in.), and metatarsus I is almost straight with five ventral spines. Female *C. theveneti* have never been described and are presumed to be similar to, but smaller than, *C. longitarsis*. What makes this situation especially interesting is that the traits that define *C. theveneti* are present on the immature male *C. longitarsis*. This presents the possibility that there is only a single dimorphic species in which some males reach sexual maturity while simultaneously preserving an array of juvenile features. *C. longitarsis* demonstrates some degree of color variation, with individuals from the coast tending to be more silvery gray, and those in the Sierra, browner.

C. longitarsis is widespread along the California coast from Monterey County to southern Humboldt County. Populations are also found across the northern Central Valley into the Sierra Nevada, with numerous records from between Sierra County and Tulare County. *C. theveneti* has only been reported from Mendocino County on the coast and Mariposa County in the southern Sierra.

NATURAL HISTORY: Nearly everything we know about the biology of velveteen tarantulas is based on Bentzien's (1976) work on *C. longitarsis*. They live from sea level to 2,300 m (7,550 ft) in elevation in oak woodlands, forest clearings, and open grasslands. In coastal areas, *C. longitarsis* burrows can be common in banks, hillsides, and road cuts with varying amounts of sun. In the drier, interior regions, they're found on the north slopes of steep canyon walls or amid the roots of thick vegetation near water. While *Calisoga* will often dig their own burrows, they also adopt empty rodent holes and hollow rotting tree roots. As nocturnal hunters, *Calisoga* wait just inside the burrow entrance for a large insect or other potential prey item to pass by. Once within range, the spider lunges forward and, using its front legs, scoops the prey back toward its fangs. After their prey is subdued, velveteen tarantulas prefer to eat deep in the safety of their retreats.

Generally, male *Calisoga* begin searching for females after the first fall rains, although there is evidence of an additional spring pulse in mating activity in some Sierran populations. An adult male initiates courtship by quivering his legs and body as he enters the female's burrow and gently touches her with his fore legs. If receptive, she will raise her fore legs and spread her fangs. The male then uses the tibial spurs on his first pair of legs to keep her fangs parted while he pushes her into a more vertical mating position. A gravid female will wait until summer to begin constructing her lone egg sac, which is attached to the roof near the back of her burrow. About two months later, the spiderlings hatch and disperse.

ANTRODIAETIDAE — Folding-door Spiders, Trapdoor Spiders, Turret Spiders

Pl. 2

IDENTIFICATION: Antrodiaetids are small to medium-size mygalomorphs, 6 to 25 mm (.24 to .98 in.) in length, with one to four sclerotized patches on the dorsal abdominal surface. Members of the family Antrodiaetidae have posterior lateral spinnerets that are quite long and capped by a fingerlike terminal segment. The anterior portions of their carapaces are distinctly elevated, and the thoracic groove shape varies. Antrodiaetidae range in color from grayish yellow to dark mahogany brown. Because of this family's secretive nature, the individuals most commonly encountered are adult males wandering in search of females.

SIMILAR FAMILIES: Antrodiaetids are superficially similar to the members of several Pacific coast mygalomorph families, including the wafer-lid trapdoor spiders (Euctenizidae, p. 39) and the cork-lid trapdoor spiders (Ctenizidae, p. 43). However, neither the wafer-lid nor the cork-lid trapdoor spiders have abdominal sclerites. The only other Pacific coast mygalomorph family sharing this feature is Mecicobothriidae (p. 47), diminutive spiders with exceptionally long posterior lateral spinnerets whose terminal portions are narrow, tapered, and pseudosegmented. The posterior lateral spinnerets of the generally larger antrodiaetids are shorter and distally rounded. The mecicobothriids also build small sheet webs connected by trail-like silken tunnels, very different from the turrets and trapdoors of the antrodiaetids.

PACIFIC COAST FAUNA: Two genera containing 22 described species. *Antrodiaetus* (pl. 2) is the more widespread of the two genera, and with 12 described species, it is also the most diverse mygalomorph genus in the Pacific Northwest. Members of the genus *Antrodiaetus* can be recognized by the presence of a fairly deep longitudinal thoracic groove. Most species have only two pairs of spinnerets, and if three pairs are present, the anterior laterals are extremely small and unsegmented. Along the Pacific coast, *Antrodiaetus* range from Central California to southeastern Alaska, reaching their greatest diversity in Oregon and Washington. *Antrodiaetus* live throughout the region, from coastal and mountain forests to dry Great Basin scrublands. Nearly all of Oregon and Washington is home to at least one species of *Antrodiaetus,* and in many areas, two or even three species can be found in close proximity. Most *Antrodiaetus* burrows have short, flexible collars that stand erect when opened, but when closed, they create a false double-door effect. Bits of dirt and plant material are normally integrated into the burrow entrance's weave, camouflaging it against the surrounding soil.

Antrodiaetus riversi (pl. 2) and *Antrodiaetus gertschi* are especially distinct, enough so that until recently they were placed in the genus *Atypoides.* Adult males of these two species have large, blunt, tusk-like apophyses on their chelicerae, and both sexes have three pairs of spinnerets. Known as the California Turret Spider, *A. riversi* is a California endemic found along the coast from Monterey County almost to the Oregon border. Additional populations live in the mountains and foothills of the central Sierra Nevada from Mariposa County north to El Dorado County and in the Sutter Buttes, a very small, isolated mountain range in the northern Central Valley. The spider's common name comes from the rigid, silk-lined turret that marks the entrance to its burrow. Up to three inches in height, the turret has an outer coating of pine needles, moss flakes, and other soil debris. Genetic and biogeographic research by Ramirez and Chi (2004) and Starrett and Hedin (2007) has revealed that the California Turret Spider actually represents a complex of at least five morphologically similar but genetically distinct species that have been isolated by periods of glaciation and geologic upheaval.

A. gertschi lives in the Sierra Nevada and the Cascade Range from Sierra County, California, to northern Klamath County, Oregon. The thick, flexible collar around the entrance to its burrow folds out, flowerlike, when open. Coyle (1971) revised *Antrodiaetus* (including *A. riversi* and *A. gertschi* under their previous placement in *Atypoides*) and included descriptions and natural history summaries for most of the described species. Additional information on the *Antrodiaetus* of southwestern Oregon can be found in Cokendolpher et al. (2005).

The other regional antrodiaetid genus is *Aliatypus* (pl. 2), which is regionally represented by 10 species, all of which are endemic to California. Unlike *Antrodiaetus, Aliatypus* cap their burrows with thin, flexible trapdoors. They further differ in the shapes of their thoracic grooves. While *Antrodiaetus* has a longitudinal furrow, in *Aliatypus* the thoracic groove can be a deep pit, a shallow depression, or absent altogether. *Aliatypus* also have three distinct pairs of spinnerets with two visibly segmented anterior laterals. Adult male *Aliatypus* are easily recognized because their palpal patellae are nearly as long as their palpal tibiae, giving them exceptionally long, slender pedipalps. *Aliatypus* are distributed widely across California with populations on either side of the Central Valley. They live in habitats as diverse as the interior range foothills, the high-elevation red fir forests of the Sierra Nevada, and the cool coastal redwood groves. The thin, cryptic trapdoors built by *Aliatypus* are similar to those constructed by members of the genera *Aptostichus* and *Promyrmekiaphila* (Euctenizidae, p. 39), but unlike euctenizid burrows, there is occasionally a narrow rim of silk ringing the burrow's inner rim, helping to make a tight fit between the trapdoor lid and the entrance. Additional information on *Aliatypus* can be found in Coyle's (1974) revision of the genus.

NATURAL HISTORY: Antrodiaetidae is by far the most thoroughly studied of North America's mygalomorph families. To prevent desiccation, antrodiaetids generally build their burrows in cool, north-facing slopes or in deeply shaded ravines, often near tree roots for added soil support. *Antrodiaetus* are most common in humid woodlands, although a few species have adapted to drier climes. One exceptional species, *A. montanus,* lives in the juniper and sage scrub of the northern Great Basin. To compensate for the natural aridity of this region, the spiders build comparatively long burrows, often at the base of rocky outcrops where water is more likely to penetrate into the soil. Additionally, they line the inside of the burrow with a thick layer of silk, trapping even more of the cool, damp air. While not formally tested, it has been proposed that the trapdoor-type cover used by *Aliatypus* does a better job of sealing the burrow against moisture loss than the folding door used by most *Antrodiaetus*. This adaptation may be what has allowed *Aliatypus* to expand into warmer, more arid parts of Southern California.

In most species of Antrodiaetidae, the adult males leave their burrows and begin searching for females in the late summer or fall, depending on the seasonal rains. They move about at night, taking cover during the day in old burrows and under fallen debris. Very little is known about the specific courtship behaviors of most antrodiaetids, but the *Antrodiaetus riversi* male uses his cheliceral apophyses to hold the female's fangs apart and position her during mating. In other species of *Antrodiaetus*, the males have spine clusters and unusual modifications to their front legs that may be used in a similar manner. *Aliatypus* males lack these features; however, Coyle and Icenogle (1994) proposed that the exceptionally long palps of *Aliatypus* may permit the male to reach the female's genital opening without raising her body into a vertical position. This would allow the pair to mate in the safety of her burrow rather than at its entrance as most other mygalomorphs do. *Antrodiaetus* females attach their lens-shaped egg sacs to the walls of their burrows, while *Aliatypus* females suspend their pendulous egg sacs just above the bottom of their retreats.

Because of the narrow environmental requirements and limited dispersal abilities, antrodiaetid burrows are often found in dense aggregations peppered across the broader landscape. In some areas, it's not uncommon to find two or more species of trapdoor spider belonging to different genera or even families living in the same assembly. Such colonies normally contain numerous adult females surrounded by the scattered burrows of their young. Adult males are known to wander widely in search of females, which helps prevent excessive inbreeding within colony sites. Like other mygalomorphs, antrodiaetids take several years to reach maturity. While males perish soon after reaching adulthood, females can live for many years, producing batches of young annually. Vincent (1993) suggested that wild *Antrodiaetus riversi* females are capable of living 16 years or more.

Like all Pacific coast mygalomorphs, antrodiaetids are sedentary nocturnal hunters. Waiting at the edge of the burrow, they detect passing prey through vibrations in the soil. Some species of *Aliatypus* even incorporate small tabs of silk and soil into their trapdoors that act as triggers, the lightest brush signaling that prey is close at hand.

EUCTENIZIDAE — Wafer-lid Trapdoor Spiders

Pls. 3, 4

IDENTIFICATION: Euctenizidae represents a diverse group of small to medium-size mygalomorphs, many of whom have distinctive abdominal patterns or build unusual burrow entrances. The major morphological features used to describe this family include a lack of abdominal sclerites, the presence of fairly dense scopulae on the tarsi and metatarsi of their anterior legs, and teeth only on their cheliceral promargins. All of the North American genera were previously housed in the family Cyrtaucheniidae, but in their major revisionary work, Bond et al. (2012b) split them from the South American and Old World genera and placed them in the newly established family Euctenizidae

SIMILAR FAMILIES: Two families have members that are likely to be mistaken for euctenizids: the cork-lid trapdoor spiders (Ctenizidae, p. 43) and the folding-door spiders (Antrodiaetidae, p. 35). Antrodiaetids are superficially similar in size and shape to the euctenizids, and one genus, *Aliatypus*, also makes a thin, wafer-type trapdoor. However, unlike antrodiaetids, euctenizids lack abdominal sclerites. Structurally, the cork-lid trapdoor spiders are most similar to the wafer-lid trapdoor spiders, but the former make thick-lidded trapdoors, are generally bulkier, lack abdominal patterning, have shorter spinnerets, and have teeth on both cheliceral margins. Additionally, female wafer-lid trapdoor spiders possess scopulae and only have a few long, slender spines on the tarsi and metatarsi of their fore legs. Female cork-lid trapdoor spiders lack scopulae and have numerous stout, thornlike spines on their anterior legs.

PACIFIC COAST FAUNA: Three genera with 41 described species. *Aptostichus* (pl. 3) is by far the most diverse of the region's mygalomorph genera, with 37 described species. Although Southern California has the greatest diversity of *Aptostichus*, they can be found along almost the entirety of the California coast and inland north to at least Tehama County. As a genus, *Aptostichus* is quite variable in appearance. Many species have irregular mottled bands across their abdomens, while in others they are faint or lacking altogether. Coloration in this genus ranges from pale yellow tan to reddish brown, often with a faint purplish wash on the abdomen. Additional features include a diagnostic series of spines on the dorsal surface of the adult male's cymbium and small but distinctive cuspules concentrated on the posterior portion of the female's endites. All of the currently recognized species are described in detail in Bond (2012).

Spiders in the genus *Promyrmekiaphila* (pl. 4) are easily recognized by their tiger-stripe abdominal pattern: broad, dark chevrons over a lighter gray or brown background. Unlike *Aptostichus*, female *Promyrmekiaphila* have cuspules across the entire surface of their endites and males lack spines on their cymbia but do have patches of long, thin spines on the ventral surfaces of tibiae I. *Promyrmekiaphila* is endemic to Central and Northern California with two described species. *Promyrmekiaphila clathrata* is found along the Coast Ranges from San Benito County to southern Mendocino County and Glenn County where it lives in the rich, dark soils of damp oak woodlands and redwood forests. *P. winnemem* is known only from Shasta County and Tehama County at the northwestern apex of the Central Valley. Research by Stockman and Bond (2007) has revealed significant genetic and ecological differences between the different *P. clathrata* populations, and future studies will likely divide the spider into as many as five different species. Both *P. clathrata* and *P. winnemem* are discussed and illustrated in Stockman and Bond's (2008) revision of the genus.

The third Pacific coast euctenizid genus is *Apomastus* (pl. 3). Endemic to Southern California, two species are found in and around the Los Angeles Basin: *A. schlingeri* to the north and *A. kristenae* to the south and east. The males can be separated by the spination pattern of tibia I, while the females are morphologically indistinguishable. Unlike other Euctenizidae genera that build thin, wafer-type trapdoors to cover their burrows, *Apomastus* construct collars or flexible turrets that can extend up to several centimeters from the ground. They are superficially similar to those built by the California Turret

Spider (*Antrodiaetus riversi,* Antrodiaetidae, p. 36); however, these two genera are separated by a distribution gap of several hundred miles. Inside the collar the silk is quite loose and, if disturbed, will collapse on itself, obscuring the spider hidden within. *Apomastus* are generally deep mahogany brown with unmarked abdomens. Among Pacific coast euctenizids, a uniformly dark coloration, a recurved thoracic groove, and a lack of cymbial spines are diagnostic of *Apomastus* males. Females are similarly colored and have straight, transverse thoracic grooves, unlike the procurved grooves of *Aptostichus* females. A more detailed exploration of the morphology and systematics of *Apomastus* can be found in Bond (2004).

NATURAL HISTORY: Until recently, little had been published on the natural history of this family. The wafer-lid trapdoor spiders reach their greatest North American diversity in Southern California where *Aptostichus* can be found in a wide variety of habitats, from coastal and desert dunes to chaparral, montane forests, and dry inland foothills. For the most part, *Promyrmekiaphila* and *Apomastus* are restricted to more humid, shaded environments.

Promyrmekiaphila and *Aptostichus* both cover their burrows with thin wafer-type trapdoors, often incorporating bits of vegetation and soil into their weaves. Excluding a few species of *Aptostichus,* the spiders in both of these genera generally construct branched burrows with smaller, dead-end tunnels extending off of the main shaft. *Promyrmekiaphila* tend to build their burrows on mild to moderate slopes and hillsides, occasionally reaching depths of 30 cm (11.8 in.) or more. Once identified, the silk and soil turrets capping *Apomastus* burrows are quite distinctive. They are nearly always found on damp north-facing stream banks and ravine walls, although there is a population from Riverside County that has adapted to a drier chaparral habitat. Like all trapdoor spiders, female *Apomastus* can reproduce year after year, and it's not uncommon to find the offspring from several broods living together in their mother's burrow.

Mygalomorphs exhibit an exceptional diversity in Southern California. Their sedentary lifestyles, limited dispersal abilities, and clustered assemblages have both promoted speciation and left them highly susceptible to localized extinctions. Because of their highly fragmented population structure, identifying species of wafer-lid trapdoor spiders solely by their morphology has been ineffective at expressing the family's true regional diversity. One species, *Aptostichus simus,* is resident in coastal dunes from northern Mexico to Point Conception, Santa Barbara County, California, with separate populations on California's Channel Islands and around Monterey Bay. This spider has been the subject of several studies focusing on the morphological, ecological, and genetic variation found across its range. Bond et al. (2001) found that despite being essentially indistinguishable in form and habitat, the four main populations (in San Diego, in the Los Angeles Basin, on Santa Rosa Island, and along Monterey Bay) have a great deal of genetic divergence. This is indicative of a cryptic species complex in which despite their outward similarities, each group is on its own independent evolutionary trajectory and has been reproductively isolated for between 1.5 and 6 million years. A more comprehensive approach using morphology, genetics, and ecological similarity to help clarify species boundaries was developed by Bond and Stockman (2008). This combined data technique shows a great deal of promise in illuminating the actual diversity of these spiders and provides an important link between systematics and ecology.

Unfortunately, *Apomastus* distribution around the Los Angeles Basin is in one of the most urbanized and densely populated regions of the state. Research by Bond et al. (2006) found that populations historically living within the basin itself have, for the most part, gone extinct and that today the genus is limited mainly to wooded ravines in undeveloped areas along the margins of its previous distribution. What makes these losses particularly concerning is that like many mygalomorphs, *Apomastus* aggregations are highly structured, with little gene flow between populations. *Apomastus* colonies within the Los Angeles Basin were adapted to a drier coastal sage scrub habitat than those currently found in the higher-elevation chaparral and oak woodlands. As research on other trapdoor spider genera has revealed, populations adapted to different habitats often demonstrate differences in morphology, behavior, or other life history traits; these adaptations are now lost with the extinction of these basin populations.

CTENIZIDAE **Cork-lid Trapdoor Spiders**

Pl. 3

IDENTIFICATION: The most obvious feature regarding these robust spiders isn't morphological but architectural. Their thick-lidded trapdoors are unique among the Pacific coast's spider fauna and in many cases are the only visible evidence of these spiders' presence. The West Coast's Ctenizidae are moderately large spiders, 15 to 28 mm (.6 to 1.1 in.) in length, although there is a great deal of variation between populations (W. Icenogle, pers. comm., 2008). They are mostly dark mahogany brown to black while the abdomens of adult males have an unmarked dusky reddish or purplish hue. The cork-lid trapdoor spiders lack abdominal sclerites, their spinnerets are short and stout, both margins of their cheliceral furrows are toothed, and their thoracic furrows are deeply etched and strongly procurved. Their burrowing is aided by their heavily spined fore legs and file-like rastella on their chelicerae.

SIMILAR FAMILIES: Members of the family Euctenizidae (p. 39), also known as the wafer-lid trapdoor spiders, may be mistaken for ctenizids. Euctenizids, however, never build the thick-lidded trapdoors characteristic of the Pacific coast's ctenizids. On wafer-lid trapdoor spiders, the posterior lateral spinnerets are relatively long with multiple segments usually visible when viewed from above. Euctenizids are also generally less bulky with longer, thinner legs, and their abdomens are often patterned with either light and dark bands or chevrons.

PACIFIC COAST FAUNA: Two genera, two regional species. Along the Pacific coast, ctenizids are restricted to Southern and Central California. The more common of the two species is the California Trapdoor Spider, *Bothriocyrtum californicum* (pl. 3). Populations of this spider show considerable variation in size and color. Adult females are usually light mahogany to dark chestnut brown with darker legs and chelicerae. Adult males are dark brown to black with a grayish-pink to brick-red abdomen. There is also a population on the Channel Islands whose members are distinctly brassy in color. The California Trapdoor Spider's range extends from northern Baja California, Mexico, north through Santa Barbara County on the coast and to Kern County inland. It lives in a variety of habitats, including oak scrub, chaparral, dry grasslands, and deserts, with colonies known from the Anza Borrego Desert north to Death Valley.

The second regional species of cork-lid trapdoor spider is *Hebestatis theveneti* (pl. 3). It is similar to the California Trapdoor Spider but is generally slightly smaller in size and distinctly darker in color. The female is almost uniformly black, and the male's abdomen is more purplish than red (W. Icenogle, pers. comm., 2008). These spiders are restricted to steep banks in damp oak scrub habitats from San Diego County to Santa Barbara County on the coast and along the southern Sierra foothills from Kern County north to Mariposa County. Recently, a colony was discovered in southern San Benito County in Central California (M. Hedin, pers. comm., 2009). Whether these spiders are the descendants of a long-isolated population or are part of a contiguous but thinly distributed collection of coastal and near-coastal *Hebestatis* remains to be seen.

The female *H. theveneti* has a shallow depression on the dorsal surface of the third tibiae, while the same leg segment on *B. californicum* is unmodified. Additionally, the female *B. californicum* has a cluster of short spines on the distal end of tibia IV, a feature the female *H. theveneti* lacks. The male *B. californicum* has a distinct knob on the underside of each palpal tibia and a bundle of spines on the distal underside of each tibia I. The male *H. theveneti* lacks the palpal knob, and rather than a group of spines under the distal end of tibia I, it has two large spines resting on a small apophysis. Despite having been found at numerous locations, with specimens in the collections of several museums and universities, the male *H. theveneti* has yet to be formally described in the scientific literature.

NATURAL HISTORY: Superficially similar, these two species demonstrate several ecological differences. The California Trapdoor Spider is much more tolerant of arid habitats. Its burrow is commonly found on a flat to moderately steep hillside or south-facing slope. Up to a foot long, the burrow helps protect the spider from digging predators and maintains more consistent temperature and humidity levels in hot, dry areas. When excavating a new burrow or enlarging an existing one, the California Trapdoor Spider either flicks the dirt away from the burrow or mixes it with long strands of webbing that can be pulled out, much like a sailor pulling an anchor from the sea. The inner walls are then swathed with a thick layer of silk, and the trapdoor is built with alternating layers of silk and damp soil. Once the trapdoor is completed, the spider repeatedly pulls the still malleable lid into the entrance until a tight fit is made. Fang holes inside the trapdoor allow the spider to hold it shut with surprising force. The California Trapdoor Spider has even been reported placing small pieces of moss on top of the trapdoor, making an already cryptic entrance even more difficult to find (Passmore, 1933).

H. theveneti is more humidity dependent than *B. californicum*. Its burrow is a shorter, horizontal structure found in steep, shaded banks, often amid the roots of poison oak and other scrub foliage. Rarely more than a few inches in length, the burrow is often very well camouflaged with moss growing directly on top of the trapdoor.

Cork-lid trapdoor spiders are mainly nocturnal hunters, using vibrations to help them target their prey. Once an unsuspecting insect comes within range, the trapdoor spider lunges from its burrow's entrance, taking care to always leave at least its rear legs and part of its abdomen under the trapdoor, preventing it from closing and sealing the spider out. In less than a second the spider pulls the captured prey back into its burrow.

Adult females never willingly leave the safety of their burrows, and adult males only vacate theirs when looking for mates. As with many of California's mygalomorphs, the mating season is closely tied to fall and winter rains. With the passing of the first major showers, adult male cork-lid trapdoor spiders can be found, usually at night, wandering in search of receptive females. The courtship of these spiders has not been described; however, the presence of crushed male bodies at the bottom of some females' burrows hints at the difficulties they face. After mating, the female lays her eggs in a sac at the bottom of her burrow. When the spiderlings hatch out, they stay in the mother's burrow for several months, feeding on insects that she captures for them. Because of their limited dispersal abilities and long life spans, it is not uncommon to find an aggregation of trapdoor spiders of different ages in the same vicinity.

When observers look in trapdoor spider burrows, they fairly commonly find not living spiders, but large, hairy cocoons belonging to spider wasps (family Pompilidae). While most spider wasps dig their own nurseries, at least one species, *Pedinaspis planatus,* takes advantage of the California Trapdoor

Spiders' already well-constructed burrows. When the female wasp finds an active spider burrow, she'll attempt to pry open the lid and, if unsuccessful, will chew and tear her way through to reach the spider inside. California Trapdoor Spiders are also parasitized by a species of small-headed fly (*Ocnaea smithi*, Acroceridae). After the fly pupates to adulthood, all that is left are the host spider's crumpled remains and the fly's cellophane-like pupal sac. An excellent series of photographs illustrating the life cycle of trapdoor spiders and their parasites can be found in Jenks (1938). This article also does an inspiring job of showing how an interested naturalist, using diligence and patience, uncovered numerous new aspects of these spiders' and their parasites' biology. With so little known regarding even the most basic aspects of many spiders' life histories, a curious and inventive naturalist could do the same or even more today.

DIPLURIDAE

Pl. 4

IDENTIFICATION: These tiny mygalomorphs, 3 to 5.5 mm (.12 to .22 in.) in length, lack abdominal sclerites and have only two pairs of spinnerets, and although the posterior lateral spinnerets are quite long, they're not pseudosegmented.

SIMILAR FAMILIES: The only spiders likely to be mistaken for a diplurid are members of the family Mecicobothriidae (p. 47). Superficially similar, mecicobothriids have abdominal sclerites and pseudosegmented posterior lateral spinnerets. Additionally, other than *Hexura rothi*, all regional mecicobothriids have three pairs of spinnerets.

PACIFIC COAST FAUNA: One genus with one regional species. The Pacific coast's only diplurid is *Microhexura idahoana* (pl. 4), an exceptionally small mygalomorph found in mountain ranges across the Pacific Northwest. Populations are found in the Cascade Range from central Oregon to Mt. Rainier in west-central Washington. There are additional records from the lower Monashee Mountains of northeastern Washington and from the Blue Mountains of northeastern Oregon and southeastern Washington. On the male, the legs and carapace are light brown with irregular dark lines extending from the thoracic furrow to the lateral edges of the carapace. The abdomen varies from grayish tan to dark purplish brown, occasionally with pairs of faint lateral bars. The female is a bit larger, is slightly lighter in color, and demonstrates greater variability in abdominal pattern. Females from Mt. Rainier usually show several pale gray dashes on the abdomen, while in other populations, these marks are reduced or absent. *M. idahoana* is closely related to the federally endangered Spruce-fir Moss Spider, *M. montivaga,* a rare denizen of the southern Appalachian Mountains of Tennessee and North Carolina. A more detailed description of *M. idahoana* can be found in Coyle (1981).

NATURAL HISTORY: The diminutive *M. idahoana* lives in the duff and moss of conifer forests between 600 and 2,300 m (2,000 to 7,500 ft) in elevation. Its webs consist of a series of small sheets and connecting tubes, usually placed under a rotting log or mat of moss in a shaded, rocky area. This species is extremely humidity dependent, and even just a few hours of exposure to dry air can be fatal (Coyle, 1981).

Females begin laying their eggs in June and when disturbed carry their spherical egg sacs in their fangs. After dispersing from their natal webs, young spiders take several years to reach maturity, with penultimate males completing the final molt between November and March. Nothing is known about the courtship behaviors or dietary range of this species. Its ability to survive in these northern mountains is due in no small part to its exceptional tolerance to cold. Even in the middle of winter, juveniles can be quite active, and numerous specimens have been collected wandering atop deep snowbanks or crawling about on tree trunks above the snow line (R. Crawford, pers. comm., 2008).

MECICOBOTHRIIDAE

Pl. 4

IDENTIFICATION: These small mygalomorphs are identified by the presence of both elongate, pseudosegmented posterior lateral spinnerets and abdominal sclerites. Males also have an unusual palpal structure in which the tarsus extends cymbium-like over the bulb. Unlike other mygalomorphs, the Mecicobothriidae thoracic groove is longitudinal and shallow.

SIMILAR FAMILIES: Mecicobothriids share several features in common with members of the family Dipluridae (p. 46), including their diminutive size, elongate spinnerets, and web design. However, diplurids lack abdominal sclerites and pseudosegmented spinnerets. The only other western mygalomorph family with abdominal sclerites is Antrodiaetidae (p. 35), but those spiders build either trapdoors or turrets and have relatively short spinnerets.

PACIFIC COAST FAUNA: Three genera with four species, all of which are described and illustrated in Gertsch and Platnick's (1979) revision of the North American fauna. The largest (13 to 18 mm, or .5 to .7 in., in length) and most widespread of the region's mecicobothriids is *Megahexura fulva* (pl. 4). This California endemic is fairly common in forested regions throughout the Coast Ranges of Southern California north to San Francisco Bay and in the Sierra Nevada north to the Lake Tahoe region. It has a diagnostic pair of pleurites extending backward from the posterolateral edge of its tan carapace, dark reddish-brown chelicerae, and a pale abdomen with reddish-purple speckling on the top and sides. *M. fulva* has two abdominal sclerites, one on its dorsoanterior margin and another, more difficult to see, directly above its pedicel. The posterior lateral spinnerets have only three segments, the last of which is pseudosegmented.

The smallest North American mecicobothriid is *Hexurella rupicola* (pl. 4), the only known *Hexurella* from the Pacific coast. Like *M. fulva*, this tiny spider (2.5 to 4.5 mm, or .10 to .18 in.) has two abdominal sclerites but is quickly differentiated by its small size, orangish-yellow fore parts, reddish-brown abdomen, and four segmented posterior lateral spinnerets. *H. rupicola* has been found living under rocks in the dry Southern California foothills of San Diego, western Riverside, and eastern Orange County (W. Icenogle, pers. comm., 2008).

Members of the genus *Hexura* (pl. 4) are intermediate in size, with adults ranging from 8 to 11 mm (.3 to .4 in.) in length. They are easily identified by the presence of only a single, broadly fused abdominal sclerite on the dorsoanterior portion of the abdomen. Two species have been described, both from forested regions of the Pacific Northwest. *Hexura rothi* has only two pairs of spinnerets and is found from the Central Oregon Coast Ranges south through the Klamath Mountains into Del Norte County, California (R.J. Adams, pers. obs., 2008). *H. picea* has three pairs of spinnerets and ranges from the central Oregon coast inland to the eastern Cascade Range and north to Puget Sound, Washington. Both species are tan to mahogany brown with a darker purplish-brown abdomen.

NATURAL HISTORY: Mecicobothriids use their long, flexible spinnerets to build small sheet webs with tubular retreats. *H. rothi* often constructs its web where a slight incline, such as the accumulated humus at the base of a fallen tree, allows it to burrow deep into the leaf litter and moss. It then spins a horizontal tube with a small sheet web extending out from its entrance. The larger *M. fulva* builds its retreats among rocks, among tree roots, or in the crevices of shaded ravines. *Hexurella* species construct their webs under rocks or conceal them beneath a thick layer of leaf litter.

Like other mygalomorphs, mecicobothriids apparently take several years to reach adulthood. *Hexura* and *Megahexura* males achieve sexual maturity late in summer and are found through fall and occasionally into early winter, while adult females and juveniles can be found throughout the year. At this time, nothing is known about mecicobothriid courtship behaviors, but the *Hexura rothi* egg sac has a lens shape and usually contains around 80 eggs. The much smaller *Hexurella rupicola* has a small, round egg sac that holds around half a dozen eggs.

HYPOCHILIDAE — Lampshade Weavers

Pl. 5

IDENTIFICATION: The lampshade weavers are a remarkable group of long-legged, cribellate spiders, easily recognized both by their unusual physical appearance and by the distinctive architecture of their webs. This ancient lineage retains several features linking them to the earliest araneomorphs. They have two pairs of book lungs like mygalomorphs, and their chelicerae are intermediate in position between the angled alignment of the araneomorphs and the parallel condition of the mygalomorphs. The Pacific coast's lampshade weavers are eight-eyed, medium-size spiders ranging from 7.4 to 10.5 mm (.3 to .4 in.) in length, although their exceptionally long, thin legs can make them look larger. They are dorsoventrally flattened and regularly reside in the center of the web, with the body pressed against the substrate and the legs in a characteristically splayed pose. The webs of these spiders are absolutely unique, consisting of a symmetrical, lampshade-shaped cone of lacy, gray cribellate silk.

SIMILAR FAMILIES: The only family whose members are likely to be mistaken for a lampshade weaver are the cellar spiders (Pholcidae, p. 73). Although pholcids will occasionally rest with their body and legs pressed against a wall or ceiling, they have more rounded or globular bodies. Additionally, the sheet and tangle webs of pholcids never have the symmetrical structure of a lampshade weaver's web.

PACIFIC COAST FAUNA: One genus, three regional species. *Hypochilus* (pl. 5) is the only genus of lampshade weaver in North America. The Pacific coast's species are endemic to California mountain ranges. Additional species are known from the Rocky Mountains and from the southern Appalachian region. Lampshade weavers are rather uniform in their overall coloration, differing mainly in the fine details of their reproductive structures. The carapace is pale around the margins, with a large, dark floret pattern in the center, and the abdomen has a heavy coating of purplish brown reticulations and a smattering of white spots over a light yellow-gray base color. *H. kastoni* is the northernmost of California's lampshade weavers and has been found in the Klamath Mountains and Cascade Range around Siskiyou County and eastern Humboldt County. The most widespread of the Pacific coast's lampshade weavers is *H. petrunkevitchi,* with populations known from numerous locations in the Sierra Nevada, including Yosemite, Sequoia, and Kings Canyon National Parks. *H. bernardino* is the southernmost of the region's Hypochilidae and has only been recorded from the San Bernardino Mountains of San Bernardino County. Specific information on California's lampshade weavers can be found in Gertsch (1958b), Platnick (in Forster et al., 1987), and Catley (1994). Considering their mountain habitats, it is not surprising that many populations end up isolated from one another, and despite their morphological similarity, they can genetically be very divergent (Hedin, 2001).

NATURAL HISTORY: With their retained ancestral features, lampshade weavers make up one of the most unusual spider families in North America. With the greatest concentration of species, the Appalachian lampshade weavers have received the most attention from biologists. However, because of the family's conservative morphology and uniformity of web design, information on the eastern lampshade weavers can be broadly used to discuss California's species. All *Hypochilus* build lampshade-type webs whose orientation can range from vertical to horizontal. The spiders are found in moist, shaded environments, including streamside boulder fields, caves, and culverts.

Web construction begins when the spider lays down a circular disk of silk that is used as an attachment point for the narrow "top" of the lampshade. The spider then spins out a series of frame and support lines before adding the gray cribellate silk that makes up the cover of the lampshade. Additional support and tangle threads extend below the open end of the web and are thought to act as a baffle, channeling flying insects into the sticky mesh above.

Once the web is built, the spider assumes a characteristic pose with its body pressed against the substrate and its legs in contact with the web's interior rim. If a small insect or other prey item gets caught in the lampshade's webbing, the spider slowly gathers it inward, biting the entangled prey until it ceases moving. Like mygalomorphs, lampshade weavers don't wrap their prey in silk after killing it. Instead, they crush their prey into a broken mass while feeding, and they drop its remains out of the web when they're finished.

Courtship and mating are nocturnal affairs. Adult males leave the safety of their webs and wander in search of receptive females, using their long front legs as feelers as they explore the surface ahead of them. Upon encountering a female's web, the male lightly taps or plucks on the webbing with his front legs. If the female is uninclined toward the male's advances, she can respond quite aggressively, rushing at him and driving him back. When a male lampshade weaver does find a receptive female, he quickly climbs into her web. After some light touching with their fore legs, she elevates her body and allows him to crawl under her. Facing in opposite directions with their legs lightly entwined and his back pressed against her belly, the spiders mate. The *Hypochilus* egg sac is a heavily camouflaged structure, often covered with an outer coating of dirt, moss, and other bits of detritus. The female either suspends the egg sac from the roof of her web or loosely attaches it to the substrate.

Like mygalomorphs, lampshade weavers take more than a

year to mature. In the Appalachian species, *H. thorelli,* the fertilized eggs develop through winter, with spiderlings emerging late in spring. They continue to grow through fall, retreating deep into rock crevices for winter. The following spring, the spiders once again become active, reaching sexual maturity in the late summer or early fall. While males die soon after mating, the presence of adult females in the early spring shows that at least a few survive into the following year (Fergusson, 1972). While California's lampshade weavers likely have similar life histories, the specific developmental chronology and habits of these unique spiders has yet to be revealed.

FILISTATIDAE — Crevice Weavers

Pl. 5, Fig. 7

IDENTIFICATION: This is a morphologically diverse group of haplogyne, cribellate spiders. They range in length from 1.5 to 18 mm (.06 to .7 in.) and vary considerably in their color and shape. A spider in this family has eight eyes tightly clustered on a central mound, chelicerae that are fused at their bases, and an abdomen often covered with dense, velvety setae. Additionally, the filistatids exhibit tibia-patella autospasy, an uncommon defensive condition in which the leg separates from the rest of the body at the tibia-patella joint when grabbed. They are secretive spiders, and often the only sign of their presence is an open funnel of lacy gray cribellate silk extending from a narrow cranny.

SIMILAR FAMILIES: Very few araneomorphs have eight eyes centrally clustered as the crevice weavers do. The wall spiders (Oecobiidae, p. 85) might be mistaken for one of the smaller genera, but they are entelegyne and have a distinctive anal tubercle. *Calponia harrisonfordi* (Caponiidae, p. 56) has a similar eye arrangement but is colored differently than any filistatid and is ecribellate. Webs made by the ecribellate tube web weavers (Segestriidae, p. 54) might be mistaken for those made by the crevice weavers, but the entrance to a tube web weaver's web is paler, with cleanly radiating threads, in contrast to the grayer, messier, cribellate silk of the filistatid's retreat.

PACIFIC COAST FAUNA: Three genera with five described and several undescribed regional species. The genus *Kukulcania* (pl. 5) is the most widespread and conspicuous of the Pacific coast's crevice weavers, and it houses all five of the region's described species of Filistatidae. *Kukulcania* are found throughout California and are common around homes, in rock piles, in thick leaf litter, and under loose bark. They range from 5 to 18 mm (.2 to .71 in.) in length and exhibit an exceptional degree of sexual dimorphism. Compared with the males, females are larger, darker, and stockier with proportionately shorter legs. The adult male has exceptionally long, thin legs, a small abdomen, and long palps and is usually pale yellow to light brown in color. In both sexes, the carapace narrows considerably along its anterior margin, and the spider's eight eyes are arranged in two adjacent clusters of four, slightly divided along the midline. Common throughout the southern United States and Central and South America, the Southern House Spider *(K. hibernalis)* has only been collected a few times in California, presumably originating as an unnoticed stowaway in shipped cargo. *K. geophila* is both the smallest and most widespread of the Pacific coast species. It is found across California, from San Diego County north to Shasta County. A slightly larger, darker subspecies, *K. g. wawona*, has been described from Yosemite National Park, Mariposa County, California. Of the three remaining regional *Kukulcania*, *K. arizonica* and *K. utahana* are found throughout Southern and Central California, while *K. hurca* has only been reported from Imperial County, California. Brief descriptions and illustrations of the endemic western species (and one subspecies) can be found in Chamberlin and Ivie (1935, 1942b).

Several unidentified species in the genera *Filistatinella* and *Filistatoides* (fig. 7) have also been found in California. They are small spiders, less than 4 mm (.16 in.) in length. *Filistatinella* are dark brown to black, have a distinctly circular carapace, and have been collected from the lowlands and coastal areas of Southern and Central California. *Filistatoides* are mainly tropical but several specimens have been collected in Southern California. They are orange brown with purplish chevrons along the abdomen and have an elongate, pointed carapace with dark median stripes.

NATURAL HISTORY: Very little has been written regarding the natural history of the western crevice weavers. They're common around homes and other buildings as well as in leaf litter, under old logs and rocks, and behind peeling bark. Members

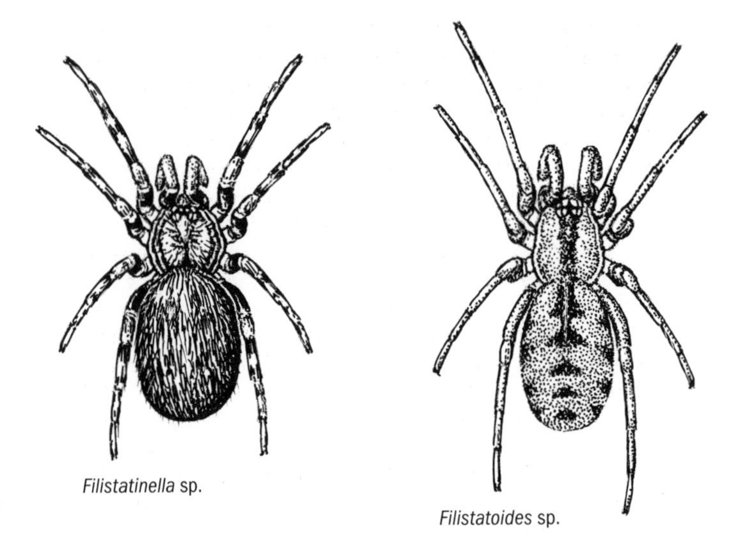

Filistatinella sp.

Filistatoides sp.

Figure 7. Small Filistatidae spiders (females).

of the genus *Kukulcania* build rather messy-looking webs of cribellate silk that radiate out from a protected retreat, either expanding like a funnel or lying like a disk, close to the substrate. Extensive networks of signal and capture lines inform the spiders about any potential prey entangled in the mesh of their webs.

Crevice weavers are comparatively long-lived spiders, taking several years to reach adulthood. Females continue to molt after reaching maturity, a feature believed unique among araneomorphs. After their final molt, sexually mature *Kukulcania* males leave their webs in search of receptive females. Almost nothing is known regarding the courtship behaviors of North America's crevice weavers. *Kukulcania* egg sacs are loosely wrapped bundles that the females guard deep in their retreats.

SEGESTRIIDAE — Tube Web Weavers

Pl. 5

IDENTIFICATION: The entrances to their webs are all most people ever see of segestriids. These six-eyed haplogyne spiders build white, wide-mouthed tube webs in protected nooks, including cracks in walls and under peeling bark. Tube web spiders have a distinctive leg arrangement in which legs I through III are directed forward while leg IV points backward. The posterior eye row is either straight or slightly recurved, and the abdomen is fairly long and cylindrical.

SIMILAR FAMILIES: While several families, including Plectreuridae (p. 70) and Filistatidae (p. 52), make tube webs, none of them have the unusual segestriid leg arrangement and they all have eight eyes. Additionally, filistatid webs are made of grayish cribellate silk, giving them a messier, lacier appearance.

PACIFIC COAST FAUNA: Two genera with five regional species. *Segestria* (pl. 5) is the most widespread of the Pacific coast genera, with three resident species. They have fairly long legs and a boldly patterned abdomen whose light background is marked by an irregular purplish brown central stripe. *Segestria pacifica* is found along the coast from Baja California, Mexico, north to British Columbia, Canada. The two additional *Segestria* species are endemic to coastal California. *S. cruzana* is known from Santa Cruz Island, part of the Channel Islands National Park, while *S. bella* has only been reported from a few locations along the coast of both Southern and Northern California. The only descriptive work done on North America's *Segestria* is the rather skeletal descriptions in Chamberlin and Ivie (1935). The genus would benefit greatly from a modern revision.

The region's second genus, *Ariadna* (pl. 5) is shorter legged than *Segestria* and has a uniformly purplish-green to brown abdomen. *Ariadna bicolor* is widespread across the eastern United States, but along the Pacific coast, it is limited to California's southernmost counties (Riverside, San Diego, and San Bernardino). *A. fidicina* is known from coastal and near-coastal areas from northern Baja California, Mexico, north to Monterey County, California. It was originally placed in a separate genus, *Citharoceps,* based on an unusual set of stridulatory files on either side of its carapace, a feature missing from *A. bicolor.* A detailed revision of the American *Ariadna* can be found in Beatty (1970).

NATURAL HISTORY: Aggregations of segestriids are common in the crevices of rock walls and cliffs, under the loose bark of trees, and beneath protected ledges. When the entire web is visible, it looks like a short, flair-lipped sock. Due to their limited dispersal abilities (like other haplogynes, segestriids don't balloon as juveniles) and the often concentrated nature of their preferred microhabitats, they regularly form dense aggregations. Along California's central coast, *A. fidicina* has done an exceptional job colonizing the large, dense stands of introduced eucalyptus common to the region.

The tube web weaver is a nocturnal hunter. It waits at the web's mouth for potential prey, monitoring the radiating signal threads with its three pairs of forward-pointing legs. When a thread is tripped, the spider lunges out, uses its heavily spined fore legs to grab the prey, and drags it back into the web retreat for feeding. In captivity, tube web weavers eat a wide variety of insects and presumably maintain such catholic tastes in the wild.

While Segestriidae courtship behaviors have yet to be described, *Segestria* mating behaviors have been recorded. The male uses his fore legs to push the female's legs up and slides himself under her carapace. He then grips the front of her abdomen with his fangs, and in this position mating occurs. Instead of building a traditional egg sac, a *Segestria* female lays her eggs in a mass at the rear of her web. Over time, she lays down a loose mesh of threads that ties them together. Adult males are rare in collections, implying that they live only a short time after maturing and mating. In contrast, adult females have been collected nearly year-round and can survive for several years in captivity, strongly suggesting that they are capable of producing multiple broods over the course of their lifetimes.

CAPONIIDAE

Pl. 6, Fig. 8

IDENTIFICATION: With the exception of *Calponia harrisonfordi,* an eight-eyed species from California's central Coast Ranges, caponiids are unique among North American spiders because they have only two eyes. They are small (2.7 to 5.4 mm, or .11 to .21 in., in length) and haplogyne and lack book lungs. Additionally, they have an especially hairy abdomen, and their posterior median spinnerets are in a transverse row with their anterior lateral spinnerets.

SIMILAR FAMILIES: With only two eyes, spiders in the genera *Orthonops* and *Tarsonops* are immediately recognizable. With

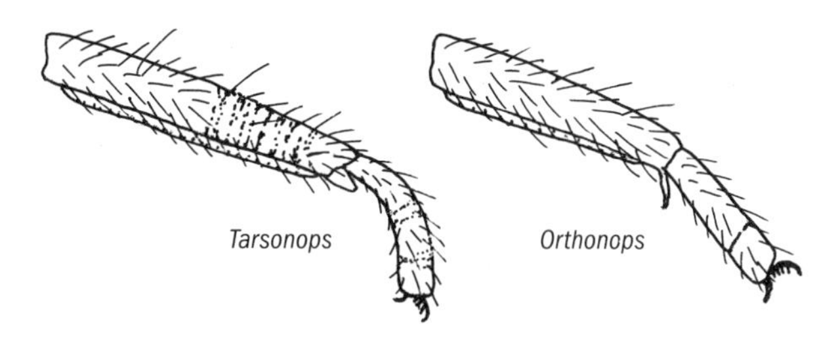

Figure 8. Structure of metatarsus and tarsus in two Caponiidae genera, *Tarsonops* and *Orthonops*.

its eight tightly clustered eyes, *C. harrisonfordi,* however, can present an identification challenge. There are several genera in the family Corinnidae (p. 146) that might be mistaken for *C. harrisonfordi* (specifically *Meriola* and *Trachelas*), but their eyes are aligned across the carapace in two distinct rows. Prodidomidae (p. 155) is a rare family whose members' eight eyes are arranged in nearly circular, procurved patterns, very different from the *C. harrisonfordi* tightly bound ocular cluster.

PACIFIC COAST FAUNA: Three genera with eight Pacific coast species. In addition to their unique eye arrangement, *Orthonops* and *Tarsonops* (pl. 6, fig. 8) have distinctive longitudinal keels and spurs on the ventral sides of metatarsi I and II and their tarsi are subsegmented. *Orthonops* is the most diverse of the Caponiidae genera, with six regional species ranging from Southern California north through the foothills and mountains of the Sierra Nevada into Central California. Ranging from 2.7 to 5.4 mm (.11 to .21 in.) in length, most species have an orange carapace, a pale gray abdomen, and dark eyes enclosed in a small, black patch. The one notable exception is *O. zebra* (pl. 6), a species endemic to the hills and canyons of Orange County and western Riverside County. Most *O. zebra* have a series of dark stripes across the abdomen, although rare individuals have been found that lack dark abdominal bands and pigmentation around the eyes. Leg I on *Orthonops* is rather stout, with unsegmented metatarsi. In his revision of *Orthonops,* Platnick (1995) provides a distribution map to the species and a key to their identification.

Tarsonops systematicus (pl. 6) is the only species in the genus known from the Pacific coast states. While it has never formally been reported from California, it is known from northern Mexico, and specimens have been collected as far north as Fresno County and Kings County (D. Ubick, pers. comm., 2009). It is around 3.0 mm (.12 in.) long and similar in shape to a small *Orthonops* but has a pale yellow carapace and exceptionally long, slender fore legs with subsegmented metatarsi. The tarsal subsegmentation is more extensive in *Tarsonops*, allowing the tarsi to more conspicuously bend. This species is described in greater detail in Chamberlin (1924).

The third regional genus is represented by a single species, *Calponia harrisonfordi* (pl. 6). Around 5 mm (.2 in.) in length, it retains the family's eight-eyed ancestral state and lacks a keel on its anterior metatarsi. Its eyes are in a tight cluster, and while the anterior medians are large and dark, the others are small and pale. Its carapace and legs are reddish orange and its abdomen is light brown. *C. harrisonfordi* is known only from the central Coast Ranges of California, with records from San Benito County to Mendocino County. This genus and species were described by Platnick (1993).

NATURAL HISTORY: Caponiids are cursorial spiders most commonly found under rocks, logs, and leaf litter and in wood rat lodges. Little is known about their natural history, although both *Orthonops* and *Tarsonops* are found mainly in arid areas. What information is available suggests that *Orthonops* are most common on warm, south-facing slopes; however, *O. giulianii* is more of a mountain forest species, living in the southern Sierra Nevada and White Mountains. *C. harrisonfordi* lives in the leaf litter of oak and bay woodlands.

Both *Orthonops* and *Calponia* have stridulatory ridges on the surfaces of their chelicerae and stridulatory picks on their palpal femora; both of these structures are used by other spiders to stridulate during courtship. The simultaneous presence of both large juvenile and adult *O. zebra* suggests that this species has at least a two-year life cycle. The egg sacs of *O. zebra* are lens shaped, wrapped in a pink papery silk, and camouflaged with dirt and other debris, while those of *C. harrisonfordi* are conical and brown. In captivity, caponiids will eat a variety of small prey items, but in the wild, at least some genera apparently specialize in hunting other spiders.

OONOPIDAE **Goblin Spiders**

Pl. 6

IDENTIFICATION: Goblin spiders are small (1 to 3 mm, or .04 to .12 in.), six-eyed, haplogyne spiders. They range in color from chalky white to dishwater brown, although the introduced *Opopaea concolor* (pl. 6) is deep reddish orange. Many species have scuta either partially or fully covering their abdomens. Their eyes are large and conspicuous, either with the posterior medians aligned with the anterior laterals (as in *Orchestina*) or in a tight cluster, with the posterior medians close to the posterior laterals (as in all other North American genera). Oonopids also have a rather unusual gait, moving in short, jerky bursts, or even hopping in the case of *Orchestina*.

SIMILAR FAMILIES: Oonopids are reasonably distinctive members of the Pacific coast's araneofauna. Although other families include small, six-eyed spiders, none have abdominal scuta. Those species of Oonopidae without scuta may be confused with members of the family Telemidae (p. 79), but goblin spiders have shorter legs, never build sheet webs, and lack sclerotized abdominal ridges.

PACIFIC COAST FAUNA: Four genera with seven described and numerous undescribed regional species. Goblin spiders in the genus *Orchestina* (pl. 6) lack scuta, have a globular abdomen, and have a distinctive eye arrangement. The posterior median eyes are in line with the anterior laterals. The fourth femur is also noticeably enlarged, a unique feature among Oonopidae that allows them to hop when disturbed. They are rather dull colored, ranging from chalky gray to brown with a purplish gray abdomen. Two native species have been described from California: *O. obscura,* from the Yosemite National Park region, and *O. moaba,* which has been collected in near-coastal leaf litter from San Diego County to Monterey County. Published records for these species show a range from 1.2 to 1.7 mm (.05 to .07 in.) in length. A third species, *O. saltitans,* native to the northeastern United States, is established in Seattle, Washington (R. Crawford, pers. comm., 2009). Additional unidentified species are known to live in both Southern California and eastern Washington (W. Icenogle, R. Crawford, pers. comm., 2009). Descriptions of the native Pacific coast *Orchestina* species can be found in Chamberlin and Ivie (1935, 1942b).

Three species of *Escaphiella* (pl. 6) are found along the Pacific coast of the United States. Females are unique in that their ventral scuta wrap around the sides of the abdomen, leaving an unprotected bare stripe along the dorsal midline. Males have a more traditional pattern, with dorsal and ventral scuta almost completely shielding the abdomen. Adult males can be separated from the other regional scute-carrying oonopid genus, *Opopaea,* by the presence of normal-size palpal patellae. *E. hespera* is dusky red to yellowish brown and has been collected under rocks and leaf litter from the chaparral and deserts of Southern California through the Coast Ranges into Sonoma County and in the Sierra foothills to Butte County (D. Ubick, pers. comm., 2009; Platnick and Dupérré, 2009a). *E. litoris* is yellowish brown with pale yellow legs and has only been regionally recorded from the deserts of eastern San Diego County, California. *E. nye* is known only from the male. Structurally similar to both *E. hespera* and *E. litoris,* a close examination of its palps is required to confirm its identity. It is known from only two locations: Nye County, Nevada, and the Inyo Mountains of Inyo County, California. More-detailed descriptions of the region's *Escaphiella* can be found in Platnick and Dupérré (2009a).

Opopaea concolor (pl. 6) is a rare, vibrantly colored, introduced species and is the only regional representative of its genus. Along the Pacific coast, this tiny vermillion spider is known only from Southern California. It has been collected in coastal scrub leaf litter (San Diego County), in termite-infested wood (Los Angeles County), and in citrus and eucalyptus groves (Riverside County) (Platnick and Dupérré, 2009b). Both sexes have ventral and dorsal scuta. On males, the palpal patellae are massive, obviously larger than the palpal femora, and the female has numerous fine striations along the lateral margins of the carapace. *Opopaea* is native to the Old World, but *O. concolor* is synanthropic and has been introduced to tropical and subtropical areas around the world. Its presence in California was confirmed by Prentice et al. (1998) and was later reviewed as part of a larger revision by Platnick and Dupérré (2009b).

There are at least two undescribed members of the genus *Oonops* (not illustrated) in Southern California. These tiny, pinkish-orange goblin spiders lack both scuta and enlarged fourth femora. They possess four pairs of ventral spines on tibiae I, and their palpal bulbs are structurally distinct from their palpal tarsi. Additional undescribed species have also been collected in California, including *Oonops*-like individuals, probably representing a new genus, that lack ventral tibial spines and have palpal bulbs that are fused to the palpal tarsi (Ubick, 2005e).

NATURAL HISTORY: North America's goblin spiders have received very little attention from researchers regarding their ecology, development, or behavior. While widespread, they are rarely collected, because of their small size and inconspicuous habits. Most are found in protected debris, including leaf litter, beneath rocks and logs, and in wood rat nests. The introduced population of *Orchestina saltitans* in Seattle appears to be restricted to the insides of homes and other buildings. Oonopidae don't build webs for hunting but do spin small silken cocoons to use as molting chambers.

For many years, our knowledge of Oonopidae distribution and systematics has been in a severe state of disarray. Most regional goblin spiders have been scarcely studied past their initial descriptions, many of which date back to the 1920s and 30s. In response to this challenge, the American Museum of Natural History in conjunction with nearly a dozen institutions worldwide initiated a Planetary Biodiversity Inventory of the world's goblin spiders. Using morphological, genetic, and ecological data, their goal is to completely revise Oonopidae and, in the process, create an Internet-accessible database with photograph-supported identification keys and detailed phylogenetic and distributional information.

DYSDERIDAE — Wood Louse Hunters

Pl. 7

IDENTIFICATION: Dysderidae is a haplogyne family represented in North America by a single introduced species, the Wood Louse Hunter, *Dysdera crocata* (pl. 7). It is a distinctive, medium-size (9 to 15 mm, or .35 to .59 in.) spider with a reddish-orange carapace and legs and an unmarked grayish-yellow abdomen. Its six eyes form a procurved arch, and its jaws are forward pointing, exceptionally large, and readily displayed when the spider is disturbed.

SIMILAR FAMILIES: Members of the family Caponiidae (p. 59) and several genera in the family Corinnidae (p. 146) share the

Wood Louse Hunter's general carapace and abdominal color pattern, but neither shares its unusual eye arrangement or its enlarged jaws. Members of the closely related family Segestriidae (p. 54) also have six eyes (in a different arrangement) but lack the Wood Louse Hunter's strongly contrasting colors. Additionally, segestriids live in tube webs, and their first three pairs of legs are all anteriorly directed.

PACIFIC COAST FAUNA: The Wood Louse Hunter (pl. 7) is fairly common in suburban and residential areas throughout California and Oregon. Less abundant in Washington, it was first recorded in the southern part of the state in 1986 and is now found as far north as Seattle (R. Crawford, pers. comm., 2008). It lives in gardens, under sod, and in other dark, moist places. *D. crocata* is highly synanthropic, and while its specific point of origin is unclear, closely related species are found along the southern Mediterranean and in North Africa.

NATURAL HISTORY: In captivity, the Wood Louse Hunter is capable of eating a variety of small arthropods; however, it specializes in hunting wood lice (terrestrial crustaceans also known as pill bugs or sow bugs). Its long jaws and chelicerae make it one of the few spiders capable of piercing the wood louse's tough exoskeleton. It is a nocturnal hunter, wandering through roots and detritus in pursuit of its prey. These cursorial habits occasionally bring it into homes, where its striking coloration and immense jaws can cause alarm. When disturbed, it often responds by rearing back with its fangs spread, rather than retreating. Fortunately the bite of a Wood Louse Hunter isn't dangerous to humans, and normally the only detrimental effect is a minor pain caused by the spider's jaws breaking the skin (Vetter and Isbister, 2006).

Courtship is a rather rapid affair, based almost entirely on tactile (and possibly chemical) signals. The male uses his front legs to caress and gently tap on the female's fore legs while holding her jaws open with his own. If the female isn't interested, she will lower her abdomen against the ground and back away from the male. If she is especially averse to mating, she may become aggressive, attacking, or even killing, the smaller male. When laying her eggs, female *D. crocata* makes a tightly woven cocoon on the underside of a rock and seals herself in with the spherical egg sac. After three to four weeks the eggs hatch, but the young stay within the cocoon for an additional month before dispersing. It takes about 18 months for the juvenile Wood Louse Hunters to reach sexual maturity, and as adults they may live for an additional two to three years.

In their native Mediterranean range, Dysderidae are often concentrated in areas with mineral-rich soils. It has been proposed by Rezác et al. (2007) that Wood Louse Hunters, while synanthropic, are not directly dependent on human dwellings. Instead, they are the indirect beneficiaries of urban soil's high calcium content. Sidewalks, building foundations, and other concrete structures leach calcium into the surrounding soils, which attracts wood lice that require calcium to grow their sturdy exoskeletons.

TROGLORAPTORIDAE

Pl. 7

IDENTIFICATION: Trogloraptorids are long-legged, six-eyed, haplogyne spiders endemic to a small region of caves and old-growth redwood forests in southwestern Oregon and northwestern California. Their most distinctive feature is their subsegmented, raptorial tarsi. Trogloraptoridae are able to fold their large claws against their heavily spined tarsi, creating a grasping structure that is presumably used for holding prey. Additionally, on trogloraptorids, the anterior and posterior lateral eyes are connected, while their posterior medians are separated. There is a membranous band connecting the anterior lateral spinnerets, and there is an unusual double row of teeth on the serrula.

SIMILAR FAMILIES: With their raptorial claws, eye and spinneret arrangements, secretive nature, and limited distribution, members of the family Trogloraptoridae are unlikely to be confused with any other North American spider.

PACIFIC COAST FAUNA: One genus representing a single described regional species, the cave-dwelling *Trogloraptor marchingtoni* (pl. 7). Ranging from 6.9 to 9.7 mm (.27 to .38 in.) in length, its exceptionally long legs can make it look significantly larger. It is pale orange with darker chelicerae and a purplish-gray abdomen that is decorated with several pale chevrons. It is only known from a small collection of caves in Josephine County, Oregon. Several juvenile trogloraptorids, likely representing a second, undescribed species, have also been found under redwood debris in Del Norte County, California. Both

Trogloraptoridae and *T. marchingtoni* are described in greater detail in Griswold et al. (2012).

NATURAL HISTORY: Very little is known about the life history of *T. marchingtoni*. It lives in the dark zone of caves where it hangs from the ceiling by a few strands of silk. Nothing is known regarding its feeding habits or courtship behaviors, and despite multiple searches, no individuals were found among the boulders and rotting logs at the mouth of one inhabited cave.

SCYTODIDAE — Spitting Spiders

Pl. 7

IDENTIFICATION: With their smooth, dome-like carapaces and unusual eye arrangement, spitting spiders are unmistakable members of North America's spider fauna. They are haplogyne, are between 3.5 and 10 mm (.14 and .39 in.) in length, and have six eyes aligned in three dyads forming a recurved arc. Spitting spiders range in color from pale yellow with grayish speckling to nearly black.

SIMILAR FAMILIES: The recluse spiders (Sicariidae, p. 65) share a number of features with the spitting spiders, including their eye arrangement, overall size, often pale coloration, and a penchant for living under rocks, in homes, and in other structures. Sicariids, however, have exceptionally flat carapaces with obvious longitudinal thoracic grooves.

PACIFIC COAST FAUNA: One genus, *Scytodes* (pl. 7), with two recognized and several undescribed or unidentified species. *Scytodes univittata* (previously known as *S. perfecta*) was introduced to the Americas (Rheims et al., 2007) and has been reported from Southern California and into the Central Valley as far north as Fresno County. It is a small yellow-and-gray spider, 3.5 to 5.5 mm (.14 to .22 in.) in length, with banded legs. The second described California species, *S. thoracica*, was presumably introduced from Europe and is now widespread throughout the eastern United States. There is at least one record from Chico, Butte County, California. However, with its synanthropic habits, it's quite possibly more common than currently recognized. Similar in size to *S. univittata*, it has a darker, more distinctly patterned carapace. Additional spitting spiders, some clearly neither *S. univittata* nor *S. thoracica*, have been found in both rural and urban areas of California from San Diego

County north to San Jose (Santa Clara County) (Prentice et al., 1998; W. Icenogle, D. Ubick, pers. comm., 2009). Whether these spiders are undescribed species, or ones that were described from elsewhere but with unrecognized introduced California populations, is uncertain. Because of their size and often synanthropic habits, spitting spiders are easily transported in luggage and freight, explaining the appearance of at least some of these predominately tropical spiders along the Pacific coast.

NATURAL HISTORY: Spitting spiders' predatory behaviors encompass a fascinating array of behavioral and morphological adaptations. Encased within their dome-like carapaces are massive, partially divided venom glands. The anterior portion of the gland secretes the spider's venom, and the posterior section produces a gluey silk. Scytodids are not visual hunters, depending mainly on touch and vibration to locate their prey. Some species build small tangle webs while others are wandering hunters, but no matter how their prey is found, the spiders' actions are fairly consistent. Pulling its body slightly up and back, a spitting spider immobilizes its prey with a zigzag-shaped spray of glue and venom from its chelicerae. This toxic, mucilaginous net both pins the prey to the substrate and initiates the paralyzing process (J. Rovner, pers. comm., 2009). The spider can then safely approach its captured prey and, biting it on the leg, further hasten its demise. The spider may also use its fore legs to roll the prey back and forth, increasingly entangling it in the sticky mesh. The spider then cuts its prey from the webbing and wraps it with silk from its spinnerets before feeding, either at the capture site or in the safety of its retreat. Scytodids also use spitting as a defensive measure, spraying their sticky venom at potential predators.

Neither the courtship nor the mating behaviors of the region's spitting spiders have been described. Like the more familiar cellar spiders (Pholcidae, p. 73), female spitting spiders wrap their eggs in a few strands of silk and carry them in their chelicerae until they hatch.

SICARIIDAE — Violin Spiders, Recluse Spiders

Pl. 7

IDENTIFICATION: While many people call any cursorial tan or brown spider a recluse, there are a few key features that you must see to confidently distinguish recluses from other, superficially similar spiders; the most important of these is their eye arrangement. Violin spiders have only six eyes arranged as three pairs in a recurved row. They are fairly small (5 to 13 mm, or .20 to .51 in., in length). They have long, thin legs without spines, only two tarsal claws, and an exceptionally large, fleshy colulus. They range from pale yellow to brownish gray or reddish brown in color, and the dusky violin-shaped cephalic mark that gives this family its common name can be conspicuous on some species but pale to absent on others.

SIMILAR FAMILIES: Violin spiders may be mistaken for the more heavily marked spitting spiders (Scytodidae, p. 64). The two families have similar eye arrangements, but spitting spiders have distinctive dome-like carapaces that house their enlarged glue and venom glands, and they are strongly patterned on both body and legs.

PACIFIC COAST FAUNA: One genus, *Loxosceles* (pl. 7), containing seven regional species, three of which are nonnative and of very rare or localized occurrence. The most widespread of the native species is the Desert Recluse *(L. deserta,* previously known as *L. unicolor),* a yellowish to tan spider with a faint or absent cephalic mark. It is commonly found in California's Mojave and Colorado Deserts north into the foothills of the southern San Joaquin Valley. Other native species include *L. russelli* from the Death Valley region, *L. martha* from the Palm Springs area, and *L. palma,* which has been collected around palm canyons and desert oases from Southern California into northern Baja California, Mexico. These spiders are very similar, so a detailed examination of their reproductive structures is often required to confirm their identification. Detailed information on North America's native violin spiders, including keys to their species-level identification, can be found in Gertsch (1958d) and Gertsch and Ennik (1983).

L. laeta is an introduced species from western South America that is established in a small region of Los Angeles, California, predominately in the Monterey Park, Alhambra, and Sierra Madre neighborhoods (Waldron et al., 1975). It lives deep underground in steam tunnels and basements and is rarely encountered by the general public. *L. laeta* is dark yellow to reddish brown with a series of striations along its cephalic region and has comparatively long palpal femora and tibiae.

Two other nonnative species have been recorded on very rare occasions along the Pacific coast. The Mediterranean Recluse, *L. rufescens,* has been collected worldwide and is a sporadic vagabond on cargo ships and in shipping containers. Although there are no established populations on the West Coast, exceptionally localized infestations have occurred, and individual specimens have been found in a number of larger western cities, including Fresno and Los Angeles in California (R. Vetter, pers. comm., 2008) and in Seattle and Spokane in Washington (G. Binford, pers. comm., 2008). The least common of the region's violin spiders is the Brown Recluse *(L. reclusa).* Despite popular belief, the Brown Recluse does not live on the Pacific coast. On fewer than a dozen occasions, this species has been transported from its core range in the south-central United States in packages and freight but has never established a viable regional population. There has also never been a confirmed Brown Recluse bite in any West Coast state. Unfortunately, the myth of the Brown Recluse is perpetuated through media and medical misinformation when any small necrotic wound is regularly, if falsely, attributed to this species. There are numerous causative agents (including bites of other spiders or ticks; bacterial, fungal, or viral infections; chemical burns; and pressure ulcers) that can result in similar injuries and either have been mistaken for Brown Recluse bites or have the serious potential for misdiagnosis (Vetter, 2008). Although the cytotoxic effects of *L. reclusa* and *L. laeta* venom have been well documented, violin spiders are quite shy, and people in the Brown Recluse's natural range have lived in homes with large infestations for years with no ill effects (Vetter and Barger, 2002). Even in those Los Angeles neighborhoods that have housed populations of *L. laeta* for at least 40 years, there has yet to be a confirmed case of envenomation from a *Loxosceles* bite.
NATURAL HISTORY: The native violin spiders are cursorial, nocturnal hunters commonly found in dry, rocky desert washes, on talus slopes, and at cave and mine openings. They are occasionally associated with wood rat middens and can be quite common around dilapidated buildings with lots of cracks and debris. Violin spiders build messy webs in small crevices that they line with strands of a microscopically unique, ribbonlike silk. They are generalist predators, feeding on small arthropods, including other recluses. They will both wander

at night in search of food and take any prey items that become entangled in their webs. Based on laboratory observations of the Desert Recluse, females normally drag their prey items back into their retreats before feeding, whereas the males often feed at the capture sites (Ennik, 1971).

Courtship behaviors have been observed in several species. In each case, it begins with the male using his front tarsi to lightly stroke the female's body and legs. The male then drums on the webbing with his palps and front legs. If receptive, the female responds with a similar drumming behavior, along with leg quivering and jerky body motions, seen in both sexes. Copulation occurs when the female allows the male to elevate her first and second pairs of legs with his fore legs. After mating, the female normally responds to additional male advances with aggression or by holding her body flat against the ground.

A *Loxosceles* egg sac has a dense, circular base and is capped by a thin silk sheet and hidden in a secluded area near the female's retreat. She weaves a canopy over the egg sac, creating a brood cell where she stays until her spiderlings have emerged. Before the offspring disperse, the female Desert Recluse will occasionally allow her early instar offspring to feed alongside her on captured prey.

DIGUETIDAE — Desertshrub Spiders

Pl. 8

IDENTIFICATION: In California's deserts and chaparral-covered foothills, a desertshrub spider's web is structurally unique. It consists of a low-slung sheet surrounded by a cobweb-like tangle with a conical retreat suspended in the center. Desertshrub spiders are haplogynes with fairly flat, oval carapaces and six eyes arranged in three dyads. The abdomen can be either white or patterned, and all of the regional species have boldly banded legs. Additionally, diguetids have three tarsal claws and chelicerae fused at their bases. While each of these anatomical features can be found in other spider families, in combination they are diagnostic of the desertshrub spiders.

SIMILAR FAMILIES: Spiders in the family Plectreuridae (p. 70) are morphologically similar to diguetids and share their preference for deserts and dry scrublands. Plectreurids, however, are cursorial hunters, never making the elaborate webs that are characteristic of desertshrub spiders. Also, plectreurids have eight eyes arranged in two straight rows.

PACIFIC COAST FAUNA: One genus with five regional species. *Diguetia* (pl. 8) is the only genus in North America, and its members are fairly common throughout the southwestern United States. *Diguetia canities* (pl. 8) is the most widespread of the Pacific coast's species. It can be identified by its abdomen's rounded posterior edge and by the presence of several sinuous bands merging at its anterior margin. *D. albolineata* is similarly marked, but the posterior edge of its abdomen has a well-defined conical tubercle. Both of these species are found across Southern California, with records of *D. canities* from as far north as Tulare County. While the published accounts of *D. albolineata* note it only from Southern California, specimens have been collected well into the dry interior hills of San Benito County in Central California (Essig Museum of Entomology, University of California at Berkeley). The three remaining species are all known only from Southern California's deserts and chaparral. Gertsch (1958a) revised the family and provides a key to the known species of *Diguetia*.

NATURAL HISTORY: Fairly common in Southern California, desertshrub spiders become progressively less abundant north through the Sierra foothills and in the dry eastern portions of the Coast Ranges. The Diguetia web consists of several sections, the most noticeable of which is a conical retreat made of dried plant material, insect parts, and old exuviae suspended in its center. A horizontal sheet spreads from the entrance of the retreat, creating a platform that the spider traverses from below. Both the sheet web and the retreat are surrounded by a mesh of tangle threads, all of which are connected to the surrounding vegetation by a series of exceptionally strong support lines. When an insect either is ensnared or tumbles onto the sheet web, the spider bites it and waits for the venom to take effect. Once the prey is immobilized, the desertshrub spider often wraps it in a few strands of silk, either to hold it in place or to pull the insect's wings closer against its body. Diguetia feeding behavior is thought to be similar to that of the earliest makers of aerial webs as they progressed from depending solely on venom for prey capture to using silk and venom together. This technique secures the bitten prey before it has time to escape and frees up the fangs for multiple bites.

Little is known regarding the specifics of *Diguetia* courtship, although both males and females possess stridulatory files on their chelicerae and have rows of small tubercles on their palpal femora. These structures are used by many spiders to create sounds during courtship. When harassed, females will stridulate in short bursts, creating noises loud enough to be heard by sensitive human ears. Over a period of several weeks, the eggs are laid in a series of three to 10 discus-shaped egg sacs that are stacked like shingles inside the female's retreat. Each sac is around 8 mm (.31 in.) in diameter and contains between 150 and 250 small, white eggs. At least one species of *Diguetia* is also an unwilling partner in one of the very few known examples of brood parasitism in arachnids. The female *D. mojavea* (pl. 8) begins building egg sacs in late summer and early fall. While the female is present, her eggs are relatively safe. Once she dies, however, two different species of desert jumping spider, *Habronattus tranquillus* and *Metaphidippus mannii* (Salticidae, p. 167), readily invade her retreat and lay their own eggs among the desertshrub spider's. The jumping spider's eggs hatch first and are distinctly larger than their desertshrub spider nest mates. Throughout winter and early spring, the adult and immature jumping spiders live in the retreat, sheltered from the environment and feeding on *Diguetia* juveniles and eggs. To lessen the impact of jumping spider predation on her young, a *D. mojavea* female lays over a thousand eggs across an extended period of time, increasing the odds that at least a few will survive. This behavior also ensures that the hatching time for the different egg sacs is asynchronous, which can be useful in the desert, where environmental conditions can change quickly and may favor the survival of one cohort of siblings over another.

PLECTREURIDAE

Pl. 8

IDENTIFICATION: Members of the family Plectreuridae are small to medium-size (5 to 12 mm, or .20 to .47 in., in length), eight-eyed, haplogyne spiders. Their carapaces and legs range from reddish orange to black, and their abdomens vary from yellowish brown to black, often with a pale heart mark on their dorsal surface. As many of the diagnostic features of this family are quite small, a detailed examination may be required to confirm familial identity of a plectreurid. Fortunately, there are only two genera in the region, and once you become familiar with the resident species, recognition of them becomes more straightforward.

While plectreurids are most abundant in the arid portions of the southwestern United States and northern Mexico, a few species have expanded north into Central California's oak and coniferous forests and through the Great Basin into south-central Washington. They make an inconspicuous tangle web among rocks, cactus, and low-growing shrubs, often with a tubelike retreat extending back into a small crevice.

SIMILAR FAMILIES: Plectreuridae is most closely related to Diguetidae (p. 68); however, the families differ in several significant ways. Diguetidae build complex aerial webs, are of different colors than Plectreuridae, and have only six eyes. Female *Kukulcania* (Filistatidae, p. 52) are superficially similar but are cribellate, and their eight eyes are gathered together in a central mound. *Calponia harrisonfordi* (Caponiidae, p. 56) of the Central California coast is also differently colored, and its eight eyes are also clustered together, not spread across its carapace. A large and dark female in the plectreurid genus *Kibramoa* can occasionally be mistaken for a female Western Black Widow (*Latrodectus hesperus,* Theridiidae, p. 95), but *Kibramoa* lack the diagnostic red hourglass mark on their undersides.

PACIFIC COAST FAUNA: Two genera with 12 regional species. The more diverse of the plectreurid genera is *Plectreurys* (pl. 8) with nine species known from Southern California to south-central Washington. These spiders have fairly short, stocky legs, and femur I is distinctly curved, is visibly thicker than femur II, and for the most part lacks dorsal spines. If dorsal spines are present on femur I, they are limited to its base. Additionally, adult males have large clasping spurs or spines at the terminal end of tibia I. Most *Plectreurys* are limited to the southwestern deserts and chaparral, but *P. castanea* is found throughout the southern Sierra Nevada and Coast Ranges and has even been collected on Santa Barbara Island, part of California's Channel Islands National Park. Another widespread species, *P. tristis,* is found in a variety of habitats, from southern deserts to cool mountain forests. It has been recorded from the Sierra foothills north into Central California. It has also been found sparingly throughout the Great Basin, with a small, apparently isolated population in Kittitas County, Washington (R. Crawford, pers. comm., 2009). All *Plectreurys* are quite similar in appearance, with a chestnut-brown or black carapace and legs and a dark-gray to dusky-orange abdomen. An illustrated key to the *Plectreurys* can be found in Gertsch's (1958c) revision of the family Plectreuridae.

The second regional plectreurid genus is *Kibramoa* (pl. 8), with three Pacific coast species. *Kibramoa* have longer, thinner legs than *Plectreurys,* with numerous spines along the dorsal surface of femur I, which is also relatively straight, lacking the obvious curve found on *Plectreurys.* Additionally, *Kibramoa* males lack a clasping spur on tibia I. *Kibramoa madrona* is the most common species of this genus, with numerous records from Central California's Coast Ranges. In both sexes of *K. madrona,* the carapace and legs range from yellow to orange brown, often with darker femora, and the abdomen is gray. *K. suprenans* is more reddish brown with a dark-gray to black abdomen and has been found across Southern California, both along the coast and inland north to Kern County. The third Pacific coast species, *K. guapa,* is quite similar to *K. suprenans* but has bright-orange or reddish legs and has only been recorded from San Diego County and northern Baja California, Mexico. Gertsch (1958c) revised the family and provides a key to its species.

NATURAL HISTORY: The family Plectreuridae is endemic to the western United States and northern Mexico. While its greatest diversity is in the region's deserts and chaparral, a few species have radiated into the more mesic forests of Southern and Central California, both along the coast and in the Sierra Nevada. *P. tristis* demonstrates the greatest environmental tolerance and can be found in Southern California's deserts, its mountain forests, and throughout the Great Basin's scrublands. With a population in Washington, it's also the northernmost of the region's Plectreuridae.

These are sedentary spiders that build small tangle webs with tubelike retreats under rocks, bark, and other debris. They often hide during the day, but at night they wait at their webs' entrances, ready to capture any insects that stumble in. Upon reaching maturity, adult males leave their webs and, wandering in search of receptive females, occasionally enter houses and other buildings. After mating, female plectreurids guard their eggs from within lacy, cocoon-like brood cells.

PHOLCIDAE — Cellar Spiders

Pls. 9, 10

IDENTIFICATION: Pholcids are small to medium-size spiders with exceptionally long, thin legs, pseudosegmented tarsi, and three tarsal claws. Except for the very small, six-eyed *Spermophora senoculata,* all Pacific coast cellar spiders have eight eyes arranged as two tightly clustered triads with the small, but well-developed, anterior median eyes between them. The abdomens of these spiders range from oblong to globose and vary from pale gray to heavily mottled. On many species, the males have spurs or other sexual modifications to either or both their clypei and chelicerae.

The common name "daddy longlegs" can be confusing. It has been applied not only to spiders in the family Pholcidae (order Araneae), but also to harvestmen (order Opiliones). These relatives of spiders look similar, but on harvestmen the cephalothorax and abdomen are broadly attached, appearing as a single unit.

SIMILAR FAMILIES: The only spiders likely to be confused with cellar spiders are the lampshade weavers (Hypochilidae, p. 49). However, on the West Coast, hypochilids are limited to California's high-elevation montane boulder fields and sheltered recesses. The lampshade weavers are dorsoventrally flattened, and their distinct "lampshade" web is markedly different from the sheets and tangles of cellar spiders' webs. The lampshade weavers also have eight eyes, but theirs are arranged in two clear rows.

PACIFIC COAST FAUNA: Eight genera with 20 regional species. The two most conspicuous of the region's cellar spider species are the introduced Long-bodied Cellar Spider (*Pholcus phalangioides,* pl. 9) and the Marbled Cellar Spider (*Holocnemus pluchei,* pl. 9). Originally from Central Europe, *P. phalangioides* is now found in homes and buildings throughout the region. It is pale yellowish gray with dusky markings on its carapace and oblong abdomen. There's a persistent urban legend that this is one of the most venomous spiders in the world; however, there is no toxicological evidence to support this claim, and there's never been a reported case of detrimental effects from the bite of this spider.

H. pluchei is a Mediterranean native that was first noted in California in the mid-1970s (Porter and Jacob, 1990). It is now widespread throughout the drier, warmer portions of the state north through southern Oregon. The Marbled Cellar Spider is particularly common in the Central Valley, Sierra foothills, and desert regions, but it is easily transported and could turn up nearly anywhere. Its body is brown with gray-and-black patterning on the abdomen, and there is a broad black band running the length of its underside. This species is more common in exposed, outdoor areas such as cattle guards, bushes, and fencerows than the more reclusive Long-bodied Cellar Spider.

There are several additional nonnative Pholcidae with limited Pacific coast distributions. The Tailed Cellar Spider (*Crossopriza lyoni,* pl. 9) is a common synanthropic species in tropical and subtropical areas around the world. It is unique among California's pholcids in that its abdomen is posteriorly pointed, giving it a roughly triangular shape. It's 4.0 to 6.0 mm (0.16 to 0.24 in.) long and has been recorded from several locations in Imperial County, California (W. Icenogle, pers. comm., 2008). Huber et al. (1999) provides an overview of the status and distribution of this species throughout the New World.

One of the largest pholcids in the world is *Artema atlanta* (pl. 10). Often over a centimeter in length, it is widespread throughout the warmer parts of Eurasia and has spread to tropical areas around the world. Small populations of *A. atlanta* have been found in Imperial County and San Diego County in Southern California. Its abdomen is globular and beige with large brown spots, and the male has massive black spurs on the outer edges of his chelicerae. Once fairly common under bridges along the New River south of the Salton Sea, *A. atlanta* appears to be undergoing displacement by the much more numerous Marbled Cellar Spider (R.J. Adams, pers. obs., 2009). Its presence is further discussed in Brignoli (1981).

On the opposite end of the size spectrum is the tiny house spider *Spermophora senoculata* (pl. 10). Native to the Asia-Pacific region, this minute (1 to 2 mm, or .04 to .08 in.) pale gray spider is the only cellar spider on the Pacific coast with only six eyes. It has a globular white to pinkish abdomen and is usually found in closets, garages, and other dark recesses.

S. senoculata has occasionally been reported from both California and Oregon; however, its small size, reclusive nature, and synanthropic habits almost guarantee this spider is more widespread than currently recognized.

Of the three native cellar spider genera, the most diverse is *Psilochorus* (pl. 10) with 12 Pacific coast species. These small (under 4 mm, or .16 in.), round-bodied cellar spiders are common throughout California but show their greatest diversity in the southern half of the state. The *Psilochorus* abdomen is typically bluish gray, and the carapace ranges from buffy white to reddish brown and is often decorated with a dark Y-shaped mark. These spiders are especially common in drier regions, where they make small tangle webs under rocks, logs, and cow droppings. *P. simoni* and *P. hesperus* are the only regional *Psilochorus* species whose distribution extends north of California along the Pacific coast, and only *P. hesperus* is commonly found as far north as Washington. The males in this genus have narrow palpal femora, each of which has a pointed spur on its ventral margin. A detailed revision of this genus, including illustrations of all of the known North American species, can be found in Slowik (2009).

Physocyclus californicus (pl. 10) is the only member of this genus commonly found on the West Coast and has been collected at a number of areas across Southern California. It is similar in appearance to *Psilochorus* but larger (usually longer than 4 mm, or .16 in.), and the male possesses enlarged palpal femora without ventral spurs. A second species, *P. globosus,* is a very rare member of the Pacific coast spider fauna, with a single record from Alameda County, California. Additional information on this genus can be found in Brignoli (1981).

The Short-legged Cellar Spider (*Pholcophora americana,* pl. 10) is the only representative of its genus in the region and is common in moist forests throughout the western United States and southwestern Canada. It ranges from 2.5 to 4 mm (.10 to .16 in.) in length and has a reddish-brown cephalothorax with a black Y-shaped mark across its carapace and a gray abdomen. Although its legs are shorter in relation to its body size than those of other cellar spiders, *P. americana* still possess the thin legs and pseudosegmented tarsi characteristic of the family. Gertsch (1982) summarizes this spider's North American distribution and illustrates the fine details of its anatomy.

NATURAL HISTORY: The native pholcid genera (*Pholcophora, Psilochorus,* and *Physocyclus*) are small spiders commonly found in thick leaf litter, under rocks and logs, in caves and cellars, and in other dark spaces. Their webs are small, messy tangles where the spiders feed on tiny invertebrates, including collembolans, flies, and ants.

The majority of the Pacific coast's cellar spiders are introduced and can be found around homes and other structures. By far the most well known is the Long-bodied Cellar Spider. It is especially common in garages, basements, and other dark places, where it hangs upside down from its loose tangle web. Both this species and the Marbled Cellar Spider are well-documented araneophages and have been recorded eating a wide variety of other spiders (including their own species) as well as harvestmen and assorted insects. Both species have also been observed invading the webs of other spiders, attacking and eating the hosts, their eggs, and any previously captured insects. The Long-bodied Cellar Spider has even been found using aggressive mimicry, plucking the web of another spider in imitation of a prey item. It then attacks the host spider when it closes in to investigate. When subduing entangled prey, both Long-bodied and Marbled Cellar Spiders begin by throwing silk over their targets with their hind legs, only coming in to bite once the prey have been safely immobilized.

During courtship, the male Marbled Cellar Spider signals his presence to a female by performing jerky body movements in the female's web, while the male Long-bodied Cellar Spider has been observed caressing the female's legs and body with his front legs before mating. Nearly all species of cellar spiders have sexually dimorphic spurs, spines, or hairs on either their chelicerae or clypei or both. These structures are extremely important when you are distinguishing closely related species. They are used by the males during copulation. Usually there are grooves or pockets on a female's epigynal plate where the male inserts his cheliceral spurs. This alters the plate's internal structure and improves his chances for successful insemination. Unlike many other spiders whose egg sacs are complex, multilayered structures, cellar spiders make egg sacs that are held together by a loose mesh of silk threads, leaving the eggs (and developing spiderlings) clearly visible. A female cellar spider will carry her single egg sac in her jaws until the eggs hatch. The female Marbled Cellar Spider builds a distinctive, spherical nursery web, where she resides until the young hatch. The newly emerged spiderlings then use the nursery web as a scaffold to which they cling for their first molt.

Several of the larger cellar spiders have unusual defensive behaviors. When disturbed, the Long-bodied Cellar Spider spins and shakes so quickly that its body becomes a blur. The Marbled Cellar Spider performs similar behaviors, including an up-and-down bouncing motion. It is thought that these actions make it difficult for potential predators to focus on the spider's overall body structure, therefore making it more difficult to attack.

LEPTONETIDAE

Pl. 11

IDENTIFICATION: Leptonetids are diminutive (1 to 3 mm, or .04 to .12 in.) haplogyne spiders with a distinctive eye arrangement. Although there are eyeless cave-dwelling species in the southeastern United States, all of the Pacific coast leptonetids have six eyes, either clustered together in a tight group or with the posterior median eyes resting noticeably behind the other four. In living spiders, the legs are uniquely iridescent, contrasting with their pale-gray to dark-brown bodies. This is one of the few families that exhibits patella-tibia autospasy, a defensive condition in which the leg breaks off at the patella-tibia joint when grabbed or ensnared. Leptonetids build small sheet webs in damp, dark areas, including in caves, under leaf litter, and in rotting logs.
SIMILAR FAMILIES: Members of the family Telemidae (p. 79) share numerous morphological and ecological features with Leptonetidae. However, telemids have a different eye arrangement and a unique sclerotized zigzag-shaped ridge on the dorsoanterior edge of the abdomen. *Spermophora senoculata* is a small, pale, six-eyed cellar spider (Pholcidae, p. 73) that also might be mistaken for a leptonetid, although its eyes are arranged as two widely separated triads.
PACIFIC COAST SPECIES: Eleven species in two genera. The most diverse of the Pacific coast's leptonetid genera is *Calileptoneta* (pl. 11) with nine described species distributed from Southern California to southern Oregon. Members of this genus can be readily identified by the placement of their posterior median

eyes well behind the other four. Given their preference for cool, humid areas, it's not surprising that most of the Pacific coast's leptonetids are known from the foothills and forests between San Francisco Bay and southern Oregon. Only two species are known from the entire southern half of the state, *C. ubicki* and *C. oasa,* which live in Monterey County and western Riverside County, respectively. Also removed from the *Calileptoneta* coastal stronghold are a few unidentifiable females and juveniles that have been collected from caves in California's Sierra Nevada. When all of the known *Calileptoneta* locations are plotted on a map, including both identified and unidentified specimens, an interesting series of wide distribution gaps appears. These are almost certainly an artifact of the spider's small size, cryptic habits, and brief seasonal abundance in otherwise dry, unsuitable areas. An increased exploration of Sierran caves and Southern California's oak woodlands and riparian corridors would likely turn up additional undescribed species. Ledford (2004) provides a detailed summary of all known *Calileptoneta* species, including a key to their identification and notes regarding their natural history and biogeography.

Another regional Leptonetidae genus, *Archoleptoneta* (pl. 11), is represented by two tiny species, 1.1 to 1.8 mm (.04 to .07 in.) in length. Unlike *Calileptoneta, Archoleptoneta* have posterior median eyes that are contiguous with the others, creating a tight ocular cluster. The spiders are light brown in color and have minute cribella and calamistra, features only discovered many years after the genus was first described, and features missing from *Calileptoneta* (Ledford and Griswold, 2010). *Archoleptoneta schusteri* has been found across the central coast of California, with records from Monterey County north to Lake County along with a second population in the Sierra Nevada from Madera County south into Fresno County. *A. gertschi* is known from a small area in California's Sierra Nevada between Placer County and Calaveras County. Additional unidentified *Archoleptoneta* specimens have been collected in the near-coastal mountains of the southern and northwestern areas of California and in the Sierra Nevada. An illustrated key and detailed descriptions of the known *Archoleptoneta* can be found in Ledford and Griswold (2010).

NATURAL HISTORY: Leptonetids are easily overlooked because of their minute size and cryptic habits. Generally preferring dark, moist places, they usually build their tiny sheet webs under rocks, deep in leaf litter, and in rotting logs. Although *A. gertschi* has been found in caves, along the Pacific coast that niche is predominately filled by spiders in the family Telemidae. Both the leptonetids and telemids share an unusual series of minute glands along their legs known as Emerit's glands. These microscopic glands are thought to secrete noxious compounds to discourage potential predators. When disturbed, leptonetids fall to the ground and lie on their backs with their legs folded over their sterna. The Emerit's glands are concentrated on the dorsal surfaces of the spiders' legs in a position that would effectively shroud a spider in a cloud of chemical repellents.

In his revision of *Calileptoneta*, Ledford (2004) provided the first detailed look at the natural history of the Pacific coast's Leptonetidae. *Calileptoneta* are most often seen hanging beneath their small, rectangular sheet webs, and when populations are dense, these webs may overlap along their edges. Courtship has been described for *C. ubicki* and is a rather simple affair in which the male signals to the female by lightly plucking on her web with his palps. He then slowly approaches her and, suspended below her web, copulates with her. After mating, the male spends several days sharing the female's web, possibly guarding her from other males. A *Calileptoneta* egg sac usually contains around 16 eggs and is camouflaged with an outer coating of dirt. An *Archoleptoneta* egg sac is disk shaped, contains four to eight eggs, and is attached to the underside of a rock or wood surface near the female's web (Ledford and Griswold, 2010).

TELEMIDAE

Pl. 11

IDENTIFICATION: Telemids are minute (1 to 1.7 mm, or .04 to .07 in.) haplogyne spiders with a diagnostic zigzag or W-shaped sclerotized abdominal ridge just above the pedicels. Telemid cephalothoraxes range from pale gray to bright orange, and their abdomens are blue green or gray and often clothed in fairly long, diffuse setae. Adult females have a distinctive brush of short, stout setae on the dorsal surfaces of their palpal tarsi. While each of the currently described North American species has six eyes, numerous eyeless cave-dwelling species still await description, making Telemidae among the most diverse of the Pacific coast's troglobitic spider families.

SIMILAR FAMILIES: Although superficially similar spiders can be found in cool, dark habitats, only telemids have a zigzag-shaped abdominal ridge. Members of the genus *Archoleptoneta* (Leptonetidae, p. 77) are the spiders most likely to be mistaken for a telemid. However, telemid eyes (when present) are arranged in three distinct pairs, while in *Archoleptoneta* the six eyes are set in a tight ocular cluster.

PACIFIC COAST SPECIES: One genus with three described and numerous undescribed regional species. All of the Pacific coast's described Telemidae are six-eyed members of the genus *Usofila* (pl. 11), although eyeless specimens awaiting description exist in the collections of various museums and private individuals. Telemids are found from Central California to Alaska and east to Colorado. However, their greatest diversity is almost certainly closely tied to California's Sierra Nevada cave systems. Marx (1891) recorded the first *Usofila* species with his description of *U. gracilis,* a tiny spider living in Alabaster Cave in El Dorado County, California. *U. pacifica* was later described by Banks (1894) from Olympia, Washington, and has since been collected from damp leaf litter in several locations throughout western Washington, including on the San Juan Islands. *U. oregona* (Chamberlin and Ivie, 1942b) was collected from a small area of west-central Oregon between the cities of Eugene and Corvallis. While both male and female *U. gracilis* and *U. oregona* have been described, only the female *U. pacifica* has been depicted in the published literature.

NATURAL HISTORY: Telemidae live in dark, damp areas, including under leaf litter and rocks and in rotting logs, but it is in caves where the true breadth of their diversity is revealed. They hang upside down from small sheet webs, where they capture and feed on tiny soil- and cave-dwelling arthropods. Both sexes also have an unusual collection of integumentary glands on the dorsal surfaces of their tibiae. Known as Emerit's glands, they produce repugnatorial secretions that may help deter predators. Courtship and mating behaviors have not been documented, but males and females will live in the same web for extended periods. Egg sacs are roughly oval in shape and are placed along the outer margins of the female's web. Interestingly, those species that live outside of caves, where there is an abundance of food and potential web building sites but also predators, produce comparatively large numbers of fairly small eggs. Cave-dwelling species that must contend with limited food supplies and areas for dispersal, though three is less predation pressure, produce only a few, distinctly larger, eggs.

MYSMENIDAE

Pl. 11

IDENTIFICATION: Mysmenids are minute, eight-eyed, entelegyne spiders with three claws on each tarsus. They are found in leaf litter, wood rat middens, and other dark, humid places. Members of the family Mysmenidae can be distinguished from other tiny, cryptic spiders by the presence of a distinctive sclerotized spot on the ventral surface of each femur I. Additionally, adult males have spine-like clasping spurs on metatarsi I and cymbia that twist near their tips.

SIMILAR FAMILIES: On the Pacific coast, *Gertschanapis shantzi* (Anapidae, p. 82) is the spider most likely to be confused for a mysmenid. The male *G. shantzi* lacks clasping spurs on metatarsi I, has a smaller and less turret-like ocular region, and has a different eye arrangement. Neither sex of *G. shantzi* has a sclerotized spot on the underside of femur I.

PACIFIC COAST FAUNA: One genus with one regional species. Originally described and illustrated by Gertsch (1960a), *Trogloneta paradoxa* (pl. 11) was initially placed in the pantropical family Symphytognathidae. It was later moved into Mysmenidae by Forster and Platnick (1977). This tiny (approximately 1 mm, or .04 in., in length) spider is a relatively rare but widespread member of the region's araneofauna and has been collected widely, if sporadically, from Southern California through western Washington. In addition to the familial features listed above, a *T. paradoxa* male has a tall, cone-like ocular turret and a high clypeus that together create a concave "facial" region when viewed from the side. Its abdomen is black with yellowish mottling along its sides and dorsal surfaces and is attached to the pedicel near its ventral midpoint, giving it a distinctly vertical orientation. The female is similarly colored but has a dramatically smaller ocular turret while maintaining the concave facial area. Compared with males, females have rounder abdomens that are not nearly as vertically inclined.

In both sexes, only the anterior median eyes are dark while the other six are pearly white.

NATURAL HISTORY: Beyond its preference for leaf litter and dark, humid recesses, almost nothing is known about the natural history of *T. paradoxa*. Like other mysmenids, it is presumed to build a tiny tangle web, but its exact structure is unknown.

ANAPIDAE

Pl. 11

IDENTIFICATION: North America's only "true" anapid is the tiny, eight-eyed *Gertschanapis shantzi*. The female is unique among Pacific coast spiders in that her palpi have been reduced to minute nubs on the anterior face of the endites. The male's abdomen is reddish brown, covered by a conspicuous scute, and sculpted with numerous grooves and small sclerotized pits along the sides. Both sexes also have enlarged spur-like tubercles on the ventral surfaces of their first and second femora. The unique web is a small, fine-meshed, horizontal orb in which extensions of the radial threads are strung above the mesh and connect to vertical support threads near its hub.

SIMILAR FAMILIES: Levi (1957a) described a small, six-eyed spider from a leaf litter sample collected in Mendocino County, California. He named it *Archerius mendocino* and placed it in the family Theridiidae (p. 93). Oi (1960) transferred it to the Old World anapid genus *Comaroma*, placing a second genus and species into the Pacific coast's Anapidae fauna. However, the presence of a setal comb on tarsus IV and the spider's palpal structure identifies it as a member of the family Theridiidae (p. 93), but until additional revisionary work is published, the rules of zoological nomenclature leave it suspended within Anapidae.

The family Mysmenidae (p. 81) is made up of tiny, superficially similar spiders with only a single representative, *Trogloneta paradoxa*, on the Pacific coast. Found in many of the same areas as *G. shantzi*, they can be immediately separated from them because the male *T. paradoxa* has a more vertically aligned abdomen, a dusky yellow-and-black carapace, and a distinctly conical ocular region. He also lacks an abdominal scute and clasping spurs on metatarsus I. The female *T. paradoxa* has well-developed palps, while these features are reduced to vestigial bumps on *G. shantzi*.

PACIFIC COAST FAUNA: *G. shantzi* (pl. 11) ranges from 1 to 1.5 mm (.04 to .06 in.) in length and is found in humid forests and cool, shaded microhabitats along the coast and in the Sierra Nevada, from Southern California to northern Oregon. The most detailed description available of *G. shantzi* can be found in Platnick and Forster (1990).

NATURAL HISTORY: Very little is known about the life cycle or natural history of G. shantzi. Its tiny orb webs have been found in fern groves, pack rat middens, and the hollows of old tree stumps. Collection records indicate that adults live year-round but are most frequently encountered from March through October.

ULOBORIDAE — Hackled Band Orb Weavers

Pl. 12

IDENTIFICATION: Uloboridae are small, cribellate, eight-eyed, entelegyne spiders whose Pacific coast representatives are unique, both in their morphologies and in their web designs. Hackled band orb weavers have exceptionally long calamistra that stretch across more than half the length of their dorsally concave fourth metatarsi. They also have multiple rows of trichobothria on femora II through IV and nondivided cribella. There are two kinds of webs made by the Pacific coast's Uloboridae: a horizontal or sloping orb web with a stabilimentum along the midline, and a vertically aligned triangle.

SIMILAR FAMILIES: There are several small spiders in the ecribellate families Araneidae (p. 110) and Tetragnathidae (p. 126) that might be mistaken for a hackled band orb weaver. Collectively, their long calamistra, dorsally depressed fourth metatarsi, and entire cribella distinguish the hackled band orb weavers from all other North American cribellate spiders.

PACIFIC COAST FAUNA: Two genera containing three regional species. The triangle web spiders, genus *Hyptiotes* (pl. 12), are found in woodlands across the western United States and are especially widespread along the Pacific coast. These inconspicuous spiders range from 2.1 to 4.4 mm (.08 to .17 in.) in length and have trapezoidal carapaces that narrow anteriorly, posterior lateral eyes on prominent tubercles, and minute anterior lateral eyes. Both sexes have a dense coating of setae on the abdomen and cephalothorax and are often exceptionally well camouflaged, with yellowish-brown, gray, and nearly black individuals regularly found. Female *Hyptiotes* have strongly arched abdomens with two pairs of lateral tubercles, while the males' abdomens are smaller and not as conspicuously sculptured. *Hyptiotes gertschi* is the most common of the region's triangle spiders and has been collected from San Diego County, California, to southeastern Alaska, reaching its greatest density in the Coast Ranges of northwestern California through northern Washington. The Pacific coast's second triangle web spider, *H. tehama*, differs mainly in the fine details of its reproductive structures. This uncommon species has only been regionally collected at a few locations in Tehama County and Siskiyou County in Northern California. Muma and Gertsch (1964) includes identification keys and illustrations of both species.

The feather-legged spiders, genus *Uloborus* (pl. 12), are represented along the Pacific coast by a single species, *Uloborus diversus*. This spider builds a small horizontal to near-vertical orb web with a stabilimentum of thick, opaque silk across its hub. Almost unmistakable, it has exceptionally long front legs with conspicuous brushes of setae below the distal portions of tibiae I and a tubercle on each side of the abdomen. It ranges from 2.1 to 4.7 mm (.08 to .18 in.) in length. Normally, it hangs beneath the web's stabilimentum, a position that obscures its outline when viewed from above. Its resting pose is characteristic, with legs I and II extended forward and III and IV held close to the body. Like other regional hackled band orb weavers, *U. diversus* is rather variable in color, although in many areas the predominant coloration is light orange and tan with a pale longitudinal carapace stripe. Especially common in Southern and Central California, *U. diversus* becomes progressively rarer farther north, with only a few scattered records from Oregon and Washington. Detailed keys and illustrations to North America's *Uloborus* can be found in Muma and Gertsch (1964).

NATURAL HISTORY: The hackled band orb weavers exhibit several highly unusual traits, including a complete loss of venom glands and the use of dry cribellate silk to make symmetrical orb and triangle webs. Because they lack venom glands, uloborids have developed other methods to subdue struggling prey. When an insect is trapped in a feather-legged spider's web, the spider shakes the webbing, further tangling it. After the prey is thoroughly ensnared, the spider wraps it in thick layers of silk.

This process applies such intense compression forces that the insect's cuticle often breaks, along with its leg joints, antennae, and eyes. Once the insect is enshrouded, the spider soaks the outside of the bundle with digestive juices, further breaking down the captured insect's membranes. It then feeds on the prey's bodily fluids as they leach through the wrapping. In at least some hackled band orb weavers, the digestive enzymes also act on the silk, thus allowing the spider to regain a portion of the proteins lost through the capture and immobilization process (Weng et al., 2006).

The web of triangle web spiders consists of four radial threads crossed by numerous sticky lines of cribellate silk. Where the radial lines come together, a bridge line is strung to a nearby twig. Hanging upside down, the triangle web spider holds the web taut through the bridge line. When a small insect crashes into it, the spider releases the tension on the bridge line, relaxing the web and entangling the prey.

Little has been published regarding courtship behaviors of hackled band orb weavers, although male triangle web spiders are known to run a thread to the female's web and strum it as part of their reproductive repertoire (B. Opell, pers. comm., 2009). The oval egg sacs of the triangle web spiders are covered in an outer layer of dusky, grayish silk and are placed flush against the sides of small twigs where they look like little more than slight swellings on the bark. In contrast, the egg sacs of the feather-legged spider, *U. diversus,* are elongate and misshapen. Looking like prey remains or dried leaves, they are hung along the perimeter of the female's web.

OECOBIIDAE — Wall Spiders

Pl. 12, Fig. 9

IDENTIFICATION: Members of the family Oecobiidae are small (1 to 4 mm, or .04 to .16 in., in length) cribellate spiders with distinctive enlarged anal tubercles (fig. 9). Their eight eyes are arranged in a compact cluster, and their flat, rounded carapaces range in color from pale yellow to dusky brown. They can be exceptionally common around homes and are often found on walls and ceilings.

SIMILAR FAMILIES: With their enlarged anal tubercles, distinctive shape, compact eye arrangement, and often synanthropic

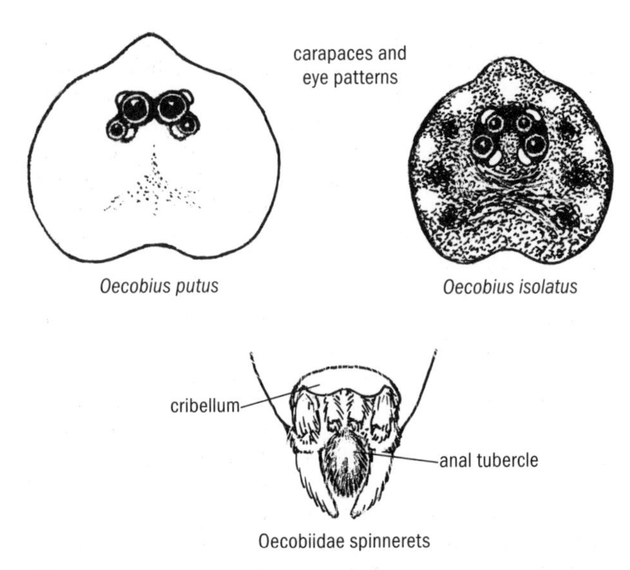

Figure 9. Characteristic features of Oecobiidae spiders (after Shear, 1970, in part).

habits, wall spiders are unlikely to be mistaken for any other North American spider family.

PACIFIC COAST FAUNA: One genus, *Oecobius* (pl. 12), containing three regional species. The most widespread Pacific coast wall spider is *Oecobius navus* (previously misidentified in the literature as the Algerian species *O. annulipes*). Its carapace is straw yellow with three distinct dashes on either side of its midline and a dark central patch extending backward from its eyes. Its abdomen is heavily mottled and its legs are banded. *O. navus* is exceptionally common both in buildings and outdoors in Southern and Central California. Farther north, it becomes increasingly dependent on homes and other structures. All Washington records are from buildings around Seattle (R. Crawford, pers. comm., 2008); however, due to its small size and synanthropic habits, it is likely more widespread than currently appreciated. *O. putus* (see fig. 9) is a desert species introduced to North America from North Africa and Central Asia and is found from Southern California to western Texas.

Its carapace is unmarked except for a dusky patch around its eyes and a faint marginal border. The native species, *O. isolatus* (see fig. 9), is only known regionally from Southern California and is the least synanthropic of the Pacific coast species. Its carapace is dark brown with a black band encapsulating the eyes and several indistinct pale spots along its margin. A detailed review of North America's wall spiders can be found in Shear (1970).

NATURAL HISTORY: The vast majority of the Oecobius research has focused on the synanthropic *O. navus*, including Glatz's (1967) detailed life history study. Wall spiders mainly build two styles of web. The most common is a double sheet woven across a small depression on a flat surface. The upper sheet is pulled taut by radiating signal lines giving it a roughly star shape. The spider rests on the smaller, lower sheet. This arrangement provides it with protection from above while allowing it to exit out the web's open sides. The second style is often built along a seam and consists of a short tube web with openings on either end.

Although ants are the main food of wall spiders, they have also been observed eating mites and even indulging in acts of cannibalism. When hunting, *O. navus* circles its prey while simultaneously throwing out a ribbon of silk from its enlarged posterior lateral spinnerets. This creates an adhesive mesh, entangling the prey and sticking it to the substrate. Once the prey is immobilized, the wall spider closes in and bites its quarry. It then cuts its prey loose and either drags it back to its retreat or feeds at the capture site. Wall spiders will also use silk as a defensive measure. Glatz (1967) observed a specimen of *O. cellariorum* whose hind leg was being bitten by an ant. In response, the spider flung silk over the ant's face, hindering any further attack and giving the spider time to drop its leg and escape. Additionally, he saw sexually disinterested females throw silk at overly enthusiastic males.

Courtship in wall spiders consists of the male wiggling his abdomen and laying down bands of silk. If receptive, the female will let the male approach and build a small tubelike mating web on top of her retreat. The spiders signal to one another by twitching and tugging on the threads of their combined structure. Only once the female enters the male's web will mating occur. This behavior may repeat itself several times, although

Glatz (1967) did observe several females attack and even eat the males during or immediately after mating. Several days later, the female lays between three and 10 eggs on a small sheet that is loosely tied together with a few strands of silk. Normally, the female builds five or more egg sacs before abandoning them, secured in the back of her retreat.

MIMETIDAE — Pirate Spiders

Pl. 13

IDENTIFICATION: Mimetids are small (2.5 to 7 mm, or .1 to .28 in.), eight-eyed entelegynes that specialize in preying on other spiders. The tibiae and metatarsi on their fore legs are armed with a diagnostic row of long spines separated by a series of short, stiff setae. Pirate spiders also often rest in a characteristic pose, with legs I and II extended forward, III held close to the body, and IV projecting backward.

SIMILAR FAMILIES: Pirate spiders are most likely to be mistaken for either orb web weavers (Araneidae, p. 110) or cobweb weavers (Theridiidae, p. 93), both of which are among their preferred prey items. However, neither family shares the exceptional Mimetidae anterior tibial and metatarsal spine arrangement.

PACIFIC COAST FAUNA: Three genera with four described and several undescribed or unidentified regional species. The most widespread of the Pacific coast's pirate spider genera are *Mimetus* (pl. 13) and *Reo* (not illustrated). Both genera have fairly short clypei and long chelicerae with a spine on each chelicera's anterior medial face. Confidently differentiating them generally requires examining their reproductive structures as described in Lew and Mott (2005). Two species of *Mimetus* have been described from the Pacific coast. Both are pale yellowish with numerous dark lines or wide bands extending backward from their eyes, and their abdomens are heavily mottled with large irregular foliate patterns. *Mimetus hesperus* is found throughout the western United States and is the only known *Mimetus* in Oregon and Washington. *M. notius* is well known from pine forests in the southeastern United States. There is also a smattering of western records, including a specimen collected in 1941 from the Laguna Mountains in San Diego County, California (D. Mott, pers. comm., 2009). An additional undescribed species of *Mimetus* is known from the San Bernardino Mountains in San Bernardino County, California (D. Mott, pers. comm., 2009). While both regional *Mimetus* species are described in Chamberlin (1923), the genus would benefit greatly from a thorough and modern revision.

The genus *Reo* is regionally represented by a single species, *Reo eutypus*. Compared with the area's *Mimetus* species, *R. eutypus* is generally darker but otherwise is exceptionally similar. It has been found in forested areas across Southern and Central California, both along the coast and in the Sierra Nevada. *R. eutypus* is described in Chamberlin and Ivie (1935) under its original placement in *Mimetus*.

Ero canionis (pl. 13) is the only identified species of *Ero* along the Pacific coast, with scattered records from central and northern Washington. *Ero* has a higher clypeus than either *Mimetus* or *Reo* and relatively short chelicerae without anterior median spines. *E. canionis* is a fairly dark spider with a large, pale-golden or tan patch on the posterior portion of its abdomen and is described in Chamberlin and Ivie (1935). An additional unidentified species of *Ero* has been collected in the Inglenook Fen area of Mendocino County, California (Griswold, 1977).

NATURAL HISTORY: Pirate spiders specialize in preying on other spiders, although occasionally they will also eat insects and spider eggs. Cobweb weavers (Theridiidae, p. 93), orb web weavers (Araneidae, p. 110), and mesh web weavers (Dictynidae, p. 198) make up the majority of most mimetids' diets. When hunting, a pirate spider regularly begins by lightly plucking at its prey's web, falsely signaling either a courting male or a captured insect. When the web's builder comes out to investigate, the pirate spider clasps its quarry between its heavily spined fore legs. Once the prey is pinned, the mimetid lunges in for the bite, its venom being especially lethal to other spiders. If it is unable to immediately subdue its prey, the pirate spider might grab a leg and tear it from its owner, feeding on the amputated limb before continuing its pursuit. This not only provides the pirate spider with a quick bit of nourishment, but may also be a deliberate technique for disabling larger spiders (Lawler, 1972). Once it captures prey, the pirate spider normally wraps it in silk and carries it to a protected area of the web for feeding. Klook (2001) showed that at times, insects make up a significant portion of a pirate spider's diet. While studying *M. notius*, he observed it chase spiders from their webs only to feed on the insects already killed, wrapped, and left behind.

Lawler (1972) studied the courtship and mating behaviors of *Reo* (then *Mimetus*) *eutypus*. Hanging upside down from some silk threads, the male slowly approaches the female, plucking the web strands to signal his presence. If the female moves toward the male, he rapidly retreats, returning a few moments later to try again. If receptive, she will eventually sit in a courtship position with her front legs curled close to her body (quite different from her open-legged attack pose). For the next five to 10 minutes the male indulges in repeated bouts of thread plucking until he gets close enough to engage her in a quick embrace. Mating itself only takes a few seconds, after which the male quickly withdraws.

Pirate spider egg sacs are distinctive. Whether suspended from a cord or encapsulated within a sphere-like net, they are covered with a loose mesh of thick, curly, coppery brown or golden silk threads. Hidden in a sheltered recess such as a gopher hole or hollow tree, *Reo eutypus* attaches its orange egg sac to the roof with a thick band of golden-yellow silk. That same material is then woven into a basketlike lattice around the outside of the egg sac. While *M. hesperus* egg sacs are very similar to those made by *R. eutypus*, *M. notius* surrounds its fluffy white egg sac with a heavy weave of curly brown silk that is then enclosed in a spherical web up to 30 mm (1.2 in.) across. This unusual structure may act as a barrier, preventing potential predators from reaching the much smaller (6 to 8 mm, or .24 to .31 in., across) egg sac suspended within.

NESTICIDAE — Cave Cobweb Spiders

Pl. 13

IDENTIFICATION: Nesticids are eight-eyed (or eyeless), three-clawed entelegynes. The anterior margins of their labia are slightly thickened, a condition referred to as rebordered. There is also a ventral comb of serrated setae on tarsus IV, and an adult male has a large, rigid extension off the base of his palpal cymbium (a paracymbium). Cave cobweb spiders are small (2.5 to 4 mm, .1 to .16 in., in length) and make distinctive "split-foot" tangle webs, so called because they split into two branches at the base, in damp, protected areas, including inside caves, within the interstices of thick leaf litter, and under rotting logs. The carapace and legs of our regional species range from pale yellow to orange, and the abdomen is usually gray with variable dark markings.

SIMILAR FAMILIES: Morphologically and evolutionarily, Nesticidae are closest to the cobweb weavers (Theridiidae, p. 93). While the two families share aspects of web design, setal combs on the fourth tarsi, and a lack of leg spines, cave cobweb spiders have thinly rebordered labia, while the labia on theridiids are smooth. Further distinguishing the two families, male theridiids have small, inconspicuous paracymbial structures. Because some Nesticidae have adapted to life in caves, they might be mistaken for members of the family Telemidae (p. 79). Telemidae are extremely small (under 2 mm, or .08 in., in length) haplogynes, and most have six eyes arranged in three pairs. Because both families also contain eyeless troglobitic species, confidently differentiating them may require looking at their reproductive structures (haplogyne versus entelegyne) or checking for the W-shaped sclerotized ridge on the telemid's abdomen.

PACIFIC COAST FAUNA: Two genera containing four regional species. Confidently differentiating these two genera requires looking at the spider's reproductive structures. *Nesticus* (pl. 13) males have large, curved paracymbia with several terminal branches, and the females have prominent, broad epigyna. The paracymbia of a male *Eidmannella* (not illustrated) are distinctly shorter without multiple terminal branches, and the female's epigynum consists of a small, narrow, inverted T-shaped structure just above the genital groove. Detailed illustrations of these structures can found in both Gertsch (1984) and Paquin and Hedin (2005).

The more diverse of the region's two genera is *Nesticus*, with three described Pacific coast species. *Nesticus silvestrii* is widespread, with records from Central California to southern British Columbia. Its legs and carapace vary from yellow to orangish brown, often with dark markings around the eyes and along the midline and margins of the carapace. Its abdomen is dusky gray and is normally (but not always) marked with dark chevrons on its dorsal surface and pale spots on its sides. It has been collected from deep leaf litter, boulder fields, and caves, both along the coast and in the Sierra Nevada. In response to these different habitats, *N. silvestrii* demonstrates a fair degree of morphological plasticity, with cave-dwelling populations showing a reduction in eye size, reduced pigmentation, and longer legs than their surface-dwelling kin. Despite these adaptations, there aren't any appreciable differences in their reproductive structures, so at this time they are still considered members of a single, morphologically variable species. A similar situation exists within the highly diverse Appalachian *Nesticus*. In contrast, Hedin (1997) showed that even among morphologically indistinguishable cave-dwelling nesticids, a great deal of genetic diversity exists between populations, revealing either incipient speciation or even cryptic species within different cave systems. Future research may reveal a comparable situation, on a smaller scale, among the West Coast's *Nesticus*.

The other two regional *Nesticus* species are both endemic to small cave systems in the Sierra Nevada of Northern California. *N. sodanus* is quite similar to *N. silvestrii,* differing mainly in the fine details of its reproductive structures, and is known only from Soda Springs Cave in Plumas County. The third species, *N. potterius,* is eyeless and pale yellow and has an unmarked whitish abdomen. It has only been found in Potter Creek Cave in Shasta County. Each of these species is described and illustrated in Gertsch's (1984) partial revision of the New World Nesticidae.

The second genus, *Eidmannella* (not illustrated), is represented along the Pacific coast by a single species, *Eidmannella pallida*. Common in caves and cool, moist soil debris throughout the southeastern United States and eastern Mexico, it has been collected several times in Southern California, and there is at least one record from Benton County, Oregon (Gertsch, 1984). Much like *N. silvestrii, E. pallida* appears to be a morphologically flexible species, becoming paler as populations evolve to survive deep inside caves. A surface-dwelling individual normally has banded legs and a yellow carapace with dusky markings around the eyes and midline. The abdomen is gray with dark lateral markings and pale middorsal stripes. This species is reviewed in Gertsch (1984).

NATURAL HISTORY: These cryptic spiders make their webs deep in caves, in moist pockets of leaf litter, and in the shaded recesses of canyon walls, often near creeks or streams. All nesticids make split-foot snares. Elastic threads are stretched from a rock shelf or other feature to the substrate below, where each thread splits into two sticky branches. When a small insect touches one of these adhesive strands, it is pulled up and further entangled in the webbing above.

Almost nothing is known regarding these spiders' courtship behaviors. *Eidmannella* egg sacs are about 2 mm (.08 in.) in diameter, relatively large in comparison with the size of the spider, and consist of only a few eggs wrapped in a thin white sheet. Females carefully guard their egg sacs and have been observed carrying them in their chelicerae and attaching them to their spinnerets.

Because some species are found both in caves and in cool, damp areas outside of caves, an interesting snapshot can be seen, as one population, capable of easy movement and a ready supply of prey, diverges from the more nutritionally constrained but environmentally stable cave dwellers. Given enough time, complete isolation can occur, not only between the spiders in different cave systems, but also between the cave and soil dwellers. A situation like this has led to an amazing array of nesticids in the Cumberland Plateau and southern Appalachian Mountains of the southeastern United States (Hedin, 1997). At least 30 species are known from this cave-riddled area. These same spiders are also in a very precarious situation, as many times a species is found in only a single cave or isolated cave cluster, making it highly susceptible to human disturbances. There is a wealth of unexplored caves in the Sierra Nevada, and recent limited surveys have already revealed a plethora of new arthropod species. If they are conscientiously studied, the Pacific coast's Nesticidae could reveal important information about the region's biogeography and the evolutionary processes that correspond to the successful colonization of cave environments.

THERIDIIDAE Cobweb Weavers, Comb-footed Spiders

Pls. 14–17, Fig. 10

IDENTIFICATION: Theridiidae houses some of the most familiar spiders on the Pacific coast, including the Western Black Widow (*Latrodectus hesperus*) and the cosmopolitan Common House Spider (*Parasteatoda tepidariorum*). Like other members of the superfamily Araneoidea, the cobweb weavers are eight-eyed (with one very rare exception), three clawed entelegynes. Most species have an unusual row of curved, serrated setae on the ventral surface of the fourth tarsi, although this feature may be hard to see or even absent on some males. Theridiids have very few or no leg spines, and the anterior margins of their labia lack rebordered edges. They are regularly found hanging upside down near the top of their tangle webs. As a family, the cobweb weavers vary from .8 to 14 mm (.03 to .55 in.) in length, and while most are assorted shades of brown, black, gray, and white, a few species are boldly patterned with red or even metallic silver.

SIMILAR FAMILIES: Theridiids share numerous features with other widespread spider families, including the orb web weavers (Araneidae, p. 110), the sheet web weavers (Linyphiidae, p. 133), and the long-jawed orb weavers (Tetragnathidae, p. 126). However, all of these families have numerous spines on their legs and lack tarsal combs. Additionally, nearly all of them make webs with an obvious architectural pattern, be it a sheet or an orb. The spiders most likely to be mistaken for theridiids are the cave cobweb spiders (Nesticidae, p. 90). Like cobweb weavers, they have combs of serrated setae on their fourth tarsi and few if any spines on their legs. However, nesticids make split-foot snare webs and have rebordered labia, and adult males have large, rather complex paracymbia. On male theridiids, the paracymbia are small and inconspicuous.

PACIFIC COAST FAUNA: Twenty-nine genera containing 94 regional species plus one enigmatic species currently misplaced in the family Anapidae (p. 82). While some species of cobweb weaver are easily recognized, confirming the identity of most requires a detailed examination, especially with the smaller, more cryptic taxa. Levi (2005c) provides a highly informative key to North America's Theridiidae genera, although since its publication, several of the more species-rich genera have undergone dramatic revisions. Additional resources may be needed to identify these species based on their current taxonomic positions. A major distinguishing feature among the different cobweb weavers is the size and shape of the colulus, a nonfunctional remnant of the cribellum located between the anterior lateral spinnerets. In some genera, including *Latrodectus, Steatoda,* and *Robertus,* it is a large, fleshy lobe, while in others it's either been reduced to a simple pair of setae (as in *Anelosimus*) or is lacking altogether (as in *Theridion* and *Parasteatoda*).

The most well-known of the Pacific coast cobweb weavers are the widow spiders, genus *Latrodectus* (pl. 14). As one of the very few spider genera whose venom presents a genuine risk to humans, *Latrodectus* inspires both fear and fascination. Two species are found along the West Coast: the widespread, native Western Black Widow (*L. hesperus,* pl. 14) and the introduced Brown Widow (*Latrodectus geometricus,* pl. 14), which has become a prevalent component of Southern California's synanthropic spider fauna. Widows are large spiders, up to 12 mm (.48 in.) in length, and both sexes are easily recognized by the orange or red hourglass-shaped mark on the underside of the abdomen. The adult female Western Black Widow is lustrous black, while the immature female has tan and white spots and stripes across the abdomen. The adult male ranges from beige to dark brown with broad, pale dorsal and lateral abdominal stripes. Widespread throughout the contiguous Pacific coast states, the Western Black Widow can be found in a diversity of habitats, from Southern California's deserts to the Cascade Range of northern Washington. It is commonly found in protected recesses, including rodent burrows, woodpiles, and sheds, along with its large, teardrop-shaped egg sacs.

First recorded in Los Angeles County in 2003, the Brown Widow is now one of the most common spiders around homes and other buildings in Southern California, with concentrated populations in San Diego County, Riverside County, and Los Angeles County (Vincent et al., 2008). Brown Widows are extremely variable in color and pattern, and exceptionally dark individuals can regularly be found along with pale tan ones. Brown Widow egg sacs are also distinctive, as they are covered with an array of conspicuous silk spikes. While there is some controversy over their point of origin, Brown Widows are now cosmopolitan in distribution, with populations established in warm areas throughout the world. Anecdotal reports suggest that Brown Widows may even be displacing Western Black Widows in parts of Southern California. Both species are discussed in Levi's (1959) review of the North American widow spiders, although at that time the Western Black Widow, *L. hesperus,* was synonymized within the Southern Black Widow, *L. mactans.*

While the widow spiders are notorious for their powerful venoms, they are reclusive by nature and only bite when threatened or physically disturbed, such as when a person inadvertently grabs one, triggering a defensive response. Their venom is neurotoxic, and often the result of a widow bite is redness, swelling, and some stiffness at the bite location. In more serious cases, cramping can spread to the major muscle groups, accompanied by headaches, nausea, and difficulty breathing. Any time a widow bite is suspected, medical attention should be sought immediately, and if possible, the spider or its remains should be brought to the hospital for identification. Although lab tests have shown that, milliliter for milliliter, the venom of the Brown Widow is more toxic than that of most American black widows, one study of confirmed Brown Widow bites in Africa found that the only symptoms expressed were mild pain and redness around the bite site (Müller, 1993). However, this does not dismiss the Brown Widow's venomous potential, as demonstrated by at least one severe reaction that required intensive hospitalization (Goddard et al., 2008). This shows that other factors are at work beyond pure toxicity, including individual sensitivity and the quantity of venom administered during the bite.

Spiders in the genus *Steatoda* (pl. 14, fig. 10) are exceptionally well represented across the western United States, with 14 Pacific coast species. These include the False Black Widow (*Steatoda grossa,* pl. 14), a common, synanthropic species that, while native to Europe, is now found in homes from Southern California to northern Washington. The female False Black Widow is fairly large (5.9 to 10.5 mm, or .23 to .41 in., in length) and superficially similar to the Western Black Widow, *L. hesperus.* While both are large, black theridiids, the False Black Widow's abdomen is slightly more oblong with a brownish or purplish sheen and never has a red hourglass mark on the ventral surface. The female False Black Widow can show a light gray band across the abdomen's dorsoanterior margin along with a similarly colored middorsal stripe. The male is generally smaller (4.1 to 7.2 mm, or .16 to .28 in., in length) and has a reddish-orange cephalothorax and legs. The abdomen is dark gray to reddish brown with light banding that may be broken into yellowish spots. False Black Widow webs are similar to those of other large theridiids and can be found in basements,

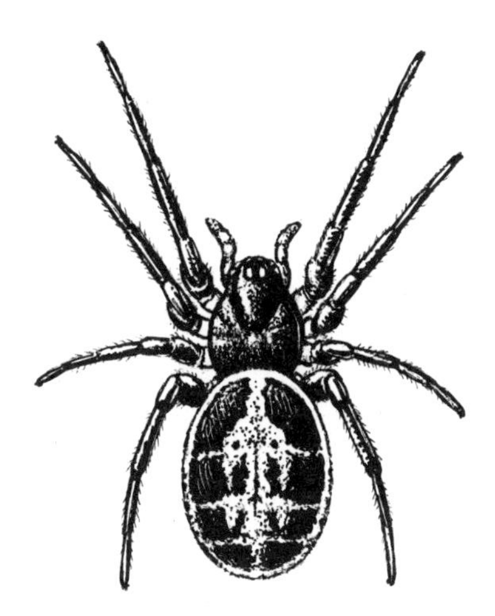

Figure 10. *Steatoda nobilis* (female).

cupboards, and other protected spots. A second large (7.8 to 14 mm, or .31 to .55 in., in length), introduced species, *S. nobilis* (see fig. 10) has recently been discovered in Ventura County, California, living in many of the same habitat niches as the Brown Widow (*L. geometricus*). Both sexes of *S. nobilis* can be recognized by their distinctive, ornate abdominal pattern.

Representatives of the 13 native *Steatoda* species are found in a wide range of habitats, from deserts to montane forests. Most range in color from brick red to purplish black and almost always show a pale band across the abdomen's anterior margin, normally accompanied by numerous white or grayish yellow triangular and oval spots. In some localized populations, the white has spread across the entire dorsal abdominal surface, leaving only two symmetrical rows of purplish spots. In both sexes of the native *Steatoda,* the abdomen is rather globular, and they regularly live in close proximity, even within the same web. *Steatoda* egg sacs are spherical, and in many species, the protective wrapping is thin enough that the eggs are visible within. In contrast, *S. grossa* egg sacs are covered with a dense layer of fluffy silk and look like small cotton balls. The Pacific coast's native *Steatoda* range from 2.4 to 9 mm (.09 to .35 in.) in length. A revision of the North and Central American species can be found in Levi (1957b), while Gertsch (1960b) provides a detailed look at the variation within the "*Steatoda fulva*" group, a morphologically variable complex of small, ground-dwelling spiders of the arid southwestern United States and northern Mexico. Vetter (2012) documents the presence of *S. nobilis* in California and provides a detailed description of the species.

Very similar to *Steatoda* is the genus *Crustulina* (pl. 15), which is represented on the Pacific coast by a single, widespread species, *Crustulina sticta*. In both sexes the abdomen is fairly globular and dark purplish brown, occasionally with light stripes along its anterior and lateral margins and down its dorsal midline. *C. sticta* is distinguished by its granulated carapace and sternum. The reddish-brown carapace is covered with dozens of tiny triangular tubercles that under magnification look like minute, forward-pointing shark's teeth. Both sexes run between 1.9 and 2.7 mm (.07 and .11 in.) in length. *C. sticta* prefers moist conditions and has been collected in leaf litter and under rocks and logs throughout the Coast Ranges, from Southern California to northern Washington, with additional records from the Sierra Nevada (Levi, 1957b).

Two species of *Tidarren* (pl. 15) are found along the Pacific coast, *Tidarren sisyphoides* and the generally smaller *T. haemorrhoidale*. Females are small to moderately large theridiids that range from 2.4 to 8.6 mm (.09 to .34 in.) in length and have globular, mottled abdomens similar to that of the Common House Spider, *Parasteatoda tepidariorum*. The adult *Tidarren* male is extremely small, averaging just over 1 mm (.04 in.) in length, and in *T. sisyphoides* has a body mass that is only 1 percent that of the much larger female. A *Tidarren* male has only a single, comparatively massive palp that makes up around 10 percent of its total body mass. Although it develops two palps as other spiders do, it purposefully amputates one just after its penultimate molt by entangling it in a silk scaffold and then moving in circles until it breaks off. Research by Ramos et al. (2004) showed that males with only one palp rather than two are able to move farther and faster, a condition that would improve their ability to find and compete for a mate. This is especially important in *Tidarren* because males are only able to reproduce once. Large numbers of males will gather around the periphery of an immature female's web waiting for her to finish her final molt. Courtship mainly consists of the males tapping on the female's hind legs, abdomen, and sternum. If the female is receptive, she lowers her abdomen and sits very still, allowing the male to begin mating. Within a few minutes of inserting his palp into her epigynum, the male *Tidarren* dies. The female then removes the expired male, and while many species will eat the dead male, *T. sisyphoides* simply discards his body from her web (Knoflach and Benjamin, 2003). She may then allow a second mating, but limited observations strongly suggest that third matings are quite rare. The web of *T. sisyphoides* is unusual among theridiids in that it consists of a dome-shaped sheet web protected from above by a dense tangle and from below by a horizontal sheet web. Suspended within the tangle is the spider's retreat, often a rolled leaf that opens onto the underside of the dome sheet. Only *T. sisyphoides* is regionally common, with California records from coastal and near-coastal areas from San Diego County to Santa Clara County. Levi (1957c) reported *T. haemorrhoidale* (under the name *T. fordum*) from San Diego County.

Argyrodes elevatus (pl. 16), also known as the Dewdrop Spider, and *Neospintharus furcatus* (pl. 16) are both small (2.3 to 4 mm, or .09 to .16 in.) kleptoparasites that make their homes in the webs of larger species. They feed off insects too small to garner the host's attention or on the remains of its abandoned meals. The Dewdrop Spider is aptly named because the female's silvery, triangular abdomen looks like a drop of dew hanging off the host spider's web. The male has a more oblong abdomen but is similarly patterned with silver sides and a wide, dark stripe down the dorsal midline. The male also has a highly modified clypeus that juts out past the ocular region, giving the impression that the chelicerae are regressed. In California the Dewdrop Spider is known to parasitize the garden spiders (*Argiope,* Araneidae, p. 111) and has been regionally reported only from a small strip of the Southern California Coast, from San Diego County north to Ventura County. *N. furcatus* is similar in size to the Dewdrop Spider but is not as silvery, and both sexes have a more distinctly triangular abdomen. On the male, the abdomen is concave along its posterior margin, giving it a shark-fin look. *N. furcatus* varies from pale gray to dark brown with scattered silver spots on the abdomen. Like the Dewdrop Spider, the male *N. furcatus* has modified cephalic features, but in this case there are several setae-tipped tubercles extending off the clypeus and between the eyes. Regionally *N. furcatus* has only been reported from Orange County, California (host unknown). Both *A. elevatus* and *N. furcatus* (under its previous placement in *Argyrodes*) are reviewed in Exline and Levi (1962).

Rhomphaea fictilium (pl. 16) is easily one of the oddest looking spiders in North America. The female's abdomen is exceptionally long and wormlike. Tan overall, it is often marked with darker stripes and silvery speckles. *R. fictilium* also has unusually long, thin legs for a spider its size (the female ranges from 5.6 to 12 mm, or .22 to .47 in., in length). The male is similarly colored but generally smaller (3 to 6.6 mm, or .12 to .26 in., in length) with a shorter, more horizontal abdomen and long, slender palps. *R. fictilium* is generally not kleptoparasitic, but instead is an araneophagic predator, capturing and feeding on other spiders in their own webs. While widespread throughout the southeastern United States, along the Pacific coast *R. fictilium* is most commonly found in moist, near-shore coniferous forests, often with a dense undergrowth of ferns (D. Ubick, pers. comm., 2010) and has been collected from Monterey County, California, to southern British Columbia. *R. fictilium* was formerly housed in the genus *Argyrodes* and is discussed in greater detail in Exline and Levi's (1962) revision.

For many years several similar spiders were placed together in the genus *Achaearanea*. Each of these species lacks a colulus and, in females, has an abdomen that in profile is higher than it is long (Levi, 2005c). Recent revisionary work has split the genus apart based on fine details of the spiders' reproductive structures, and all of the North American spiders previously lumped into *Achaearanea* have been reassigned (Yoshida, 2008). The most familiar of these species is the cosmopolitan Common House Spider (*Parasteatoda tepidariorum*, pl. 14). It is found in garages, sheds, and undisturbed nooks throughout the region but is especially common in the larger cities of Southern and Central California. It is a medium-size spider (the female is normally between 5 and 6 mm, or .2 and .24 in., long) and highly variable in color, ranging from pale tan to nearly black, often with numerous gray chevrons along its abdomen's dorsoposterior surface. In profile, the female's spinnerets are quite far back along the ventral surface of her abdomen. This is in contrast to most of the other species previously classified as *Achaearanea*, in which the female's spinnerets are near the abdomen's ventral midpoint. The male *P. tepidariorum* is smaller (3.8 to 4.7 mm, or .15 to .19 in., in length) with a less bulbous abdomen and can occasionally be found inhabiting a female's web. The egg sacs of the Common House Spider are round with a beige, papery outer layer and are suspended within the female's web.

Nearly all of North America's spiders previously placed in *Achaearanea* are now in the genus *Cryptachaea* (pl. 16) with four representatives along the Pacific coast: *Cryptachaea fresno* is known from several locations in California's Sierra Nevada; *C. canionis* is widespread throughout the western United States; *C. porteri*, a species normally associated with the southeastern United States and Central America, has been collected in San Benito County, California (R.J. Adams, pers. coll., 2007); and *C. blattea* is a cosmopolitan spider with records from both California and Washington. Another former *Achaearanea* species, *Hentziectypus schullei* (not illustrated), is part of a predominantly Mexican and Central American genus and has been found in Imperial County, California. The Pacific coast's *Cryptachaea* and *Hentziectypus* are all diminutive spiders, with females ranging from 1.4 to 3.5 mm (.05 to .14 in.) in length. They build small cobwebs in thick bushes, under rocks, and beneath loose bark. Because of their small size and highly variable color patterns, the confirmation of a specimen's identity will nearly always require a detailed examination of its reproductive structures as outlined in Yoshida (2008) and Levi (1955, 1963b).

Theridion (pl. 17) is another genus that has undergone numerous revisions. Twenty-three species live along the Pacific coast, and it is not uncommon for multiple species to live in the same area. They are small spiders with females ranging from 1.3 to 4.8 mm (.05 to .19 in.) in length, and there is a great deal of diversity in color and pattern both within and between the different species. Features uniting the genus include the female's nearly spherical abdomen (except for *Theridion positivum*, whose abdomen is wider than it is long) and the lack of a colulus. *Theridion* range from white to dark brown with pink spots, and nearly all species have a patchy dorsal abdominal stripe. Paler species often build their cobwebs in foliage, while darker species are more likely to be found near the ground. The most comprehensive key to the region's *Theridion* is Levi (1963c).

Numerous spiders have been moved in and out of *Ther-*

idion, occasionally multiple times, illustrating the complexity involved in interpreting their systematic relationships. The other genera share many features with *Theridion,* and any attempt to positively identify a given individual will require a close examination of its reproductive structures, combined with a review of the relevant literature. *Ohlertidion ohlerti* (not illustrated) is a holarctic species that has been collected widely across Washington and Oregon and has been found in Yosemite National Park, California. *Neottiura bimaculata* (pl. 17), described in Levi (1956), and *Simitidion simile* (not illustrated) are also holarctic in distribution, with numerous records of both species from throughout Washington. Two species of *Platnickina* (not illustrated) have been found along the Pacific coast. *Platnickina alabamensis* has been found several times in San Diego County, California, and there are multiple records of *P. tincta* from Washington and Oregon. *Canalidion montanum* (not illustrated) is found across Eurasia and North America, with several records from both Oregon and Washington. *Rugathodes sexpunctatus* (pl. 17) is extremely common in the coastal and near-coastal areas from Southern California through northern Washington. One of the most distinctive of the species formerly placed in *Theridion* is *Wamba crispulus* (pl. 17), a small spider (the female spans 1.4 to 2.6 mm, or .05 to .1 in., in length) whose abdomen is wider than it is long, giving it a broadly oval shape when viewed from above. The dorsal surface of the female's abdomen is white with a large, conspicuous black patch on its anterior half. It is common in the southeastern United States, with numerous coastal records from Central California to southern Oregon. Identifying many of these spiders requires using the keys in Levi (1957e, 1963c) and then cross-checking those names against their current taxonomic placement in the online World Spider Catalog (Platnick, 2013).

Another distinctive species previously placed in *Theridion* is the introduced Red House Spider (*Nesticodes rufipes,* pl. 17). The female ranges from 4.2 to 5.3 mm (.16 to .21 in.) in length and has an orange-red cephalothorax and legs and a spherical, dusky-gray abdomen. The male is similarly colored but noticeably smaller (2.8 to 3.7 mm, or .11 to .15 in., in length) with a much smaller abdomen. Common in warm, tropical regions worldwide, *N. rufipes* has only a tenuous hold along the Pacific coast. Often found in shipments of fresh fruit, the Red House

Spider has never spread beyond a few scattered colonies and lone individuals. One population was found in the basement of the old California Academy of Sciences building in San Francisco, California, but disappeared when the building was demolished (D. Ubick, pers. comm., 2010). Other specimens have been found in fruit-packing plants as far north as Seattle, Washington. This species is reviewed in greater detail in Levi (1957e).

The genera *Euryopis* and *Dipoena* are especially interesting because they have given up hunting with webs and have become stalking and ambush predators. They wait on the ground and low foliage to hunt passing ants, sometimes many times their own size. Their chelicerae lack teeth and are armed with exceptionally long, flat fangs. Four species of *Euryopis* (pl. 15) have been found along the Pacific coast. They are small to medium-size spiders with females ranging from 1.3 to 4.7 mm (.05 to .18 in.) in length, and they have a distinctive subtriangular abdomen that comes to a point above the spinnerets. The most widespread *Euryopis* on the Pacific coast is *Euryopis formosa* (pl. 15). It is found across most of California (except for the desert regions) and through Oregon and Washington. It is a dark spider whose abdomen is black along the sides and below and is decorated dorsally by a large, dark, roughly triangular patch narrowly outlined in white. In Southern California's deserts, it is replaced by the much paler *E. californica* (pl. 15). This species has a pale yellow to brown cephalothorax and legs, and its abdomen is almost entirely grayish white with a large, black triangular patch on its dorsal surface. The two remaining regional species are known from only a few scattered records, *E. argentea* from northwestern Oregon and *E. spinigera* from several widely dispersed localities across California. While little has been written about any of the Pacific coast's species, the closely related *E. coki* has been observed hunting around ant mounds in Idaho. When an ant walks by, the spider flings strands of sticky silk over it and attaches it to the ground. Once the ant is entangled, the spider dispatches it with a bite to the leg. The spider then drags its prey away from the mound by a silk sling attached to its spinnerets. At times, up to a dozen spiders may be found gathered around a single ant mound (Porter and Eastmond, 1982). The American *Euryopis* are revised and illustrated in Levi (1954, 1963a).

Eleven described and several undescribed species of *Dipoena* (pl. 15) are known from the Pacific coast. Like *Euryopis,* they hunt ants on the ground and in foliage rather than building a web. Regionally the females range from 1.5 to 4.3 mm (.06 to .17 in.) in length, and in contrast to *Euryopis,* they have rounded, more globular abdomens. Males have unusual elevated carapaces with the ocular region raised quite high above the chelicerae, often crowned by a deep semicircular groove with numerous radiating furrows. Some species are boldly patterned, others are unmarked, and they vary in color from solid black to pale yellow with pinkish-gray abdomens. *Dipoena* are found across the region in cool, coastal forests, deserts, and sagebrush steppes. Additional information and identification keys can be found in Levi (1953, 1963a).

The genus *Enoplognatha* (pl. 16) is regionally represented by seven species, four native and three introduced. While all of the indigenous species and *Enoplognatha thoracica* (native to western Europe and the Mediterranean region) are fairly dark or marbled and live close to the ground, *E. ovata* (pl. 16) and *E. latimana* are pale yellow, often with carmine-red patches and black dots on the dorsal abdominal surface, and they make their webs in bushes and shrubs. *Enoplognatha* are small to medium-size theridiids, ranging from 2.9 to 7 mm (.11 to .28 in.) in length. This genus can be separated from similar cobweb weavers in the genera *Cryptachaea* and *Theridion* by its generally more ovoid abdomen and by the presence of a distinct colulus. Native to Europe, *E. ovata* has been introduced along both the Atlantic and Pacific coasts of North America and is common from Northern California through western Washington and southern British Columbia. It is a polymorphic species with three recognized major color patterns, although intermediates are occasionally reported. In the *lineata* form, the abdomen is entirely yellow with two longitudinal rows of black spots. In the *redimita* form, the spots are encompassed within a pair of red dorsolateral stripes. The least common form, *ovata,* has a broad, red shield covering most of the abdomen's dorsal surface. Each of these color morphs occurs throughout the range of the species and can even be found in individuals from the same brood. Because there is a sex-linked aspect to the genes that control *E. ovata* color patterns, adult males only display the *lineata* form. *E. latimana* is very similar to *E. ovata* and has also been introduced to the Pacific Northwest from Europe, although adult females demonstrate only the *lineata* and *redimita* morphs.

The most common of our native *Enoplognatha* is the Marbled Cobweb Spider (*E. marmorata,* pl. 16). Found widely across North America, it has been collected in wooded areas from Southern California to northern Washington. Like other dark *Enoplognatha,* it generally makes its web in a shaded area near the ground. The male and female are similar in size, with published records ranging from 3.9 to 7 mm (.15 to .28 in.) in length. Although there is quite a bit of variability in the depth of the colors and pattern, it usually has a heavy, spotted, silver-tan abdomen with numerous irregular dark-brown to black patches. Levi (1957e) reviewed the North American species of *Enoplognatha* but didn't include *E. latimana,* which was later described by Hippa and Oksala (1982). Since Levi (1957e) was published, the native *E. intrepida* has been collected from several locations in eastern and central Washington (Crawford, 1988).

Only a single species of *Robertus, R. vigerens* (pl. 17), is known from the Pacific coast states. It is a small (2.9 to 4.5 mm, or .11 to .18 in.), rather plain spider with a reddish-brown cephalothorax and pearly white eyes, although the anterior medians are slightly duskier than the rest. The abdomen is dark gray with minute light splotches, ovoid, and slightly flattened, and its colulus is large and conspicuous. The male has an unusual series of stridulatory ridges on the posterior margin of the carapace that align with a collection of stout setae on the anterior face of the abdomen. Little is known about this enigmatic spider other than that it lives in leaf litter, in moss mats, and under stones and boards. *R. vigerens* has been collected from San Francisco County, California, north through central and western Washington and into British Columbia. Kaston (1946) reviewed and illustrated the reproductive structures of the North American *Robertus* species under their previous generic name, *Ctenium.*

Anelosimus analyticus (pl. 17) is the only representative of its primarily tropical genus along the Pacific coast. Its abdomen is more ovoid than spherical and is grayish yellow with a broad, wavy, dark-gray to black band along the dorsal midline. It is a relatively small spider, with the female ranging from 3.2 to 4.4

mm (.13 to .17 in.) in length. The male is similarly patterned but generally darker and smaller, running between 2.5 and 2.8 mm (.1 and .11 in.) in length. The colulus is highly reduced and is represented by a pair of setae between the anterior lateral spinnerets. *A. analyticus* has been reported along the coast of Southern California in both San Diego County and Orange County. A detailed description and illustrations of its reproductive structures can be found in Levi (1956).

In addition to the spiders listed above, there are several genera whose members are extremely small, rare, or unlikely to be seen by anyone except the most diligent naturalist. In many cases nothing is known about these spiders beyond their original descriptions. Species in the genus *Thymoites* are minute denizens of leaf litter, low plants, and loose bark. They range from 1 to 2.5 mm (.04 to .1 in.) in size, and four species are known from the Pacific coast states. Like *Theridion,* they lack coluli and have rather spherical abdomens that vary from solid black to yellow with dark spots. *Thymoites* are smaller than most adult *Theridion* and have comparatively shorter legs. Levi (1957e) reviews and illustrates the North American species under their old generic name, *Paidisca.*

Theridula opulenta (not illustrated) is a tiny (1.6 to 2.6 mm, or .06 to .1 in., in length) spider with an exceptionally broad abdomen laterally tipped with small, dark tubercles. An individual from northern regions is normally bright red with a yellow dorsal abdominal patch and a black carapace. One from the southern United States is generally entirely black with several large white abdominal spots, and intermediates between these two very different color morphs are not uncommon across the range of the species. It lives in foliage where it constructs small cobwebs between the leaves. Levi (1966) reported a female specimen from Crater Lake National Park, Klamath County, Oregon.

Chrosiothes iviei (not illustrated) is a minute spider (around 2 mm, or .08 in., in length) that was described in Levi (1964). It has a nearly spherical abdomen with an entirely black center whose dark coloration extends up around the pedicel. It has been found in Imperial County and Orange County in Southern California. *Chrysso* (not illustrated) is represented regionally by two rare species, *Chrysso pelyx* and *C. nordica.* They are small, pale spiders with distinct tubercles on the dorsoposterior margin of the abdomen and have been collected sporadically across a large swath of the northwestern United States. Both species were illustrated and described by Levi (1957d) under the genus *Arctachaea,* a name that is still often used for these species. *Coleosoma normale* (not illustrated) is a tiny spider (around 1.5 mm, or .06 in., in length), and the female has a pale, globular abdomen with two parallel black stripes running the length of its dorsal surface. It has been reported from the south-eastern corner of California and is described in Bryant (1944).

Officially listed as a member of the family Anapidae (p. 82), *"Comaroma" mendocino* is a tiny (1 to 1.5 mm, or .04 to .06 in., in length), six-eyed spider with an abdominal scute that has been collected in Mendocino County, California. It was described by Levi (1957a), who placed it in *Archerius,* a new theridiid genus. Oi (1960) moved it into the otherwise Eurasian anapid genus *Comaroma*. However, the comb of serrated setae on tarsus IV and palpal structure identify it as a genuine theridiid. Until a review of the anatomy of *"Comaroma" mendocino* along with a formal re-designation is published, the rules of zoological nomenclature dictate that this species remain in the family Anapidae under its current name.

NATURAL HISTORY: Despite the cobweb weavers' great diversity and familiar presence, the life histories of only a few genera and species have been explored in any depth, with the majority of studies focusing on only the most synanthropic or medically important groups. As their common name implies, most theridiids build cobwebs. While tangle webs may appear to be a chaotic snarl of threads, they are architecturally complex structures whose features are designed to capture specific types of prey. There is also exceptional diversity between the webs built by different theridiid groups. Many of our most familiar genera, including Latrodectus, Steatoda, and Parasteatoda, build a tangle of stout support lines with numerous gumfoot threads. A gumfoot line is highly elastic and has a coating of sticky silk around its base, which is lightly glued to the substrate. When a small arthropod touches a gumfoot thread, it breaks the thread's attachment to the ground. The prey is then pulled into the tangle portion of the web, where it becomes further ensnared before the host spider subdues it with a bite and a wrapping of silk. Other theridiids, including those in the genus Theridion, don't use gumfoot lines. Instead, they lay an extensive network of viscid threads throughout their webs to catch small flying insects. A third major type of cobweb uses no viscid silk. This style is a highly complicated tangle that acts as a knockdown trap, forcing flying insects onto a sheet web near the spider's retreat. It is used by several genera, including Tidarren and Coleosoma.

Unlike orb web weavers (Araneidae, p. 110), which tear down and rebuild their webs on a daily basis, theridiids leave their webs up for extended periods, adding or replacing damaged portions as needed. While the majority of research on theridiid webs has focused on their efficiency at capturing prey, Blackledge et al. (2003) demonstrated that their three-dimensional structure also provides protection from several genera of mud dauber wasps (Sphecidae) that specialize in capturing spiders as hosts for their parasitic larvae. The cobweb's complex arrangement creates a tangled net around the spider's retreat that both hinders the attacking wasp and provides a longer warning period, allowing the spider to escape. While foraging diversification certainly was a major aspect in the evolution of three-dimensional cobwebs from ancestral two-dimensional orb webs, the added security they afford may have also played a role in their development.

Another function of webs is that of pheromonal beacon. When male spiders reach sexual maturity, they leave their own webs in search of receptive females. Male Western Black Widows, *Latrodectus hesperus,* use airborne pheromones carried off the females' webs to guide them across unfamiliar and dangerous territory. Research by Kasumovic and Andrade (2004) showed that male Western Black Widows were able to recognize the pheromones of conspecific females and could even distinguish between the webs of females originating from different regions, in this case preferring those from a British Columbia population (where the field tests were performed) over those descended from spiders collected in Arizona. While airborne pheromones help guide males to the females' webs, courtship is stimulated by the presence of a separate contact pheromone. This pheromone is thought to be less specific than the airborne pheromones, because males placed in the webs of different widow species will still initiate courtship behaviors (Ross and Smith, 1979).

There has only been limited research on the courtship behaviors of theridiids across the different genera, with the most intensive studies focused on the widow spiders. What has been published appears fairly consistent when compared with other Araneoidea families. When encountering a female, the male announces his presence by plucking at her web with his pedipalps and fore legs and rapidly vibrating his abdomen. If the female is receptive, she responds in kind. It may take several attempts on the male's part to initiate the female's interest, and if she is not receptive, she will either drive him from her web or kill and eat him. If the male is accepted, he slowly approaches the female, often cutting the surrounding threads as he works his way toward her. A Western Black Widow male then strokes the female's legs and abdomen before climbing on top of her. Additionally, he may throw several strands of silk over her, forming what is referred to in the arachnological literature as a bridal veil. The bridal veil is not strong enough to physically restrain the female but is thought to be coated with pheromones that keep her in a relaxed state, and despite popular belief, the males are not always eaten by the females after mating.

As in other large, diverse families, theridiid egg sacs are quite variable and in some cases are useful in identifying the spider that made them. The egg sacs of both *Latrodectus* and *Parasteatoda* are covered with a sheet of buffy, parchment-like silk, while those of *Steatoda grossa* are enveloped in a thick layer of loose, fluffy silk. In some species the egg sacs are clearly visible, hanging in the matrix of the female's web, while other species secrete them away in their retreats. Although post-hatching maternal care is unusual in spiders, *Theridion* mothers regurgitate food to their newly hatched spiderlings, providing them with additional nutrients before they balloon away (Kaston, 1965).

Most theridiids are considered generalist predators, feeding on almost any insects that stumble or fly into their webs. Some genera, however, are specialists. *Euryopis* and *Dipoena* capture ants by sticking them to the ground with swaths of thrown silk, and *Rhomphaea fictilium* is an araneophage that hunts other spiders in their own webs. The presence of spiders can have a striking effect on the behavior of their prey. Like many desert-dwelling theridiids, Western Black Widows are well-known predators of large ants. A male or juvenile will often build a web in the vegetation around the periphery of an ant mound, capturing workers along their foraging trails.

In response to this predation pressure, the ants will close up their nest entrance for several days, cutting off food supplies to the spiders and encouraging them to leave (MacKay, 1982). Another common spider of the desert Southwest, *Steatoda fulva*, builds its web directly above the entrance to a harvester ant nest. By constructing its web during the hottest portion of the day when ant activity is at its lowest, the spider is able to quickly trap numerous individuals in the late afternoon when foraging resumes. The ants normally respond by closing off the affected entrance and opening a new one several feet away a few days later (Hölldobler, 1970).

ARANEIDAE — Orb Web Weavers

Pls. 18–22, Fig. 11

IDENTIFICATION: Araneids are among the most familiar and morphologically diverse of North America's spider families. With their characteristic web design, orb web weavers are what come to mind when most people think of spiders. They range from 1.5 to 30 mm (.06 to 1.18 in.) in length, and their abdomens can be smooth, lobed, or spiny. While many species are cryptically colored, others are brightly decorated. With one rare exception, the region's araneids build vertically inclined orb webs, some of which include stabilimenta across their hubs. Like other members of the superfamily Araneoidea, including the long-jawed orb weavers (Tetragnathidae, p. 126) and the cobweb weavers (Theridiidae, p. 93), orb web weavers are eight-eyed, three-clawed entelegynes. The anatomical features that unite Araneidae are subtle and best seen under a microscope. Fortunately, the webs are characteristic enough that the familial identification of most individuals is straightforward. The challenge comes when identifying a spider away from its web, such as one taken in a sweep net, or a wandering male. Orb web weavers' legs are adorned with numerous heavy spines, their clypei are fairly low (less than three times the diameter of the anterior median eyes), and their endites are either squarish or slightly rectangular. Females usually have discernible scapes, and the males normally have squat palpal tibiae that are wider than they are long, with uneven distal margins. A detailed key to North America's araneid genera can be found in Levi (2005a).

SIMILAR FAMILIES: The spiders most likely to be mistaken for orb web weavers are the long-jawed orb weavers (Tetragnathidae, p. 126). While tetragnathids tend to build either horizontal or inclined webs, this is not an absolute, and vertical webs are fairly common among some genera. Distinguishing these two families often entails closely examining the spider's reproductive features. Male tetragnathid have long, narrow palpal tibiae, and females lack well-defined epigynal scapes, unlike most Araneidae. The exception are spiders in the araneid genera *Zygiella* and *Parazygiella*. In these genera, the males have elongate, cone-shaped palpal tibiae and the females have flat or swollen transverse epigyna without scapes. Fortunately, *Zygiella* and *Parazygiella* generally build characteristic, easily recognized webs (see below under Pacific Coast Fauna). *Hypsosinga* is the only other regional araneid genus whose females lack epigynal scapes. It is represented along the Pacific coast by a pair of uncommon, fairly distinctive spider species. In addition to their characteristic reproductive structures, tetragnathids have elongate endites that often widen distally, in contrast to the more square endites of the araneids. Fortunately, with a good hand lens, this feature can often be seen when a spider is resting in the hub of its web.

PACIFIC COAST FAUNA: Seventeen genera with 55 regional species. Among the largest and most charismatic of the region's araneids are the diurnal garden orb weavers, genus *Argiope* (pl. 18). Four species have been recorded along the Pacific coast, three of which are regularly found in suburban gardens and grassy fields. Each species of garden orb weaver has characteristic black, yellow, and white markings on its abdomen, and the posterior median eyes are located behind the posterior laterals. In most other orb web weaver genera, the posterior median eyes are either in line with, or anterior to, the posterior laterals. Female garden orb weavers are large, with the three most common species ranging from 14 to 26 mm (.55 to 1.02 in.) in length. The males are less strikingly patterned and distinctly smaller (4 to 5.8 mm, or .16 to .23 in., in length). The most widespread of these attractive spiders is the Banded Garden Spider (*Argiope trifasciata*, pl. 18). It is found from Canada to southern South America and is common across large areas of the Pacific coast. The Black and Yellow Garden Spider (*A. aurantia*, pl. 18), with its contrasting black and yellow abdominal markings, is less common, with a more coastal distribution, and is known from Southern California to northern Oregon. The adults of both species include a stabilimentum of zigzagging, ribbon-like silk through the hub of the web. The Silver Garden Spider (*A. argentata,* pl. 18) has a silverish sheen across the base and edges of its abdomen, along with several large lateral lobes. It ranges from Southern to Central California with records from both Monterey and San Benito County (R.J. Adams, pers. coll. and obs., 2005). The Mexican Garden Spider, *A. blanda,* is a regionally rare species that has been collected in San Diego County, California. It is similar to the Silver Garden Spider, but the female Mexican Garden Spider is smaller, ranging from 9 to 14 mm (.35 to .55 in.) in length, and tends to have a wider white transverse band on the ventral side of the abdomen. In some cases, a close examination of the spiders' reproductive structures may be necessary to confidently differentiate these two species. Unlike the vertical stabilimenta of the Black and Yellow and Banded Garden Spiders, adult Mexican and Silver Garden Spiders construct a conspicuous X shape through the hubs of their webs. The egg sacs of the Black and Yellow and the Banded Garden Spiders are roughly spherical, covered with brown parchment-like silk, and are hidden among the leaves of nearby bushes or, not infrequently, under the protective eaves of a house. The Silver Garden Spider's egg sac is more triangular and, when suspended in a bush, looks like a dead, slightly chewed leaf. Levi (1968) provides additional descriptions, illustrations, and a key to the American *Argiope* species.

The only other Araneidae to share the garden orb weaver's unusual eye arrangement is the boldly marked *Gea heptagon* (pl. 21). This relatively small spider (2.6 to 4.5 mm, or .1 to .18 in., in length) has a roughly heptagonal abdomen that's often adorned with lateral lobes along its posterior margin and dorsal tubercles near its anterior edge. It may also have a dark, backward-pointing triangular mark on the posterior portion of the abdomen. According to Levi (1968), the regional distribution *G. heptagon* is limited to the Southern California coast, where its small, tightly woven orb webs are found close to the ground in grassy fields. Occasionally the spider will remove a large wedge-shaped section on the lower half of its web, a behavior whose purpose is still not understood. When disturbed, *G. heptagon* drops from its web, folds its legs tightly over its body, and quickly changes its coloration. It is able to turn the white portions of its body dark brown to more effectively hide in debris below its web. Its egg sac is triangular and flat and looks like a piece of dried leaf. Placed away from the main web, the egg sac is protected by a barrier web and is camouflaged by an irregular cross-hatching of dark silk threads.

With 16 known Pacific coast species, *Araneus* (pl. 19) is the most numerous and morphologically diverse of the region's araneid genera. Also known as the round-shouldered and angulate orb weavers, *Araneus* can be especially challenging to identify, because its members are often quite variable, both within and between species. United by shared details of their reproductive structures, they can be large or small, strikingly colored or drab, and although many species have conspicuous "shoulder" humps, others have smoothly rounded abdomens. Levi (1971a, 1973) divided the genus into three broad categories: the *diadematus* group, which contains our largest and most common species, and the smaller and more arboreal *pegnia* and *sturmi* groups. Pacific coast *Araneus* males range from 3.2 to 11 mm (.13 to .43 in.) in length, while females are distinctly larger and as a genus range from 4 to 18 mm (.16 to .71 in.) in length. *Araneus* includes some of our most abundant and distinctively marked fall-maturing spiders. These include the Shamrock Orb Weaver, *Araneus trifolium*; the introduced Cross Orb Weaver, *A. diadematus*; *A. andrewsi;* and the Cat-faced Spider, *A. gemmoides* (plate 19), which is known to hybridize with the very similar *A. gemma*. In contrast to the frequently encountered larger *Araneus* are the smaller, less often seen species, including, among others, *A. montereyensis, A. bispinosus,* and *A. detrimentosus* (pl. 19). Unlike the larger *Araneus* species that prefer open areas, these spiders live predominately in thickly wooded areas and in the forest canopy. For such a widespread genus, even basic life history information is lacking for many species. A concentrated study of even a single population would certainly discover new information about their lives. In addition to the references listed above, Levi (1975a) discusses several new *Araneus* records from Oregon.

There are a few genera that are similar to *Araneus* orb web weavers in their overall appearance and habits, differing mainly in their reproductive structures. Members of the genus *Neoscona* (pl. 20), also known as spotted orb weavers, can be distinguished by the presence of a longitudinal thoracic groove. In *Araneus,* the groove is indistinct and, when visible, transversely crosses the carapace. On the ventral abdominal surface, *Neoscona* have a rather sharply defined black square bordered by one or two white patches. However, this easily seen characteristic is shared by members of several other araneid genera, including *Larinioides* and some *Araneus,* making it only a secondary identification characteristic. Three species of spotted orb weavers are known from the Pacific coast region. To identify adult females, the dorsal abdominal color pattern can be useful, but it is not as reliable for the smaller, more variable juveniles or males. The most widespread of the region's *Neoscona* is the Arabesque Orb Weaver (*Neoscona arabesca,* pl. 20). This small to medium-size spider (4.2 to 12.3 mm, or .17 to .48 in.) is especially common in moist meadows and in low trees and shrubs, especially along the coast. The adult female has an oval abdomen with numerous angled black dashes bordering a paler central line. The slightly larger Western Spotted Orb Weaver (*N. oaxacensis,* pl. 20) has a longer, narrower abdomen that is mainly black with paired light dashes along the midline, along with a peppering of small white spots and broad, irregular lateral patches. Exceptionally common throughout California, it becomes progressively rarer as one moves north into southern and eastern Washington. The least common of the Pacific coast's *Neoscona* is also generally the largest. *N. crucifera* (pl. 20) is widespread across the southeastern United States and is believed to have been introduced to Southern California, with records from both San Diego and Orange County. The male ranges from 4.5 to 15 mm (.18 to .59 in.), and the female measures between 8.5 and 19.7 mm (.33 and .78 in.) in length. *N. crucifera* is usually orangish brown with a pale stripe along the dorsal abdominal midline and is more prevalent in chaparral and dry oak woodlands than the other two *Neoscona* species. Its egg sac is relatively flat or lens shaped and composed mainly of loose, fluffy silk and is either hung in the spider's retreat or is hidden inside a nearby curled leaf. A more comprehensive review of *Neoscona* anatomy and taxonomy can be found in Berman and Levi (1971), where *N. crucifera* is referred to by the now obsolete name *Neoscona hentzii.*

Three species of *Larinioides* (pl. 20) are found along the Pacific coast, each of which has a dorsoventrally flattened, oval abdomen with a large central foliate mark. On the underside of the abdomen, just anterior to the spinnerets, is a black patch bordered by two white comma-shaped marks. The region's most common species, *Larinioides patagiatus* (pl. 20), is a moderately large spider, with the female ranging from 5.5 to 11 mm (.20 to .43 in.) in length. The smaller male is between 5.8 and 6.5 mm (.23 and .26 in.) long. *L. patagiatus* can be found in a variety of cool, shaded areas, often near water, and has frequently been found along freshwater docks and piers. As numerous records from coastal California show, it becomes increasingly abundant farther north and is quite common throughout western Oregon and across most of Washington. The similar Furrow Orb Weaver (*L. cornutus,* pl. 20) is a rare and sporadic member of the western araneofauna, with less than two dozen published records from the coast of Central California to northern Washington. The Bridge Orb Weaver, *L. sclopetarius,* is a predominately Eurasian species thought to have been introduced to North America. While it is common in the northeastern United States, there are only a few regional records from western Washington. The Bridge Orb Weaver differs from our two native species by the presence of a distinct white border around the margin of its carapace and a gray cross-like pattern on its abdomen. All of these spiders make rather flimsy looking, open orb webs with widely spaced radii and viscous threads. It has been noted that under optimum conditions, *Larinioides* can become extremely numerous, with individuals of different ages building overlapping and interconnected orb webs. A key to most of the world's *Larinioides* species, including all of the North American fauna, can be found in Levi (1974a) under their earlier placement in the genus *Nuctenea.*

Spiders in the genus *Cyclosa* (pl. 21, fig. 11) are among the most common orb web weavers along the Pacific coast. Known as trash line orb weavers, they are often found in suburban gardens, woodlots, and coastal scrub. Each of the three regional species is distinctive, and for the most part, the adult female is identifiable by the shape and position of the abdominal tubercles. The smaller, darker male often requires a close examination of the reproductive structures to confirm its identity. *Cyclosa* are relatively small spiders, with Pacific coast females ranging from 3.6 to 7.9 mm (.14 to .31 in.) in length. Trash line orb weavers earn their name by hanging prey remains, assorted

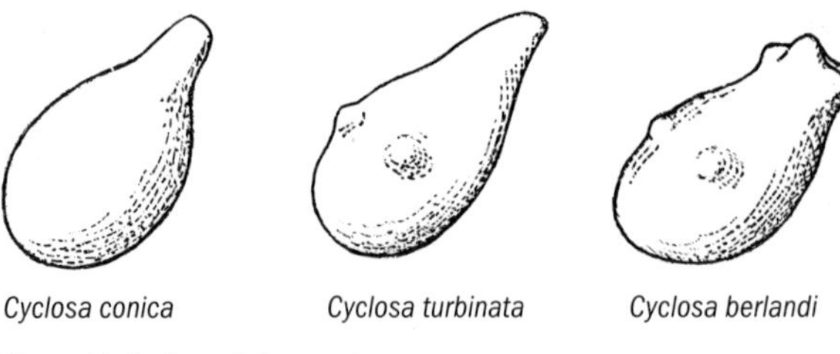

Figure 11. *Cyclosa* abdomen shapes.

debris, and their brownish-gray egg sacs in a vertical line across the hub of the web. The spider then rests in the center of this ornamented stabilimentum, hidden amid the accumulated refuse. The most widespread of the region's three species is *Cyclosa conica* (pl. 21, see fig. 11). This spider has a Holarctic distribution and is common in forested areas throughout the western United States. The female's abdominal tubercle is thick and broadly triangular. According to Levi (1977a), specimens from the Pacific Northwest show the greatest regional variation in size, and unlike the other *Cyclosa* species, the female *C. conica* attaches her elliptical, yellowish-brown egg sac to a twig or under a nearby leaf. *C. turbinata* (see fig. 11) is similar to *C. conica* but generally smaller. The posterior abdominal tubercle of the female is narrower in a dorsal view and looks more taillike than the triangular projection of *C. conica*. Additionally, *C. turbinata* often has a small pair of "shoulder" bumps near the abdomen's anterodorsal margin, a feature missing from *C. conica*. Though *C. turbinata* is widespread throughout the southeastern United States, in this region it is found almost exclusively in coastal and near-coastal areas from Southern California to Puget Sound, Washington. The third species, *C. berlandi* (see fig. 11), was only recently described by Levi (1999). With four tubercles on the posterior tip of its abdomen and two more on its dorsal surface, *C. berlandi* is the most distinctive member of the region's *Cyclosa* species and lives along the coast of California north to the San Francisco Bay Area. All three species make small orb webs in brush and tall grasses and when disturbed will rapidly vibrate up and down, blurring their outlines to any potential predators.

Eustala (pl. 21) is unique among the region's orb web weavers in that the female's epigynal scape is directed anteriorly. The abdomen is broadly triangular with rounded anterior corners. Additionally, many species have one or more small tubercles on the abdomen's posterodorsal margin. Of the four species known from the Pacific coast, the most common is *Eustala rosae* (pl. 21). This is a medium-size spider found in wooded areas across much of California, with at least one record from southwestern Oregon. The female ranges from 6.8 to 9 mm (.27 to .35 in.) in length while the smaller male is between 5 and 5.9 mm (.20 and .23 in.) long. There's a fair degree of variation in *Eustala* abdominal color patterns, even within the same species. The most prevalent pattern on the dorsal abdominal surface of *E. rosae* is a gray or tan background with a dark folium bisected by a black, longitudinal line. Additionally, there is often a small tubercle on the posterodorsal margin and two more on the abdomen's posterior surface above the spinnerets. Regarding the region's other three *Eustala* species, *E. californiensis* and *E. conchlea* are regionally limited to the coast of Southern and Central California while *E. anastera,* a predominately eastern species, has been found sporadically in all three of the contiguous Pacific coast states. *Eustala* are nocturnal spiders that rest by day tightly pressed to twigs or other structures near their webs. A detailed review of the region's species can be found in Levi (1977a).

The genus *Metepeira* (pl. 21) contains some of the most easily recognized spiders on the Pacific coast, not because of their striking color pattern, but because of the structure of their webs. They build a unique composite web in which an orb web is hung within the maze of a tangle-like barrier web. Suspended within the barrier is the spider's cuplike retreat made of stitched-together leaves. Of the seven regional species, only two, *Metepeira grandiosa* and *M. foxi,* make it as far north as Washington. In those two species, the underside of the abdomen has a black patch bisected by a narrow white line (it's broad in Southern California's *M. gosoga*), and the eye region is lighter in color than the posterior portion of the carapace. They have spherical or ovoid abdomens with ornate, dorsal foliate patterns. The Pacific coast's resident species range in size from 3.6 to 9.4 mm (.14 to .37 in.) for females and 2.6 to 6.7 mm (.1 to .26 in.) for males. *M. spinipes* (pl. 21), a common fall-maturing species along the coast north to central Oregon, is unique among the region's orb web weavers in that it is considered a communal-territorial spider. Aggregations of up to several hundred individuals build intricate, interconnected webs. While the barrier and scaffold threads attach to one another, each spider maintains the territorial integrity of its own orb. Research by Uetz and Cangialosi (1986) has shown that social web-building behaviors in *M. spinipes* have evolved in populations with a rich and steady prey base, whereas those in drier, less productive areas are more likely to be solitary. Levi (1977b) provides a solid overview of the North American fauna, and any in-depth study should also include Piel's (2001) revision of the Neotropic species.

Superficially similar to *Metepeira* is *Aculepeira packardi* (pl. 20), the only regional representative of its genus. Like *Metepeira, A. packardi* has a well-defined folium on the dorsal surface of its abdomen. The ventral side is decorated with a bold white stripe bisecting a black field that is bordered on either side by narrow white bands. Its egg-shaped abdomen is widest across its anterior portion. *A. packardi* is a medium to large spider, with the adult female spanning 5.6 to 16.5 mm (.22 to .65 in.) in length. The male is significantly smaller, ranging from 5 to 8.9 mm (.20 to .35 in.) in length. Uncommon but widespread, it has been collected across the Pacific coast, from Southern California to northern Washington. Preferring cooler, more mountainous regions over hotter valleys and deserts, it has been found in chaparral, open woodlands, and grassy fields. *A. packardi* is a nocturnal spider, and the female normally spends the day hidden in a shallow retreat of leaves attached near the edge of her web. A key to North America's *Aculepeira* can be found in Levi (1977b).

Three species of *Larinia* (pl. 20) have been found along the Pacific coast, although only one, *Larinia directa,* can be considered a regular part of the region's fauna. On all *Larinia*, the abdomen is elongate. On the underside is a black patch that is split by a longitudinal white stripe that, in turn, is occasionally divided by a thin black line. The abdomen's dorsal pattern consists of several longitudinal tan, brown, and yellowish gray stripes similar to those found on the slender crab spiders in the genus *Tibellus* (Philodromidae, p. 187). However, on *Larinia* the markings don't extend onto the carapace. Additionally, crab spiders are wandering hunters and never build orb webs. In most species of *Larinia* there is a small tubercle on the abdomen's anterior margin, giving it a slightly forward-pointing look. *L. directa* is a medium-size spider, with records of the adult female running from 4.8 to 11.7 mm (.19 to .46 in.) in length while the male spans 4.5 to 6.5 mm (.18 to .26 in.) in length. It is widespread in Mexico and has been collected in grassy fields along the coast of Southern California north to Santa Barbara County. By day *Larinia* can be found tightly pressed against plant stems or blades of grass, but at night they wait for prey, resting head down in the hubs of their webs. Two other species, *L. borealis* and *L. famulatoria,* have been found sporadically along the Pacific coast. *L. borealis* has been collected in both California and Washington, and *L. famulatoria,* normally a spider of mountainous regions east of the Great Basin, has been found on several occasions in Southern California. A key to these spiders can be found in Levi (1975b).

Two orb web weaver genera often recognized by the structure of their webs are *Zygiella* (pl. 22) and *Parazygiella* (pl. 22). Until recently, all of the North American species, introduced and native, were grouped together in the genus *Zygiella*. Wunderlich (2004) split the genus based on differences in the spiders' reproductive structures, leaving the Pacific coast with two introduced *Zygiella* and two native *Parazygiella* species. Both genera build orbs between 6 and 14 cm in diameter. A narrow wedge is often missing from the orb's upper half, and a signal thread runs from the web's hub through this missing section to the spider's hidden retreat. In both of these genera, the male's palpal tibiae are elongate and the females have only lightly sclerotized epigyna, features similar to those of members of the family Tetragnathidae (p. 126). Both genera are yellowish gray with black and silver foliate marks on the dorsal surface of the abdomen. Although many of the features used to differentiate the different species are parts of their reproductive structures, with experience you can recognize many individuals in the field by their overall coloration and pattern. Of the two regional *Zygiella,* the most widespread is *Zygiella x-notata* (pl. 22). Introduced from Europe, *Z. x-notata* is commonly found around homes and gardens along the coast of California through western Oregon and Washington. Another European native, *Z. atrica* is regionally known only from northwestern Washington (and southern British Columbia). The male *Z. atrica* can be readily identified by its exceptionally long, thin palpal tibia. Pacific coast *Zygiella* are between 5 and 8 mm (.20 to .31 in.) in length, and as in other spiders, the males are generally smaller than the females. Of our two native species, *Parazygiella carpenteri* appears to be unique within the genus in that its small orb web normally lacks a missing wedge. On average, it is smaller than the other species within the *Zygiella-Parazygiella* complex, averaging 5.5 mm (.22 in.) in length with little difference between males and females (Gertsch, 1964). While it is predominantly a spider of the Sierra Nevada, there are also records from both the California coast and southeastern Washington. It is most common on large conifers, where it places its web across a crevice in the tree's bark. *Parazygiella dispar* is a Holarctic species with a preponderance of records from the immediate coast of southern British Columbia to Monterey County, California. Limited measurements imply that this is a fairly large spider, with the male around 7 mm (.28 in.) and the female around 10 mm (.39 in.) in length, although according to Levi (1974b), individuals from California are distinctly larger than those from Alaska. All of these species are reviewed under the genus *Zygiella* by Gertsch (1964) and Levi (1974b).

The only representative of the predominantly tropical genus *Eriophora* in the western United States is *Eriophora edax* (pl. 21). This is a moderately large spider, with the female 12 to 16 mm (.47 to .63 in.) in length and the male 8 to 12 mm (.31 to .47 in.) long. *E. edax* is easily recognized by its broadly triangular abdomen, which is ventrally marked by a crisply defined black trapezoidal patch surrounded by a yellow or white border. It often has a white stripe and two tubercles along its dorsal abdominal midline: a larger one near its anterior margin, and a smaller, more pronounced one near the abdomen's posterior tip. The female's epigynal scape is long and swordlike, regularly stretching nearly halfway across her abdomen, and can often be seen while she rests on her web. *E. edax* is generally reddish brown above with orange or grayish yellow patches along the abdomen's sides. An entirely nocturnal spider, it has been found in numerous locations throughout Southern and Central California north to Contra Costa County. Detailed illustrations of its reproductive structures and measurements can be found in Levi (1971b).

Two species of the diminutive orb web weaver *Hypsosinga*

live along the Pacific coast: *Hypsosinga pygmaea* (pl. 22) in Washington and *H. funebris* (pl. 22) in Southern California. They range from 2.2 to 5.1 mm (.09 to .2 in.) in length and are exceptionally variable in their overall coloration. Regionally, *Hypsosinga* have an orange carapace with black patches around the eyes and an abdomen that varies from pale orange with dusky stripes to boldly marked in black and white. The northern species *H. pygmaea* is quite rare in the western United States, with only a few records from Washington. Its abdomen is generally black with five sharply contrasting white stripes: two lateral, two ventral, and one dorsal. The abdomen on *H. funebris* varies from orange to black. It normally has a broad white folium on its dorsal abdominal surface that connects with lateral stripes along its anterior margin, leaving two irregular black bands converging at the abdomen's posterior edge. Along the Pacific coast, *H. funebris* is known along the coast of California north to Monterey County. Both species are diurnal and have been collected in wet meadows and weedy fields. The genus was first revised by Levi (1972), in which both of our regional species were described in great detail but under different names (*H. variabilis* and *H. singaeformis*, respectively). Later, Levi (1975a) recognized that both of these spiders were identical to species described earlier, thus requiring that the previous names be dropped, a move that was not without some controversy (Crawford, 1988).

Araniella displicata (pl. 22) is one of the most colorful of the region's small orb web weavers. Also known as the Six-spotted Orb Weaver, it is found widely throughout the Pacific coast from Southern California to northern Canada. It builds small orb webs among the leaves of bushes and deciduous trees, often placing its entire web across the curved surface of a single broad leaf. It has three (rarely four) small black spots on either side of its abdomen. Many individuals are chalky white with a yellowish wash, while others are brushed with swatches of brown, green, pink, and red. *A. displicata* ranges from 4 to 7.2 mm (.16 to .28 in.) in length, and like other spiders, the female is significantly larger than the male. The spiders mature over summer and are diurnally active, and their egg sacs are loose, cottony balls attached to leaves or twigs. Additional information and a key to North America's *Araniella* species can be found in Levi (1974a).

Without a doubt, the female Spiny-backed Orb Weaver (*Gasteracantha cancriformis,* pl. 22) is one of the most distinctive spiders in North America. It has a flat, broad, white or yellow abdomen in which the dorsal surface is dotted with numerous black pits and is ringed by six large, red (occasionally black) spines. The two most posterior spines are separated from the bulk of the abdomen by a deep longitudinal groove. The spider's cephalothorax and legs are black, and the adult female ranges from 5.8 to 8.6 mm (.23 to .34 in.) in length. The male is only 1.9 to 2.7 mm (.07 to .11 in.) long, grayish, and rather beetle-like. Although it lacks spines, it does have a prominent longitudinal groove crossing the posterior margin of its compressed, gray abdomen. The Spiny-backed Orb Weaver is an uncommon resident of Southern California, with published records from San Diego County and Orange County along the coast and from San Bernardino County inland. Its large orb web, around 30 cm in diameter, is often built at an angle. The spider spends most of the day resting in the hub, with its dark underside facing up, a behavior that may have both camouflage and thermoregulatory benefits. The Spiny-backed Orb Weaver's egg sac is a flattened, oblong mass of loosely tangled white, yellow, or greenish threads, often bisected by a dense longitudinal stripe of green silk, and is secured to the underside of a leaf or branch near the female's web. Levi (1978) discusses and illustrates *G. cancriformis* in detail, and additional life history notes may be found in Muma (1971) and Gregory (1989).

Mastophora cornigera (pl. 22) is the only bolas spider along the Pacific coast. Undoubtedly, *Mastophora* are among the oddest spiders in the Americas. Rather than waiting for an insect to land in its web, the adult or subadult female bolas spider actively hunts moths by swinging a short strand of silk with a ball of glue at the end. The adult female *Mastophora* emits pheromones that mimic those used by specific groups of female moths. When a male moth approaches, the spider swings the strand of silk and glue through the air, striking and ensnaring the duped moth. The pheromones released by each species of *Mastophora* attract only a few kinds of moth. This allows different species of bolas spiders to live in the same area without competing for food. They also adjust the ratio of pheromones over the course of the evening to capture moths that fly at different times (Haynes et al., 2002). Juvenile female and tiny male bolas spiders use pheromones as well, but they attract male moth flies (order Diptera, family Psychodidae). Rather than using balls of gluey silk, they grapple the fly with their heavily spined legs. Experimental work by Yeargan and Quate (1996, 1997) demonstrated that, like the adult female *Mastophora* using selective pheromonal signals, male bolas spiders of different species were attracting different species of moth fly. Like other members of the genus, *M. cornigera* rests by day on a leaf or twig where it looks like a small bud or bird dropping. The adult female has a bulbous, humped abdomen and branching, coral-like tubercles on the carapace. The female is moderately large (8.8 to 14 mm, or .35 to .55 in., in length) and diffusely patterned in buff, yellow, and pinkish gray. The male is minute (around 1.5 mm, .06 in., long) and also has the bulbous abdomen but lacks the carapace tubercles. Male *M. cornigera* emerge from their egg sacs as mature adults, effectively inhibiting inbreeding, as their female siblings still have several months until they became sexually mature. Each female lays a multitude of stalked, mottled egg sacs and suspends them from a network of heavy threads in the branches of a tree or under the eaves of a building. Bolas spiders are uncommon and cryptic, and in many cases, their distinctive egg sacs are the only visible evidence of their presence. Regionally, *M. cornigera* is found along the California coast from San Diego County north to the San Francisco Bay. A revision of the genus can be found in Levi (2003).

NATURAL HISTORY: As expected with such a large and diverse family, the natural history of the orb web weavers is a complicated affair and only a few species have been studied in any detail. The building of an orb web is a fairly standard procedure, although some differences can be found across the breadth of araneid genera and species. Generally, the spider begins by releasing a strand of silk from an exposed point, allowing the wind to carry it until it catches on another structure. After it is pulled taut, this bridge line provides the upper boundary of the web's framework. For added support, many species lay several additional lines of silk along this strand. Under and parallel to the bridge line, the spider lays down a slack line connected in the middle by another thread that runs to the substrate below. This creates a Y-shaped scaffold beneath the initial bridge line that becomes the centerpiece of the spider's web. The spider then adds several additional frame lines, enclosing the Y in an inverted triangle. Using the axis of the Y as the web's hub, the spider runs numerous radial threads (radii) to the frame, much like a bicycle's spokes connecting a wheel's hub to its rim. The spider uses strong, nonsticky threads for both the frame and radii. Each species tends to build webs with similar radial densities. For example, *Cyclosa conica* generally has between 40 and 50 radii in its web, whereas *Neoscona arabesca* usually has around 20. From here, the spider can take one of two routes. One route is to install a nonsticky auxiliary spiral around the web. The line loosely circles outward and provides additional support to the radii. The spider then switches to sticky silk. Beginning at the periphery of the web, the spider slowly circles inward, carefully attaching the glue-covered thread to each radial line as it goes. As it works back to the center, it removes and eats the threads from the auxiliary spiral, allowing it to recycle its proteins for later use. The alternate route is to forego the auxiliary spiral and go directly from constructing the radii to laying down the sticky silk. In some species the spider normally leaves an open zone around the hub that allows it to easily move from one side of the web to the other.

Another web feature common to several genera, including the garden spiders (genus *Argiope*), is the stabilimentum: a silk ribbon spun between two or more radii either as a vertical bar or as an X shape through the hub. Several purposes have been proposed for the stabilimentum, although most are not mutually exclusive. Experimental evidence has shown that it may act as an advertisement to flying birds, warning them of the web's presence. It is also thought to help spiders thermoregulate by blocking heat from the sun, attract insects by its UV-reflective nature (a feature shared with many flowers), and provide camouflage, especially for the trash line orb weavers (genus *Cyclosa*). For such intricate structures, orb webs are remarkably ephemeral. Most orb web weavers build a new web daily, leaving only the framework in place and either tearing down or consuming the old strands. They will also readily abandon a site that's unproductive or experiencing harsh environmental conditions.

Like most spiders, araneids are often considered generalist predators. However, because of variations in design, chemical composition, and placement, their different webs tend to capture different groups of prey. In addition to the well-known bolas spiders (genus *Mastophora*), which use pheromones to attract prey, other spiders may also make use of pheromonal mimicry. Observational evidence strongly suggests that the webs of the Banded Garden Spider *(Argiope trifasciata),* the Black and Yellow Garden Spider *(A. aurantia),* and the Shamrock Orb Weaver *(Araneus trifolium)* actively attract male New England Buckmoths (*Hemileuca lucina,* Saturniidae) (Horton, 1979). Uetz et al. (1978) found that the difference in mesh size between Black and Yellow Garden Spider and Banded Garden Spider webs dramatically affects the prey they capture, even when the webs are nearly side by side. Spiders can also differentiate between easily controlled prey and prey that are potentially dangerous. A spider may choose to ignore large, aggressive prey until they escape, or to cut them from the web if they become moderately entangled. Some orb web weavers actively attack larger, more calorically rich insects while ignoring smaller ones, even when they are significantly more numerous (Uetz and Hartsock, 1987). Many araneids capture more aggressive prey, such as grasshoppers and wasps, by initiating a wrap-and-bite strategy. This allows them to immobilize a captured insect before having to come in close and risk injury. Smaller species that feed on unaggressive insects often bite their prey first to keep it from escaping, and then begin wrapping it.

Orb web weaver courtship has only been studied in a few species, but there appear to be some general behaviors common to most members of the family. Because males are often distinctly smaller than females, they must be extremely careful to approach females' webs correctly or they risk being mistaken for prey. An araneid male normally begins by plucking at the periphery of a female's web. Based on observations of the Cross Orb Weaver, *Araneus diadematus,* and the Marbled Orb Weaver, *A. marmoreus,* a receptive female will respond by plucking a signal of her own back to the male (Blanke, 1986). The male then attaches a mating thread to the female's web. The female climbs onto the mating thread, where she hangs, ventral side up. The male approaches her in this position, lightly tapping on her mouth and sternum before mating. After copulation, the male quickly retreats from the female's web.

TETRAGNATHIDAE — Long-jawed Orb Weavers

Pls. 23, 24

IDENTIFICATION: Tetragnathidae is a morphologically diverse family of orb weavers regularly found near streams, marshes, and other waterways. Collectively, they are occasionally referred to as long-jawed orb weavers (a common name also applied specifically to members of the genus *Tetragnatha*) because the chelicerae on many species are exceptionally long and adorned with numerous stout teeth. While individual genera are fairly recognizable, the features uniting the family as a whole are subtle and often difficult to discern. Tetragnathidae is part of a larger clade of spiders called Araneoidea that houses some of our most familiar spider families, including the orb web weavers (Araneidae, p. 110) and the cobweb weavers (Theridiidae, p. 93), all of which have eight eyes, three tarsal claws, and at most a single tarsal trichobothrium. Tetragnathids have legs decorated with numerous spines and a clypeus that is fairly low (less than three times the diameter of the anterior median eyes). Both sexes have endites that are rectangular, many of which widen distally. The males' palpal tibiae are elongate and conical (except in the genus *Meta,* whose palpal tibiae are quite short), and the females' epigyna are absent or reduced to a flat plate or only a simple bulb-like protuberance. In addition to their anatomy, there are also some behavioral traits that are widespread across the family. With one rare exception, tetragnathids usually build horizontal or inclined orb webs with open hubs, although vertical and near-vertical webs are not uncommon in some genera. A key to North America's long-jawed orb weaver genera can be found in Levi (2005b).

SIMILAR FAMILIES: Tetragnathids are superficially most similar to the orb web weavers (Araneidae, p. 110); however, none of the Pacific coast araneids regularly build horizontal orb webs. Because both families contain so many structurally diverse genera, you may need to examine the spider's reproductive structures and mouthparts when facing an unfamiliar individual or one away from its web. Araneids have shorter, more squarish endites than the tetragnathids. Male araneids also have shorter palpal tibiae, and females have much more complex epigyna, often with prominent scapes. In two orb web weaver genera, *Zygiella* and *Parazygiella,* the males have elongate palpal tibiae and the females have scapeless, unsculptured epigyna. However, both genera normally build vertical orb webs, often with a conspicuous section missing from the upper half. This design is unlike the web of any tetragnathid.

Members of the genus *Uloborus* (Uloboridae, p. 83) also build horizontal orb webs, but theirs always have stabilimenta across the hubs. Additionally, *Uloborus* are cribellate and have highly visible tufts of setae at the distal ends of tibiae I, features strikingly different from any tetragnathid.

PACIFIC COAST FAUNA: Seven genera with 20 regional species. The most widespread and easily recognized species are in the genus *Tetragnatha* (pl. 23). Commonly referred to as long-jawed orb weavers, they are the namesakes of the family and are especially abundant in wet meadows, along streams and rivers, and in other moist areas. With their long legs and cylindrical bodies, they are among the most distinctive of the region's araneofauna. Adult males possess unusually long, toothed chelicerae that are used to hold the females' jaws apart during mating. Ten species have been recorded from the Pacific coast states, and while most are uncommon and thinly distributed, *Tetragnatha versicolor* and *T. laboriosa* are abundant throughout the region. These two species are also more general in their habitat preferences and can survive in forests and grasslands away from a constant source of water. Long-jawed orb weavers range in color from green to reddish brown, often with silvery speckles and stripes. They demonstrate a great deal of variation in body size, with regional species ranging from 3.8 to 13.2 mm (.15 to .52 in.) in length. They are often found either resting across the hubs of their large, open orb webs or clinging to nearby twigs or blades of grass, exquisitely camouflaged with their legs stretched out in line with their bodies. A detailed key to North America's *Tetragnatha* can be found in Levi (1981).

The second most common tetragnathid genus along the Pacific coast is *Metellina* (pl. 23). These are relatively small spiders, 3.8 to 5 mm (.15 to .2 in.) in length, that build tightly woven orb webs in humid, shaded areas. Three species are known from our region, each of which has an egg-shaped or roughly triangular abdomen that is widest across the anterior margin and is subtly banded in gray, cream, and brown. *Metellina* also have a black longitudinal bar bordered by two white stripes on the ventral side of the abdomen (a feature shared with several other genera in the family Araneidae, p. 110) and three very large teeth across the promargin of each cheliceral furrow. *Metellina curtisi* (pl. 23) and *M. mimetoides* (pl. 23) are both native to North America and found along the Pacific coast from Southern California through northern Washington, although *M. mimetoides* is much less abundant beyond northern Oregon. *M. curtisi* lives almost entirely in coastal and near-coastal regions, and while it usually has a smoothly rounded abdomen, a few individuals have small anterior humps. *M. mimetoides*, while common along the coast, also lives in the interior, with records from the Sierra Nevada and from caves across the western United States. It has a more triangular abdomen than *M. curtisi*, often with a prominent pair of anterior humps. The third regional species, *M. segmentata*, is a Eurasian spider introduced to the Vancouver area of British Columbia in the mid-1960s and has since spread across western Washington. It is now found as far south as Lewis County and appears to be replacing *M. curtisi* as the resident *Metellina* on Orcas Island in the San Juan Islands (R. Crawford, pers. com., 2009). *M. segmentata* has a smooth, egg-shaped abdomen without any noticeable protrusions. Despite these described differences, there is a great deal of overlap of shapes in these spiders, and identifying them to the species level often requires examining their reproductive structures as described in Levi (1980).

Among the most beautiful of North America's spiders is the Orchard Spider (*Leucauge venusta*, pl. 23). It ranges from 3.2 to 8 mm (.13 to .31 in.) in length and has greenish legs and an oblong silverish abdomen with variable green, black, and yellow stripes above and on the sides and orange spots below. It also has two tightly clustered rows of trichobothria along the basal third of femur IV, a feature unusual in any group of spiders. The Orchard Spider is an uncommon resident of the Southern California coast, with a single record from Mendocino County in Northern California. It builds a fairly large, tightly woven orb web in the lower branches of a tree or bush and is described in greater detail in Levi (1980).

One of the most unusual of the region's tetragnathid genera is *Pachygnatha* (pl. 24). Unlike other long-jawed orb weavers, adult *Pachygnatha* don't build webs. Instead, they are cursorial hunters, frequenting bogs and other moist places (although some juveniles are reported to still make use of webs). They are mostly between 3.3 and 6 mm (.13 to .24 in.) in length, with large jaws and a clypeus that is at least twice as high as the diameter of the anterior median eyes. The abdomen is tan to brown, often with a narrow, light folium mark and light lateral bands. The carapace is dusky yellow to chestnut brown and normally has a dark stripe along its midline. Three very rare regional species have been recorded: *Pachygnatha xanthostoma* and *P. clerckii* from Washington and *P. dorothea* from Oregon and Washington. More comprehensive identification information can be found in Levi (1980).

The smallest tetragnathid along the Pacific coast is *Glenognatha foxi* (pl. 24). Unlike other long-jawed orb weavers, this tiny spider (1 to 1.8 mm, or .04 to .07 in., in length) is limited to hot, dry regions of southeastern California. It has a distinctly spherical abdomen, a dark orange carapace, and pale yellow legs. Its abdominal pattern is variable, but the most common arrangement on the female is a pale orange dorsum with a pair of black-bordered silver spots on either side and a smattering of silver flecks along its anterior margin. The male is smaller and often entirely orangish red. It makes small, tightly woven horizontal orb webs among dry grasses and shrubs. Levi (1980) discusses in greater detail the variation found in *G. foxi*.

At the other end of the size spectrum is *Metleucauge eldorado* (pl. 24), a large, thick-legged spider (7.8 to 13 mm, or .31 to .51 in.) of the Sierra Nevada and northwestern California. It is fairly dark with an ovoid, dorsoventrally flattened abdomen decorated with a distinct silvery-gray folium dorsally and a striking black patch bordered by two sinuous white lines ventrally. This species makes large, open-hubbed orb webs in the shaded recesses of streamside boulders. It is described and illustrated in Levi (1980).

By far the rarest of the Tetragnathidae in North America is the Dolloff Cave Spider (*Meta dolloff*, pl. 24). Originally described by Levi (1980), it ranges from 11 to 14 mm (.43 to .55 in.) in length and is endemic to several small cave systems along the Central California coast. *M. dolloff* is mahogany brown with diffuse black bands on its legs. Its globular, rather vertically aligned abdomen is lightly marked with gray triangular splotches dorsally and a pair of pale longitudinal stripes ventrally. *M. dolloff* builds its web over a niche or depression in a cave wall and feeds on moths and harvestmen. Unfortunately, several of these caves are located very close to both the town of Santa Cruz and the University of California at Santa Cruz, making them highly susceptible to pollution, vandalism, and other human disturbances.

NATURAL HISTORY: As the most conspicuous, geographically widespread, and taxonomically diverse genus in the family, *Tetragnatha* has been the focus of most of the research into the natural history of North America's long-jawed orb weavers. *Tetragnatha* males initiate courtship by delicately tapping on the females' webs. Danielson-François et al. (2002) found that when a female *T. elongata* responded by plucking out arrhythmic vibrations, she wasn't interested in mating and often chased the male away. If she was receptive, she would respond by tapping out a regular rhythmic response. Both spiders would then move toward one another, their impressive jaws spread wide. After vigorous grappling, their chelicerae and fangs would become interlocked and their bodies elevated into a ventral-to-ventral position. In this stance, mating would occur. About a week after mating, *Tetragnatha* females build fluffy white egg sacs against long leaves and cover them with taut sheets of silk. They then guard the sacs until the eggs hatch several weeks later. Male spiders in several genera, including *Tetragnatha* and *Metellina*, indulge in pre- and postcopulatory mate guarding, which helps ensure that no other males get access to the female. As an added challenge, female *M. segmentata* regularly engage in sexual cannibalism. To compensate, the male will often wait at the edge of the female's web until she is feeding before he initiates courtship.

Although most tetragnathids build either horizontal or inclined orb webs, this is affected by the physical structures around them, and fully vertical orbs are not uncommon in some genera. The location of the web is also influenced by a variety of both biotic and abiotic factors. Gillespie (1987) found that *T. elongata,* a common spider of the eastern United States with a few records from the Pacific Northwest, actively chooses to build its web in a humid area with abundant prey. If the area is too dry, the spider will suffer from dehydration and quickly move to a new location. A lack of prey will also trigger the spider to move until it reaches an area where its nutritional needs are met, a behavior known as prey sampling. In addition to catching prey in their webs, *Tetragnatha* have been observed seizing small insects that wander by while they lie in wait, camouflaged against twigs or blades of grass.

PIMOIDAE

Pl. 24

IDENTIFICATION: Pimoidae is a small family of eight-eyed entelegyne spiders that is part of a larger superfamily called Araneoidea. This superfamily contains some of our most common spider families, including the cobweb weavers (Theridiidae, p. 93), the orb web weavers (Araneidae, p. 110), and the sheet web weavers (Linyphiidae, p. 133).

With one very rare exception, pimoids range from 5 to 12 mm (.2 to .47 in.) in length. Most have exceptionally long, moderately spined, reddish-brown legs that are coated with setae and demonstrate patella-tibial autospasy, an adaptation that allows the leg to break away at the patella-tibia joint when grabbed. Members of the genus *Pimoa* have dark, oval abdomens decorated with assorted light chevrons, spots, and stripes. Their cephalothoraxes have conspicuous pit-like thoracic furrows and are tan to mahogany brown with dark markings around the eyes and midline. *Pimoa* also have robust chelicerae with stridulatory files along the lateral face. With experience, you will find it fairly easy to recognize the larger Pimoidae in their messy, extensive sheet webs. The identification of an individual away from its web, however, may require focusing on the structure of its reproductive organs and mouthparts as described in Ubick (2005c).

The exception with respect to many of these features is the extremely rare *Nanoa enana*. Known only from Northern California and southern Oregon, this tiny spider is only around 1.5 mm (.06 in.) in length, has very few to no spines on its legs, and has small, weak chelicerae without stridulatory files. Its placement within Pimoidae was only possible after a detailed examination of its reproductive structures by Hormiga et al. (2005).

SIMILAR FAMILIES: Pimoids are closely related to the sheet web weavers (Linyphiidae, p. 133) and, for the most part, share many of same features, including their basic web architecture, prominent leg spination, and patella-tibia autospasy. While the clearest traits that differentiate these two families are based on the structure of their reproductive organs, other features can be useful. Overall, sheet web weavers are generally quite small, rarely exceeding 5 mm (.2 in.) in length. Additionally, the region's largest linyphiid genera, *Neriene* and *Frontinella*, build complex, fine-meshed domed or multilevel webs. Members of the genus *Calymmaria* (Hahniidae, p. 209) look like pimoids and share many of the same habitats. However, *Calymmaria* build conical, basketlike sheet webs and have significantly reduced or missing coluli. On *Pimoa*, the colulus is large and clearly visible between the anterior lateral spinnerets. Additionally, *Calymmaria* lack stridulatory files on their chelicerae and are unable to perform patella-tibia autospasy.

The highly unusual *N. enana* was not described when Ubick et al. (2005) went to press, and the familial placement of a specimen would be impossible using its key. The inclusion of *N. enana* in Pimoidae was based on the structure of its complex reproductive organs, the details of which can be found in Hormiga et al. (2005).

PACIFIC COAST FAUNA: Two genera representing 14 regional species. By far the largest and most visible of the Pacific coast's two pimoid genera is *Pimoa* (pl. 24). With 13 species known from both coastal and mountain forests, *Pimoa* are conspicuous members of the Pacific Northwest's araneofauna. Their dark coloration hides them exquisitely in the cool, moist woodlands they call home. *Pimoa* has a fascinating distribution: the entire New World contingent lives in the Pacific Northwest, while additional species are found on isolated mountain ranges in northern Spain and across Eurasia to China and Japan. This distribution implies that *Pimoa* was once much more widespread across the Northern Hemisphere and that, over time, populations have become fragmented, isolated from one another in humid, Holarctic mountain forests.

Along the Pacific coast, *Pimoa* have been collected in the redwood forests of Sonoma County, California, north through the Coast Ranges and Cascade Range into western Oregon and Washington. Additional *Pimoa* species are known from the central and northern Sierra Nevada of California and the Kettle Range and Selkirk Mountains of northeastern Washington. A key to the region's *Pimoa* species can be found in Hormiga (1994).

N. enana (not illustrated) was described by Hormiga et al. (2005). Only a few specimens have ever been collected, all of which were found in forest litter from the eastern Siskiyou Mountains and southern Cascade Range. The entire known distribution of this enigmatic species is a small area including parts of Siskiyou County, California, and Jackson County, Oregon. *N. enana* has a gray abdomen with two large pale spots near its dorsal anterior margin and numerous lighter markings on its posterior half. Its cephalothorax and legs are yellowish brown, with a dark patch around its eyes. To confidently identify this species, you must carefully examine the spider's reproductive structures and compare them with the published description, a level of detail beyond the scope of this book.

NATURAL HISTORY: Essentially all of the information published regarding the natural history of Pimoidae focuses on the genus *Pimoa*. Almost nothing is known regarding the behavior or ecology of *N. enana*. *Pimoa* build large, "messy" sheet webs in protected recesses, including inside hollow tree stumps and between streamside boulders. Their webs are chaotic-looking sheets up to a meter across with numerous threads both above and below that provide support and possibly act as baffles, funneling insects onto the sheets. Some specimens have also been collected from caves, although this appears more closely tied to their preference for shaded, humid habitats than any specific adaptation to a troglobitic lifestyle.

Pimoa are nocturnal and walk along the underside of the web like sheet web weavers. Nothing has been published regarding *Pimoa* courtship behaviors, although it is known that their egg sacs are white and coated with outer layers of dirt and debris. They are suspended beneath webs, where they are actively guarded against predators and parasites.

LINYPHIIDAE — Sheet Web Weavers, Dwarf Spiders, Money Spiders

Pls. 25, 26

IDENTIFICATION: Linyphiids are among the most common spiders in North America, with more described regional species than any other family. However, most are minute inhabitants of the leaf litter, rarely seen by the casual observer. A sheet web weaver's web consists of one or more horizontal, domed, or concave sheets supported above and below by auxiliary threads. Linyphiids are often found hanging beneath their sheet webs, a position that allows them to attack their prey from below. They range from less than 1 to 8.5 mm (.04 to .33 in.) in length, although most are under 4 mm (.16 in.) long. All are three-clawed entelegynes, and with the exception of a few rare cave-dwelling species, they have eight eyes. Linyphiids have only a few spines on their legs and exhibit patella-tibia autospasy, an adaptation that allows the leg to break off at the patella-tibia joint when entangled. Most species also have rows of thin ridges or striations on their chelicerae. Linyphiidae is often broken down into two subfamilies. The generally larger Linyphiinae regularly exhibit patterning on their abdomens. The smaller but more diverse Erigoninae, whose abdomens are normally dark, lack conspicuous patterning and may have scuta. Additionally, in many Erigoninae, the male's carapace is elevated and bizarrely sculpted with ocular turrets and other unusual growths. This said, the morphological features dividing these two subfamilies are not absolute, and several genera demonstrate a mix of structures.

SIMILAR FAMILIES: As members of the superfamily Araneoidea, Linyphiidae share numerous anatomical features with other common spider families, including the orb web weavers (Araneidae, p. 110) and the cobweb weavers (Theridiidae, p. 93). While the webs of many of our more conspicuous species are fairly diagnostic, wandering males and spiders examined away from their webs can present an identification challenge. In this situation, the presence of leg spines and epigynal scapes separates them from similar spiders in the families Theridiidae and Nesticidae (p. 90). Most linyphiids also have an elevated clypeus that is at least three times the diameter of the anterior median eyes. This feature, along with cheliceral stridulatory files and patella-tibia autospasy, differentiates them from the araneids and the long-jawed orb weavers (Tetragnathidae, p. 126). Sharing many morphological and behavioral features with the linyphiids are members of the family Pimoidae (p. 131). With the exception of *Nanoa enana,* a tiny, very rare species from Northern California and southern Oregon, pimoids are normally larger than linyphiids and they build rather messy-looking silken platforms, in contrast to the linyphiids' elegant domes and sheet webs. Additional features separating these two families include the pimoids' setae-covered legs and the fine details of their reproductive structures as described in Ubick (2005c) and Hormiga et al. (2005).

PACIFIC COAST FAUNA: Seventy-one genera and just under 300 species (as described Draney and Buckle, 2005, and the North American checklist in Buckle et al., 2001). Because of their minute size, incredible diversity, and complex identification challenges, even a cursory review of the region's erigonine genera and species is beyond the scope of this book. While a few males are illustrated to show their size and unusual carapace structures, the focus of this account is on the larger, more conspicuous linyphiine genera and species.

One of the most distinctive sheet web weavers in North America is the Bowl and Doily Spider (*Frontinella communis,* pl. 25; also known as *F. pryamitela*). A familiar sight in the eastern United States, it is a fairly common resident throughout much of California, becoming progressively rarer to the north. In Washington, it is only regularly found on the Okanogan Highlands in the northeastern portion of the state (R. Crawford, pers. comm., 2009). It is a moderate-size Linyphiidae, with the female normally ranging between 3 and 4 mm (.12 and .16 in.) in length while the male is slightly smaller. The abdomen is unusually trunk-like, with a high anterior end and a characteristic series of pale lateral stripes on a chocolate brown background. The web is a regionally unique multilevel sheet web with a bowl-like structure of nonsticky silk suspended over a flat sheet (or "doily"). Above the main body of the web is a multitude of barrier threads that disrupt the movement of small insects flying past, dropping them into the bowl. Hanging from the bowl's underside, the spider attacks it prey from below. It builds its web in an open woodland or meadow. Maturing in late spring and early summer, the adult male often lives with the female in her web, a situation that, while rare in spiders as a whole, is not unusual in some linyphiid species.

The genus *Neriene* (pl. 25) includes several of the most common of the region's sheet web weavers. Two of these, the Filmy Dome Spider (*Neriene radiata,* pl. 25) and the Sierra Dome Spider (*N. litigiosa,* pl. 25), build prominent dome-shaped sheet webs in the open spaces of small trees and bushes. In both species, males and females often inhabit the same web. Unlike the "boxy" abdomen of the Bowl and Doily Spider, the more oblong abdomens of spiders in the genus *Neriene* narrow at their anterior margins. Both species are boldly patterned and can frequently be identified by their web designs and overall appearance. The Filmy Dome Spider is Holarctic in distribution, with records from across North America, Europe, and Asia. It has an entirely black venter, and its abdomen's white lateral stripes are washed in yellow across their base. It also has a dark, broken dorsal abdominal stripe, and the female has distinctive white margins on her otherwise brown carapace. Uncommon in California, the Filmy Dome Spider has been collected at inland locations throughout Washington. The male and female are similar in size, ranging from 3.4 to 6.5 mm (.13 to .26 in.) in length. Its dome web is deep, fairly steep sided, and between 10 and 20 cm (3.9 and 7.9 in.) across.

Much more common is the similar but generally larger Sierra Dome Spider. It ranges from 5.1 to 8.5 mm (.2 to .33 in.) in length, and its shallow dome web can be up to 61 cm (24 in.) across. The Sierra Dome Spider's abdomen is generally more cylindrical than that of the Filmy Dome Spider. It is chalky white to pale green and has a thin, dark dorsal line that expands into a large splotch on its posterior margin along with numerous longitudinal and lateral stripes. The Sierra Dome Spider is found from British Columbia to northern Mexico and is common in wooded areas throughout the region. The third species, *N. digna* (pl. 25), is a small but abundant member of the coastal and near-coastal spider fauna from southern Alaska to Central California. It is found not only in wooded and brushy areas, but also as a common garden spider along the Central California coast. It is darker and slightly smaller (3.8 to 5.2 mm, or .15 to .2 in., in length) than the other two regional *Neriene* species. In both sexes the abdomen is brown and the carapace is orange brown. The female *N. digna* has wavy-edged white bands that merge to create a rough folium-shaped patch on the abdomen's dorsal surface. The adult male has a pair of white patches near the abdomen's anterolateral margins. The *N. digna* web is roughly rectangular and flat, lacking the obvious "dome" structure of the other *Neriene* species. It is built low in a bush, fern bed, or shaded grassy field. Morphological information as well as palpal and epigynal illustrations of the region's *Neriene* can be found in Helsdingen (1969).

Common but often overlooked are the hammock spiders in the genus *Pityohyphantes* (pl. 26). They are relatively large for linyphiids with Pacific coast species spanning 3.9 to 6.1 mm (.15 to .24 in.) in length. Hammock spiders are easily recognized by the presence of a wide, dark stripe on the carapace that divides into two, thin parallel bands that run into the dark eye patch. The dorsal abdominal stripes are often jagged and look like a row of dark, overlapping triangles on a light background. *Pityohyphantes* live among the woody branches of bushes and trees (especially conifers) and are regularly found near the forest canopy. Eight species have been recorded from the contiguous Pacific coast states. Descriptions and illustrations of their reproductive structures can be found in Chamberlin and Ivie (1941b, 1942b, 1943).

Another distinctive genus is *Wubana* (pl. 26) with six regional species. They are small spiders, 2 to 3 mm (.08 to .11 in.) in length, and an adult male has an unusual hornlike protuberance behind its eyes that is normally adorned with numerous long, stiff setae. The carapace and legs are yellowish to dusky orange and the abdominal pattern consists of one or more pale bands or spots on an otherwise gray background. *Wubana* have been found in all three of the contiguous Pacific coast states. They mainly prefer damp areas around streams or lakes and have been collected in leaf litter and under rotting logs. Chamberlin and Ivie (1936) revised the genus and provides a key and illustrations to the region's species.

Spiders in the genus *Microlinyphia* (pl. 25) are fairly common along the Pacific coast, especially in wet meadows, fern groves, and well-watered gardens. They range from 2.8 to 5.8 mm (.11 to .23 in.) in length and build small, relatively flat sheet webs in low vegetation. Although four species are known from the region, *Microlinyphia impigra* and *M. pusilla* are quite rare, with only a few Washington records of each. The most common of the region's *Microlinyphia, M. dana* (pl. 25), is found from Southern California to Alaska. It has an entirely grayish brown to black venter that extends over the sides of the abdomen and is bordered by a broad, pale lateral band. Its dorsal abdominal surface is beige to chocolate brown and is adorned with a pair of splotchy lateral bands. The region's second common species, *M. mandibulata* (pl. 25), is represented by two subspecies, *M. m. mandibulata,* which reaches from the eastern United States and Canada into northeastern Washington, and *M. m. punctata,* a slightly paler, more ornately marked form that lives from Southern California to northern Washington. Both forms have a reddish-orange cephalothorax and legs and a dark reddish-brown abdomen marked by a pair of white spots near the dorsoanterior margins. There are also several white patches, often melded into a wavy band across the sides of the abdomen. *M. m. punctata* is distinguished by the presence of five white spots on its ventral abdominal surface. While the Pacific coast's *Microlinyphia* are reasonably distinct, a detailed examination may be required to confirm their identification. Thorough descriptions and illustrations of *Microlinyphia* reproductive structures can be found in Helsdingen (1970).

Especially common are the many small and inconspicuous linyphiine sheet web weavers in the genera *Bathyphantes* (pl. 26), *Lepthyphantes* (pl. 26), and *Linyphantes* (not illustrated). They are found from Southern California to northern Washington and represent the most diverse of the region's linyphiine genera, with a dozen or more Pacific coast species in each genus. Most are between 1.5 and 3.5 mm (.06 and .14 in.) in length and make delicate sheet webs beneath the cover of low-lying vegetation, in leaf litter, and under rocks. Many show some degree of abdominal patterning, including chevrons, stripes, and diffuse speckling. Separating these very similar genera requires a microscopic examination of their leg spination and reproductive structures as described in Draney and Buckle (2005).

These descriptions cover only the most conspicuous of the Pacific coast's numerous linyphiid genera. Tiny and easily overlooked are the great numbers of erigonine sheet web weavers (pl. 26). Members of the subfamily Erigoninae include many of our smallest spiders, earning them the common name "dwarf spiders." They are also occasionally referred to as money spiders because in England they are said to bring good fortune to those they land on. They are especially prevalent in the damp leaf litter community, where they regularly make up a large percentage of the individual spiders present. The males of many species also have highly modified carapaces whose extraordinary sculpting includes ocular turrets, lobes, and other unusual growths, the importance of which has received only cursory study.

NATURAL HISTORY: Uniquely among the large spider families, Linyphiidae reaches its greatest diversity in the northern temperate zones instead of the tropics. While Linyphiidae is the most species-rich family in North America, the vast majority of ecological and behavioral studies have mainly focused on the

Bowl and Doily Spider, *Frontinella communis,* and the dome spider genus, *Neriene.* All these spiders build large sheet webs below a loose network of tangle threads. When a flying insect gets caught in the tangle and is forced down, the spider runs along the sheet's underside, biting the prey from below. Known prey of the larger linyphiids include flies and leafhoppers, while the smaller, leaf litter–dwelling species feed on springtails, gnats, and other minute, soil-dwelling arthropods.

Unlike most spiders, the males of several linyphiid genera spend considerable time in the females' webs, both before and after mating. The female Bowl and Doily Spider lays down pheromone-laced silk in her web that attracts males and stimulates courtship behaviors. The male's courtship-specific behaviors include rapidly flexing his abdomen and vibrating the web through his legs. If the female responds in a similar manner, courtship continues. Additional signals include leg waving, web plucking, and a display in which both sexes move toward and retreat from one another (Suter and Renkes, 1984). An act known as mounting courtship is apparently widespread among the linyphiines. In this behavior, the male goes through the motions of mating, but without actually inseminating the female. It is thought that this allows him to gauge the female's reproductive status prior to copulation. After mating, a male regularly spends up to several days cohabitating with the female, competing for food and fending off rivals. Once the initial male leaves, the female may mate with several more males before laying her first egg sac. Unlike other spiders that hang their egg sacs within the matrix of the web, the female Bowl and Doily Spider places hers in a moist, sheltered spot on the ground. The loose silk wrap is thin enough that the orange eggs are visible within. This leaves them vulnerable to water loss, so by secreting them in leaf litter or against a damp log, the female gives them greater protection against desiccation (Suter et al. 1987).

Similar courtship behaviors exist in the Sierra Dome Spider, *Neriene litigiosa.* Adult males have been found guarding penultimate females in their webs, and like Bowl and Doily Spider males, Sierra Dome Spiders males stay with newly inseminated females for an extended period. A different kind of courtship has been observed in *Lepthyphantes leprosus.* In addition to vibratory signals, some *L. leprosus* males perform an unusual web-removal exercise. After signaling his presence to the female, the male rapidly moves around her web adding new support lines. He then bites away most of the web, leaving the female stranded on a tiny strip of silk. Courtship and mating in the erigonine sheet web weavers has been examined in only a few species (Vanacker et al. 2003; Uhl and Maelfait, 2008). Some male erigonine sheet web weavers house glands in the unusual sculpting on their carapaces that are thought to produce pheromones and secretions for the females to eat during courtship and mating.

While collection records imply many linyphiids have a fairly straightforward annual life cycle, extensive research on Filmy Dome Spiders in Maryland has revealed a much more complicated picture (Wise, 1984). In winter, you can find both large and small immature Filmy Dome Spiders secreted under bark, in rock piles, and in other protected locations. The large spiders mature early in spring, reproduce, and die. Their offspring mature rapidly but are physically smaller than their parents. They lay their own egg sacs in August, the young of which represent the physically smaller of the overwintering stock. These diminutive juveniles don't mature until late the following spring, but because they've had a longer growth period, they develop into physically larger adults. Their offspring grow through summer and fall and comprise the larger of the overwintering juvenile spiders. In effect, three generations of spiders are produced every two years, with peaks of adult abundance in spring and late summer. There is, however, some plasticity in the developmental chronology of these spiders, with changes in temperature and food availability affecting their overall maturation rate.

ANYPHAENIDAE — Ghost Spiders

Pl. 27

IDENTIFICATION: Ghost spiders are small (3.3 to 6.5 mm, or .13 to .26 in., in length), eight-eyed entelegynes with unique lamelliform claw tufts. Unlike the thick, brush-like tufts of other two-clawed spiders, the setae on ghost spiders broaden distally, giving them an unusual spatulate appearance. Another distinguishing feature of the anyphaenids is the placement of the tracheal spiracle. On other spiders, it is near the spinnerets, but on ghost spiders, it is either near the midpoint of the abdomen or close to the epigastric furrow. Ghost spiders vary from white to brown, often with darker bands or other markings on the carapace, legs, and abdomen.

SIMILAR FAMILIES: No other North American spiders share the ghost spiders' unique claw tufts or unusual tracheal spiracle placement. Like anyphaenids, spiders in the families Clubionidae (p. 145) and Miturgidae (p. 143) are wandering nocturnal hunters. Western clubionids and members of the miturgid genus *Cheiracanthium* lack distinct bands or chevrons on their carapaces and abdomens, although they may show brownish heart marks. Members of the family Liocranidae can also be boldly patterned, and a close examination of the claw tufts or tracheal spiracle may be needed to confirm an unknown spider's familial identity.

PACIFIC COAST FAUNA: Three genera representing eight regional species. With six species, *Anyphaena* (pl. 27) is the most widespread and diverse of the region's ghost spider genera. These spiders are foliage and leaf litter hunters that are found from Southern California to northern Washington and range from 3.3 to 6.5 mm (.13 to .26 in.) in length. In this genus the carapace is generally pale yellow to dark orange with dark paramedian bands, and the abdomen is pale gray to orange with dark spots and transverse stripes. Because there is both a great deal of similarity between the different species and also considerable variation within species, an examination of the spider's reproductive structures is normally required to confirm an individual's specific identity. A revision of North America's *Anyphaena* and a key to their identification can be found in Platnick (1974).

Hibana incursa (pl. 27) is the only representative of its genus along the Pacific coast. Superficially similar to *Anyphaena,* its abdominal spiracle is located just posterior to the epigastric furrow, while on other ghost spiders it is near the midpoint of the abdomen. It has dark brown chelicerae that stand in bold contrast to the generally pale color of its carapace and abdomen; however, its abdominal color can darken after feeding on certain prey items. The adult *H. incursa* is around 6 mm (.24 in.) in length. It is widespread throughout the southwestern United States, extending north into Central California. *H. incursa* is reviewed in greater detail in Platnick (1974) under its previous placement within the genus *Aysha.*

The most distinctive of the Pacific coast's ghost spider genera is represented by a single species, *Lupettiana mordax* (pl. 27). This spider has unusual dark purplish-brown chelicerae that project forward from its carapace, giving it a uniquely elongate look. Its cephalothorax is brown, and its abdomen is grayish green with a row of dark chevrons along its dorsal surface. A fairly small spider, *L. mordax* ranges from 3.5 to 4 mm (.14 to .16 in.) in length. In the southeastern United States and in Central America, the chelicerae of the male are usually about 1 mm (.04 in.) long (similar to those of female). However, on approximately one in five males, the chelicerae are much longer, extending over 2 mm (.08 in.) from the carapace. It is unknown what purpose this variation serves, and there have not been enough specimens collected in California to determine if this pattern exists on the Pacific coast as well. *L. mordax* is an uncommon spider along the California coast from San Diego County to Monterey County and is discussed under its old name, *Teudis mordax,* in Platnick (1974).

NATURAL HISTORY: Ghost spiders are predominately nocturnal wandering hunters. Often found on foliage, they have also been collected in leaf litter, under rocks and boards, and in homes. *Hibana* is unusual in that its members are diurnal and are regularly seen hunting on trees and bushes. There has been little research devoted specifically to the Pacific coast's species, although it is known that *H. incursa* is a voracious predator of several aphid species and that *L. mordax* is a known prey item of the araneophagic Long-bodied Cellar Spider, *Pholcus phalangioides* (Pholcidae, p. 73) (R.J. Adams, pers. obs., 2006).

The Anyphaenidae respiratory system is quite extensive, with the trachea extending through the pedicel into the cephalothorax and legs. In related families, the trachea is restricted solely to the abdomen. The anyphaenids' respiratory design provides additional oxygen directly to the legs and is probably related to the spiders' active hunting style and rapid response to any perceived threat or disturbance. Additionally, in those species that have been examined, the trachea is distinctly larger in males than in females, suggesting that this is related to the spiders' exceptionally vigorous courtship displays. Ghost spiders do not spin elaborate webs but use silk to construct retreats and egg sacs. The egg sacs of the eastern species *Anyphaena pectorosa* are described as soft and white without a papery outer layer, and they are hidden inside folded blades of grass or in rolled-up leaves.

MITURGIDAE — Prowling Spiders

Pl. 27

IDENTIFICATION: Miturgids are two-clawed entelegynes whose eight eyes are aligned in two broad rows. As in related families, the prowling spiders' anterior lateral spinnerets are conical and almost contiguous across their bases. The detail that best distinguishes a prowling spider is the shape of its posterior lateral spinnerets, the tips of which are capped by elongate or conical segments. Miturgids also lack a clearly defined thoracic groove. Our regional prowling spiders can be concolorous or patterned, arboreal or terrestrial, and as in other families with diverse genera and a complicated taxonomic history, it is often easier to recognize the genera separately than the family as a whole.

SIMILAR FAMILIES: The sac spiders (Clubionidae, p. 145) are the spiders most likely to be confused with prowling spiders along the Pacific coast. Spiders in the clubionid genus *Clubiona* are similar to members of the miturgid genus *Cheiracanthium*. The clubionids, however, have shorter and thicker legs, conspicuous thoracic grooves, and posterior lateral spinnerets with short, rounded tips.

PACIFIC COAST FAUNA: Two genera containing four described and one undescribed Pacific coast species. The most widespread of the region's prowling spiders are the introduced long-legged sac spiders in the genus *Cheiracanthium* (pl. 27), *C. inclusum* and *C. mildei*. They are extremely similar, and differentiating them requires examining their reproductive structures. They range from 4 to 9.7 mm (.16 to .38 in.) in length, although their long legs can make them look significantly larger. Most individuals are pale yellow to light green with a dark heart mark along the abdomen's dorsal midline. Different foods, however, can temporarily turn them orange or even brown. Long-legged sac spiders are nocturnal, foliage-dwelling hunters that spend the day in silk sacs, which are often hidden in curled leaves. They are also common in homes, deciduous woodlands, suburban gardens, and vineyards. The Agrarian Sac Spider, *C. inclusum*, is the more widespread of the two species. Native to Africa, it is common in California's lowlands, becoming progressively rarer farther north, with only a scattering of records from southeastern Washington. *C. mildei* is native to the Mediterranean region and is found widely across California, with at least one record from near Seattle, Washington (Crawford, 1988). While there is overlap, *C. mildei* is generally more closely tied to homes, gardens, and suburban parklands than *C. inclusum*. Spiders in the genus *Cheiracanthium* have received a great deal of attention because their bites have been blamed for causing large necrotic wounds. Unfortunately, this misinformation has persisted for many years in the medical literature. Studies of nearly 40 verified *Cheiracanthium* bites to humans have found only one case of mild necrosis. The remaining bites induced pain and swelling that in most cases lasted from only one hour to several days (Vetter et al., 2006). Illustrations of the two spiders' diagnostic features can be found in Dondale and Redner (1982).

The second regional prowling spider genus is *Syspira* (pl. 27). These spiders are boldly patterned, medium to large (5.9 to 17.2 mm, or .23 to .68 in., in length) desert dwellers with a pair of black, irregular paramedian bands running the length of the carapace and a black bar across the anterior edge of the abdomen. The legs are long and moderately spined, and when the spiders are at rest, the legs are held splayed out from the body. *Syspira* are terrestrial hunters, hiding by day under rocks and debris. The only taxonomic work that looks at the *Syspira* of the western United States is Olmstead (1975), but because this is an unpublished thesis, the changes it proposes are not officially recognized. It is, however, the only research that addresses *Syspira* in its entirety. It identifies two previously described species as members of California's spider fauna, *Syspira tigrina* and *S. longipes*. *S. tigrina* has three thick black stripes on its carapace while *S. longipes* has only a pair of weaker, reticulated bands. Olmstead (1975) also described a third, currently unrecognized species as a widespread member of the region's *Syspira* fauna. *Syspira* can be found throughout Southern California's deserts north to Inyo County and in the arid grasslands of western Fresno County.

NATURAL HISTORY: *Cheiracanthium* and *Syspira* are both nocturnal, wandering hunters. While *Cheiracanthium* are foliage specialists, *Syspira* are entirely terrestrial. *Cheiracanthium* have received a great deal of attention due to their history as spiders of medical concern, their abundance, and their potential for pest control. They voraciously eat a wide variety of insect prey, including the eggs and caterpillars of crop-injurious moths. Adults are regularly found from late spring through early fall. A male long-legged sac spider courts a female by tapping on the outside of her retreat while simultaneously cutting away the silk from its entrance. After mating, the female either returns to her retreat or seeks out a more protected location for egg laying. Once situated, she puts down a thick layer of silk, on top of which she lays a loose mass of yellow eggs, which are then covered with a thin sheet of silk. She stays with the eggs until they hatch, dying soon after.

CLUBIONIDAE — Sac Spiders

Pl. 27

IDENTIFICATION: Clubionids are two-clawed entelegynes whose distinguishing features as a family are subtle but whose members are not difficult to recognize once you become familiar with their unassuming appearance. Their anterior lateral spinnerets are conical and nearly connected at their bases, and their posterior lateral spinnerets have short, rounded tips. Their eight eyes are aligned in two fairly straight rows, and they possess a small but clearly visible longitudinal thoracic groove on the dorsal surface of the carapace. There are tiny precoxal triangles around the outer rim of the sternum, and there are three or fewer pairs of spines on the ventral faces of the anterior tibiae. Pacific coast sac spiders range from pale yellow to mahogany brown, and on paler species a dark heart mark is often visible on the abdomen's dorsal midline.

SIMILAR FAMILIES: Clubionids have a complicated taxonomic history. Numerous now-independent families were historically assigned as subfamilies within a more broadly defined version of Clubionidae. This was due to the strong similarity between many two-clawed spiders. Those families most likely to be mistaken for sac spiders include the ghost spiders (Anyphaenidae, p. 140) and the long-legged sac spiders in the genus *Cheiracanthium* (Miturgidae, p 143). Ghost spiders tend to be more distinctly patterned than Clubionidae. They also have unusual, lamelliform setae in their claw tufts, and their tracheal spiracles are located near the middle or anterior portion of the abdomen. Long-legged sac spiders lack thoracic grooves, have longer, thinner legs, and have elongate tips to their posterior lateral spinnerets. Some members of the family Liocranidae (p. 152) are morphologically similar to clubionids but are more distinctly patterned, often with dark markings on the carapace and abdomen. Separating these two families may require looking at the spiders' endites. On clubionids, they are concave along the inner margin. On liocranids, they are either straight or convex along their borders with the labium. Additionally, except for *Hesperocranum,* none of the liocranid genera have precoxal triangles, and this genus is easily recognized by the thick strips of bristles on the undersides of its anterior tibiae.

PACIFIC COAST FAUNA: One genus containing 16 regional species. *Clubiona* (pl. 27) vary from pale yellow to a rich orange brown, and the abdomen and cephalothorax can be either similar or different in color. The abdomen is generally unpatterned except for a conspicuous heart mark on some species. *Clubiona* range from 3 to 11.5 mm (.12 to .45 in.) in length and are commonly found in woodlands, marshes, and grassy fields throughout the region. A key covering most of North America's species can be found in Edwards (1958). Several additional species, both native and introduced, are described in Roddy (1966). The Pacific coast's species are also discussed in the revisionary studies of Roddy (1973) and Dondale and Redner (1976a).

NATURAL HISTORY: Sac spiders are found in thick, deciduous foliage, grassy fields, and leaf litter. They don't make snaring webs but wander at night in search of prey. They earned the name sac spiders because they commonly weave silk retreats in curled leaves, under loose bark, and beneath rocks. Courtship and mating take place in spring and summer inside the female's retreat. In at least a few species, the female builds a special enclosure for concealing and guarding her disk-shaped egg sac. After dispersing, the young spiders overwinter either as juveniles or in their penultimate state, maturing to adulthood the following year.

CORINNIDAE — Ant-mimic Sac Spiders

Pl. 28

IDENTIFICATION: Corinnidae is a diverse family of two-clawed entelegynes that were historically housed in a more broadly defined version of the family Clubionidae (p. 145). Excepting the genus *Drassinella,* they have conical, anterior lateral spinnerets that are either contiguous or very nearly touching at their bases. Because the four Corinnidae subfamilies are exceptionally diverse, it is easier to describe them separately. Spiders in the subfamily Trachelinae lack leg spines but have numerous tiny cusps on the ventral surfaces of their anterior tibiae and metatarsi, a feature unique both within the family and among the former Clubionidae as a whole. The subfamilies Corinninae and Phrurolithinae can be recognized by the presence of four or more pairs of spines on the undersides of their anterior tibiae, precoxal triangles, and shallow to nonexistent notches in their trochanters. Members of the subfamily Castianeirinae have fewer than four pairs of spines on their anterior tibiae's ventral faces and are often boldly patterned. Males have a scute covering most of the abdomen's dorsal surface, while on the female the scute is reduced to a small patch on the abdomen's dorsoanterior edge.

SIMILAR FAMILIES: Corinnidae is the most morphologically diverse of the families formerly housed within Clubionidae. However, each of the other families has one or more features that can be used to differentiate it from the ant-mimic sac spiders. Spiders in the family Anyphaenidae (p. 140) have broad, lamelliform setae in their claw tufts and tracheal spiracles located away from their spinnerets, and they are much more likely to be found in foliage. Members of the genus *Agroeca* (Liocranidae, p. 152) could easily be mistaken for members of the Corinnidae subfamily Phrurolithinae, but they lack precoxal triangles. Sac spiders (Clubionidae, p. 145) have four or fewer pairs of spines on the ventral surfaces of their anterior tibiae. Members of the Corinnidae subfamilies Corinninae and Phrurolithinae are most likely to be mistaken for clubionids, but these corinnids have more than four spine pairs. The genus *Micaria* (Gnaphosidae, p. 156) contains numerous small ant-mimicking spiders similar to *Drassyllus* and members of the subfamily Phrurolithinae, but *Micaria* have iridescent scales scattered across their abdomens and their anterior lateral spinnerets are fairly long and cylindrical.

PACIFIC COAST FAUNA: Nine genera representing at least 32 described and numerous undescribed or unidentified regional species. Because of the exceptional structural differences between the different ant-mimic sac spider genera, the North American fauna has been divided into four subfamilies. The position of the genus *Drassinella* within Corinnidae's subfamilial arrangement is unclear and is addressed separately in this account.

The subfamily Trachelinae contains two Pacific coast genera, *Trachelas* (pl. 28) and *Meriola* (pl. 28). Instead of spines, a smattering of short, spur-like cusps decorate the legs of these spiders. The posterior eye row on *Trachelas* is strongly recurved, and the carapace narrows distinctly around its cephalic portion. On *Meriola,* the posterior eye row is nearly straight, and the carapace is much broader, especially around the cephalic region. *Trachelas pacificus* (pl. 28) is the only species of *Trachelas* found along the Pacific coast. *T. pacificus* ranges from 5.4 to 7.5 mm (.21 to .3 in.) in length and has an orange-brown cephalothorax with a dusky gray abdomen. *T. pacificus* has been found throughout most of California, from its southern border through the Coast Ranges to the San Francisco Bay Area and into the central Sierra Nevada. A more thorough description of *T. pacificus* can be found in Platnick and Shadab (1974b).

Three species of *Meriola* live along the Pacific coast. The two native species are *Meriola decepta,* which is regionally limited to Southern California, and *M. californica,* which is found in coastal and near-coastal areas from Southern California to northern Washington. The third species, *M. arcifera,* was introduced from South America and has been found at scattered locations throughout California. *Meriola* have a reddish-brown carapace and legs, and the gray abdomen is decorated with a heart mark along with several lateral bars and chevrons. *Meriola* are generally smaller than *Trachelas,* running between 2.9 and 5 mm (.11 and .2 in.) in length. The native *Meriola* are reviewed in Platnick and Shadab (1974a), while *M. arcifera* is covered in Platnick and Ewing (1995), all under their previous placement within *Trachelas.*

Members of the subfamily Castianeirinae have at most three pairs of ventral spines on their anterior tibiae, while spiders in the remaining subfamilies have between five and eight pairs of spines. Castianeirinae is regionally represented by a single genus, *Castianeira* (pl. 28). Diurnal and often boldly patterned, *Castianeira* are the most conspicuous of North America's corinnids. *Castianeira* are often black or reddish in color with contrasting red, white, or yellow abdominal spots and stripes, presumably to better mimic large ants or highly aggressive wasps, also known as velvet ants, of the family Mutillidae. Pacific coast species range from 3.5 to 9.7 mm (.14 to .38 in.) in length and have sclerotized plates, or scuta, on the abdomen. On males the scute covers nearly the entire dorsal surface, while on females it is reduced to a small patch on the dorsoanterior margin. Seven species live along the Pacific coast, reaching their greatest diversity in Southern California. Only three species are known from Oregon and Washington: the widespread *Castianeira longipalpa* and the much less common *C. walsinghami* and *C. thalia* (pl. 28). A key to the species and numerous illustrations can be found in Reiskind (1969).

The subfamily Corinninae is composed of medium-size spiders, normally greater than 5 mm (.2 in.) in length, with between four and six pairs of spines on the ventral surfaces of tibiae I. Members of the subfamily Corinninae are mainly tropical in distribution, and only two genera, each representing a single species, can be found along the Pacific coast. Both genera are limited to Southern California. *Creugas bajulus* (pl. 28) is generally between 8.6 and 9.3 mm (.34 to .37 in.) in length with a reddish-brown carapace and legs and a gray abdomen. It was described and its reproductive features were illustrated in Gertsch (1942). It has been found in both Riverside County and San Diego County. *Septentrinna steckleri* (pl. 28) is slightly smaller, ranging from 4.6 to 7 mm (.18 to .28 in.) in length. Its carapace is yellowish-brown, its legs are pale yellow, and its abdomen is a creamy, unmarked gray. It has been collected from ant colonies in the sand dunes of Imperial County. The female was described by Gertsch (1936), but it wasn't until Bonaldo (2000) that the male was described and the spider's reproductive features were illustrated.

Spiders in the subfamily Phrurolithinae and those in the genus *Drassinella* (not illustrated) are small, less than 5 mm (.2 in.) in length. The three regional Phrurolithinae genera, *Scotinella, Piabuna,* and *Phrurotimpus,* live under rocks, fallen wood, and leaf litter. The taxonomy of this subfamily is in moderate disarray with undescribed species housed in museum collections around the country. Although *Phurolithus* (representing numerous Nearctic species) and *Phruronellus* are recognized under the rules of zoological nomenclature, unpublished research by Penniman (1985) sank these genera into *Scotinella*. While this change has not been officially recognized for procedural reasons, it has been accepted by numerous researchers and appears in Ubick and Richman (2005a), the most comprehensive and widely accessible key to North America's corinnid genera. For this reason, we have chosen to consider the Nearctic *Phurolithus* and all species currently placed in *Phruronellus* synonymized within *Scotinella*.

Genera in the subfamily Phrurolithinae can be recognized by the presence of one or no dorsal spines on each femur and the complete lack of ventral or lateral spines on legs III and IV. Species in the genus *Phrurotimpus* (pl. 28) have a single spine on the dorsal surface of each femur. The carapace has wide, dark bands extending back from the eyes to the posterior rim and thinner lines along the lateral margins. Six species have been described from the woodlands of Central California through northern Washington. *Piabuna* (pl. 28) are minute spiders, less than 2 mm (.08 in.) in length, with a very unusual eye arrangement. Their anterior median eyes are large and dark (in contrast to the others, which are light) and positioned dorsally on the carapace. *Piabuna* lack dorsal femoral spines and conspicuous markings on their bodies or legs. They live in dry pine forests and arid rocky regions across the southwestern United States. There are unpublished specimen records from Southern California (Ubick and Richman, 2005a). *Scotinella* (pl. 28) have eyes that are more traditionally aligned in two clear rows. They also lack dorsal femoral spines but have two prolateral spines on femora I and are generally larger than *Piabuna*, with most records ranging from 2 to 5 mm (.08 to .2 in.) in length. Six species of *Scotinella* have been recorded from the Pacific coast, ranging from Southern California to northern Washington. Unfortunately, almost nothing is known about most of these spiders beyond their original descriptions. Because none of the three Pacific coast phrurolithine genera have been thoroughly revised, any identification attempt will require delving into the original literature and species descriptions, the most relevant of which are Chamberlin (1921) (for *Phrurotimpus*), Chamberlin and Gertsch (1930) (*Phrurotimpus* and *Scotinella*), Chamberlin and Ivie (1935) (*Piabuna* and *Phrurotimpus*), Dondale and Redner (1982) (*Phrurotimpus* and *Scotinella*), Gertsch (1941) (*Phrurotimpus* and *Scotinella*), and Schenkel (1950) (*Phrurotimpus* and *Scotinella*).

PLATES

Family Theraphosidae

Baboon Spiders, Tarantulas

PAGE 28

One genus, 18 regional species. California.

IDENTIFICATION: Small to very large, hairy spiders found predominately in California's deserts, chaparral, and grasslands. They have claw tufts, urticating hairs on their abdomens, and in adult males, two-pronged spurs on tibiae I.

Aphonopelma

PAGE 28

18 species. California.

The region's tarantulas can be divided into three groups. The summer-breeding *Aphonopelma reversum* complex, including *A. steindachneri*, are large and fairly dark and have scopulae that are less than half the length of metatarsi IV. *A. eutylenum* complex tarantulas, large fall breeders, have scopulae covering at least half the length of metatarsi IV and are generally paler brown with noticeably darker femora. The third group, the dwarf tarantulas, are endemic to the Mojave Desert area. They are comparatively small and build unusual turret-like structures around their burrow entrances.

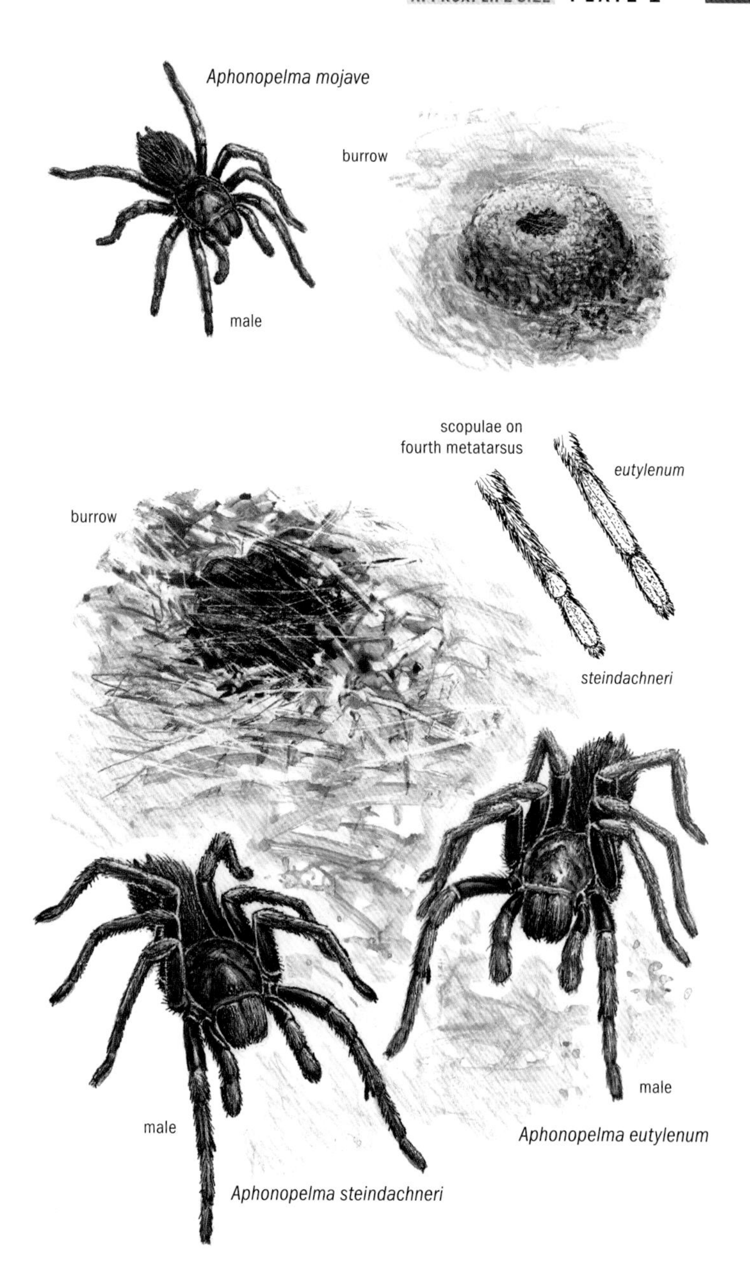
Aphonopelma mojave
burrow
male
scopulae on
fourth metatarsus
eutylenum
burrow
steindachneri
male
male
Aphonopelma eutylenum
Aphonopelma steindachneri

Family Nemesiidae

Velveteen Tarantulas, Aggressive False Tarantulas

PAGE 33

One genus, two regional species. California.

IDENTIFICATION: Fairly large mygalomorphs covered with a velvet-like pubescence that varies from silvery gray to dark brown. They have long posterior lateral spinnerets, lack claw tufts, and have aggressive threat displays.

Calisoga

PAGE 33

Two species. Central and Northern California.

Calisoga longitarsis is fairly common along the coast and in the mountains. Smaller but similar is the rarely collected *C. theveneti*. It has been collected both along the coast and in the Sierra Nevada.

Family Antrodiaetidae

Folding-Door Spiders, Trapdoor Spiders, Turret Spiders

PAGE 35

Two genera, 22 regional species. California, Oregon, Washington.

IDENTIFICATION: Small to medium-size mygalomorphs with one to four abdominal sclerites and fairly long, round-tipped spinnerets. Their burrows are capped by thin trapdoors, flexible collars, or turrets.

Antrodiaetus

PAGE 35

12 regional species. Central California to northern Washington.

Antrodiaetus have deep, longitudinal thoracic furrows and cap their burrows with thin, flexible collars. The California Turret Spider, *Antrodiaetus riversi,* constructs a distinctive turret of dirt and plant debris above its burrow's entrance. The adult males of several species have unusual tusk-like apophyses on their chelicerae.

Aliatypus

PAGE 37

10 species. California.

Aliatypus construct thin-lidded trapdoors over their burrows. When present, their thoracic furrows are represented by a deep pit or a shallow depression. Adult males are easily recognized by their exceptionally long palpal patellae.

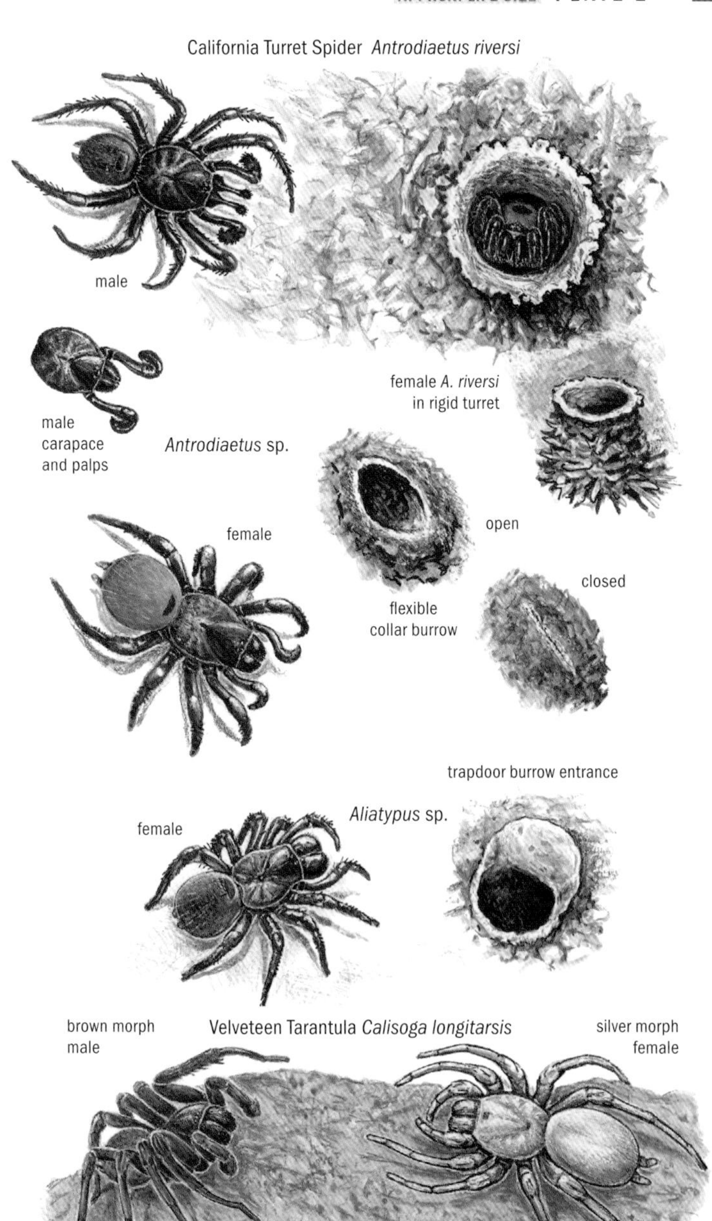
California Turret Spider *Antrodiaetus riversi*
male
male carapace and palps
Antrodiaetus sp.
female *A. riversi* in rigid turret
female
open
closed
flexible collar burrow
trapdoor burrow entrance
Aliatypus sp.
female
brown morph male
Velveteen Tarantula *Calisoga longitarsis*
silver morph female

Family Euctenizidae (in part)

Wafer-Lid Trapdoor Spiders

PAGE 39

Three genera, 41 described regional species. California.

IDENTIFICATION: Small to medium-size mygalomorphs. Many have distinctively patterned abdomens and scopulae on their anterior tarsi and metatarsi. Their burrows may be topped by thin-lidded trapdoors or flexible turrets.

Aptostichus

PAGE 39

37 described species. California.

Aptostichus vary from boldly striped to uniformly colored, and they cap their burrows with thin trapdoors. Less-conspicuous identifying features, including patterns of spination and their endite structure, are discussed in the text.

Apomastus

PAGE 40

Two species. Southern California.

Apomastus are uniformly brown with recurved thoracic furrows. They are found above the Los Angeles Basin, where they build distinctive, turret-topped burrows on moist, shaded slopes.

Family Ctenizidae

Cork-Lid Trapdoor Spiders

PAGE 43

Two genera, two regional species. California.

IDENTIFICATION: Robust spiders with deep, procurved thoracic furrows, short spinnerets, and heavily spined fore legs. Their burrows are capped by thick-lidded trapdoors.

Bothriocyrtum

PAGE 43

One species. Southern California.

Bothriocyrtum californicum is found in arid regions. The female ranges from light to dark brown, and the abdomen on the adult male varies from pinkish to dark red.

Hebestatis

PAGE 44

One species. Southern and Central California.

The *Hebestatis theveneti* female is generally smaller and darker than *B. californicum* with a distinctive depression on the dorsal face of tibia III. The male *H. theveneti* (not illustrated) has not been formally described but is known to have a purplish-tinged abdomen and a unique palpal structure. It prefers damp, shaded banks.

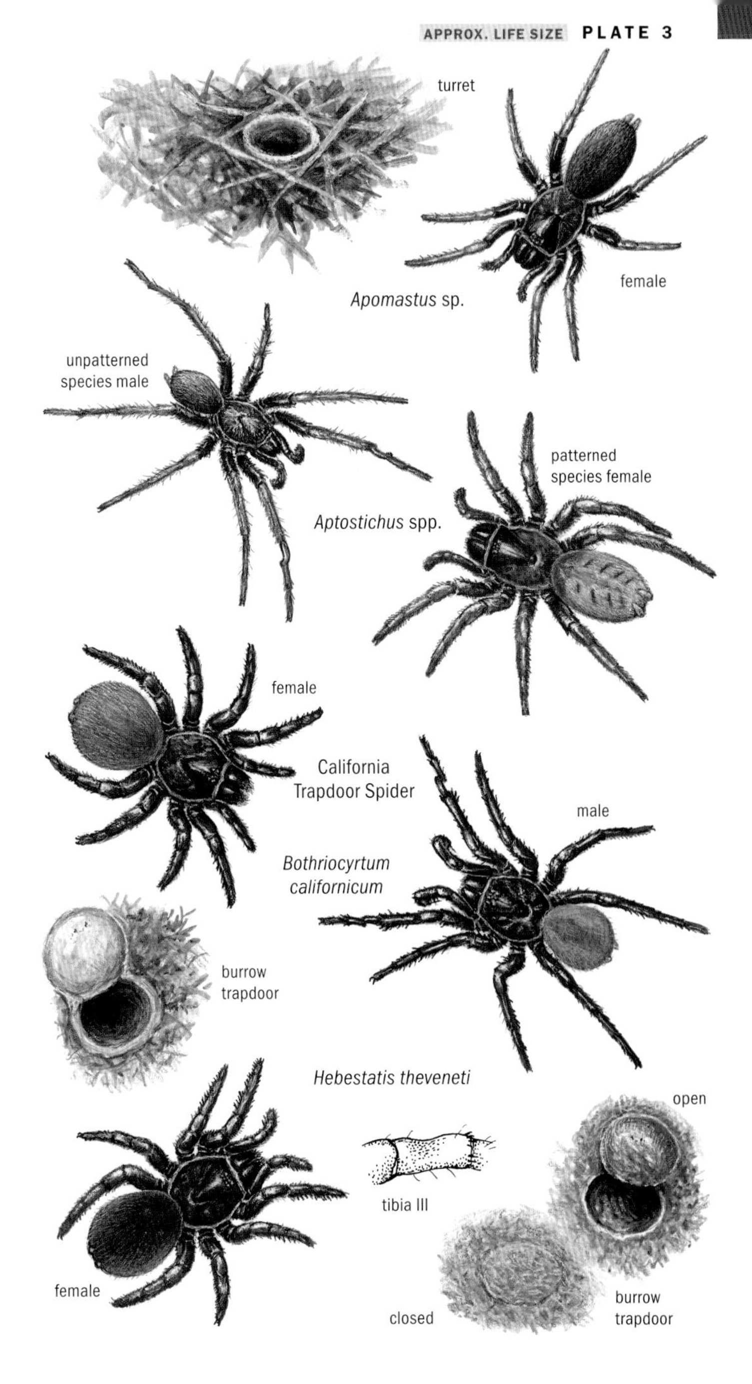
turret
female
Apomastus sp.
unpatterned
species male
patterned
species female
Aptostichus spp.
female
California
Trapdoor Spider
male
Bothriocyrtum
californicum
burrow
trapdoor
Hebestatis theveneti
open
tibia III
female
closed
burrow
trapdoor

Family Euctenizidae (cont.) PAGE 39

Promyrmekiaphila PAGE 40

Two species. Central and Northern California.

Promyrmekiaphila have distinctive broad bands on their abdomens and cover their burrows with thin-lidded trapdoors. Additional identifying features involving their spination and endite structure are discussed in the text.

Family Dipluridae PAGE 46

One genus, one regional species. Oregon, Washington.

IDENTIFICATION: Tiny mygalomorphs that lack abdominal sclerites and have only two pairs of spinnerets.

Microhexura PAGE 46

Microhexura idahoana is a minute, montane mygalomorph that constructs silken tunnels under moss mats and damp, rotting wood.

Family Mecicobothriidae PAGE 47

Three genera, four regional species. California, Oregon, Washington.

IDENTIFICATION: Small mygalomorphs with exceptionally long, flexible posterior lateral spinnerets; shallow, longitudinal thoracic grooves; and abdominal sclerites.

Megahexura PAGE 48

One species. California.

Megahexura fulva has two abdominal sclerites, one of which rests just above the pedicel and is difficult to see. It also has pleurites extending backward from the posterolateral edges of its carapace.

Hexurella PAGE 48

One species. Southern California.

Hexurella rupicola is tiny and reddish yellow and has two abdominal sclerites. It is found in Southern California's arid foothills under rocks and debris.

Hexura PAGE 48

Two species. Northern California through northern Washington.

Hexura live in shaded woodlands and have a single, broad sclerite on the abdomen. *Hexura rothi* has two pairs of spinnerets, *H. picea* has three. They build small sheet webs with connecting tunnels under rotting logs, leaf litter, and shaded slopes.

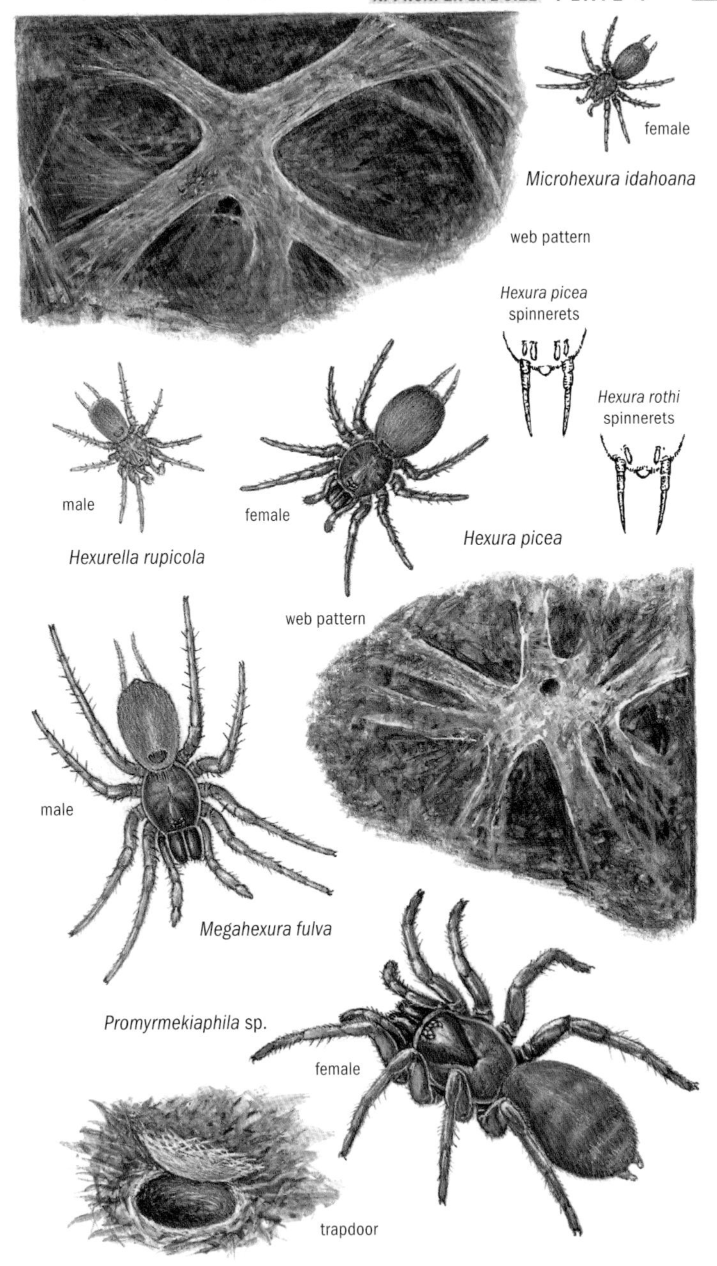
female
Microhexura idahoana
web pattern
Hexura picea
spinnerets
Hexura rothi
spinnerets
male
female
Hexurella rupicola
Hexura picea
web pattern
male
Megahexura fulva
Promyrmekiaphila sp.
female
trapdoor

Family Hypochilidae

Lampshade Weavers PAGE 49

One genus, three regional species. California.

IDENTIFICATION: Montane spiders that build distinctive lampshade-shaped webs in cool, dark areas. They are long-legged, flat-bodied spiders and have two pairs of book lungs. They often rest in a distinctive splayed position in the center of the web.

Hypochilus PAGE 50

Three species. California.

The Pacific coast's *Hypochilus* are all extremely similar to one another. Each is found in a different California mountain range.

Family Filistatidae

Crevice Weavers PAGE 52

Three genera, five described and several undescribed regional species. California.

IDENTIFICATION: Cribellate, haplogyne spiders whose eight eyes are clustered together on a central mound. If these spiders are grabbed, their legs exhibit tibia-patella autospasy. Undescribed species belonging to two small-bodied genera are known from Southern California. They are discussed and illustrated in the text (fig. 7).

Kukulcania PAGE 52

Five species. California.

Kukulcania build a tube web of cribellate silk collared by a messy net of radiating threads. *Kukulcania* are extremely sexually dimorphic. The males are distinctly paler, with thinner, longer legs than the more robust females.

Family Segestriidae

Tube Web Weavers PAGE 54

Two genera, five regional species. California, Oregon, Washington.

IDENTIFICATION: Six-eyed haplogyne spiders that build white, flare-lipped tube webs under ledges and in narrow crevices. Their legs are heavily spined and distinctively arranged. Pairs I through III extend forward while IV points backward.

(continued)

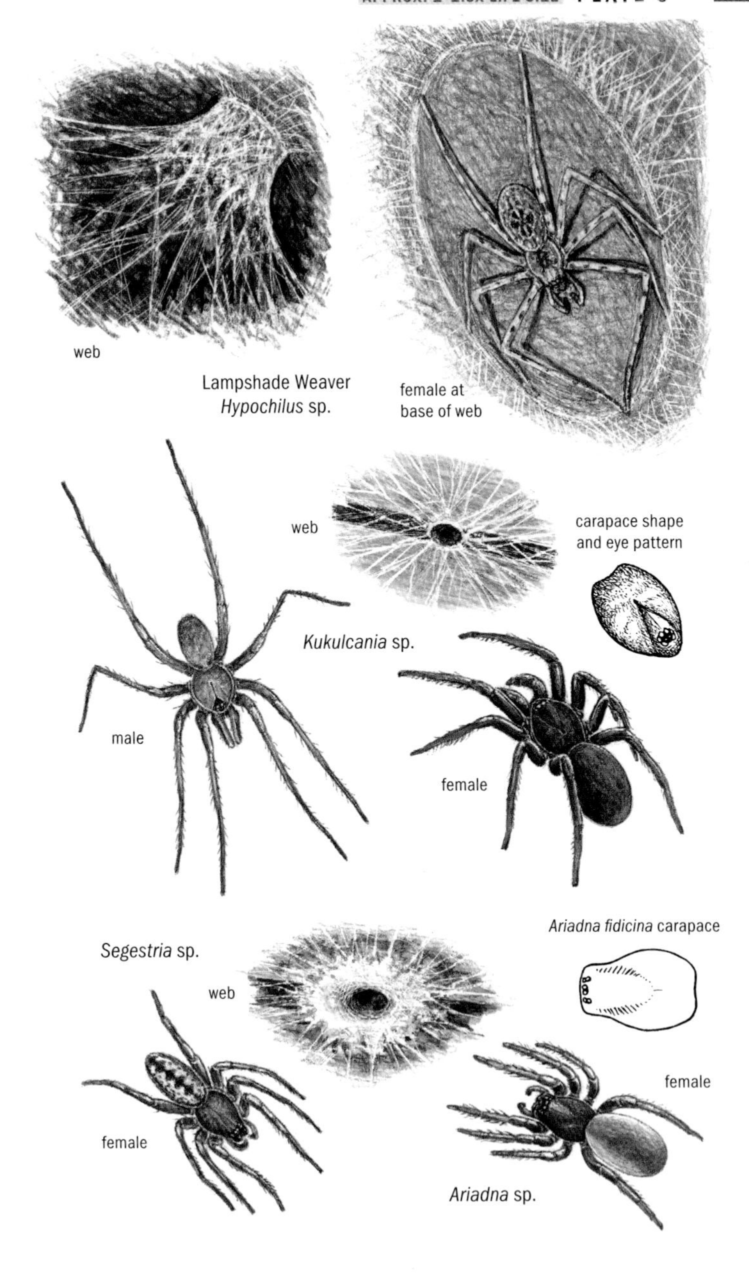
web
Lampshade Weaver
Hypochilus sp.
female at
base of web
web
carapace shape
and eye pattern
Kukulcania sp.
male
female
Ariadna fidicina carapace
Segestria sp.
web
female
female
Ariadna sp.

(continued)

Segestria

PAGE 55

Three species. Coastal California through Washington.
Segestria can be recognized by their patterned abdomens and banded legs.

Ariadna

PAGE 55

Two species. Southern and Central California.
Ariadna lack any obvious abdominal patterning and are stockier than *Segestria. Ariadna fidicina* has distinctive rows of stridulatory files across the anterior portion of its carapace.

Family Caponiidae

PAGE 56

Three genera, eight regional species. California.

IDENTIFICATION: Small haplogyne spiders that, with the exception of *Calponia harrisonfordi,* all have only two eyes. They lack book lungs and their posterior median spinnerets are in line with their anterior laterals.

Orthonops

PAGE 57 (FIG. 8).

Six species. Southern and Central California.

Except for *Orthonops zebra*, *Orthonops* have unpatterned abdomens. They have only two eyes and stout fore legs with subsegmented tarsi.

Tarsonops

PAGE 57 (FIG. 8).

One species. Southern and Central California.

Tarsonops systematicus has only two eyes and is pale yellow. Its first pair of legs is long and slender with subsegmented tarsi and metatarsi.

Calponia

PAGE 58

One species, Central California.

The eight eyes of *Calponia harrisonfordi* are arranged in a tightly clustered mound. It is a small spider of the chaparral.

Family Oonopidae

Goblin Spiders

PAGE 59

Four genera, seven described regional species. California, Washington.

IDENTIFICATION: Tiny, six-eyed, haplogyne spiders found under leaf litter and debris. Many have abdominal scuta, and they often move in short, rapid bursts. Several unidentified *Oonops* (not illustrated) and *Oonops*-like individuals have been collected in Southern California. These are described in greater detail in the text.

Orchestina

PAGE 59

Three species. California, Washington.

Orchestina are dull colored with globular abdomens and without scuta. Their fourth femurs are swollen, and their posterior median eyes are aligned with their anterior laterals.

(continued)

(continued)

Escaphiella

PAGE 60

Three species. California.

Males have large abdominal scuta and unremarkable palpal patellae. On females, the distinctive ventral scuta wrap around the sides of the abdomen, leaving a bare dorsal stripe.

Opopaea

PAGE 60

One species. Southern California.

Opopaea concolor is an introduced, vermillion-colored spider. Both sexes have extensive abdominal scuta, and the male has enlarged palpal patellae.

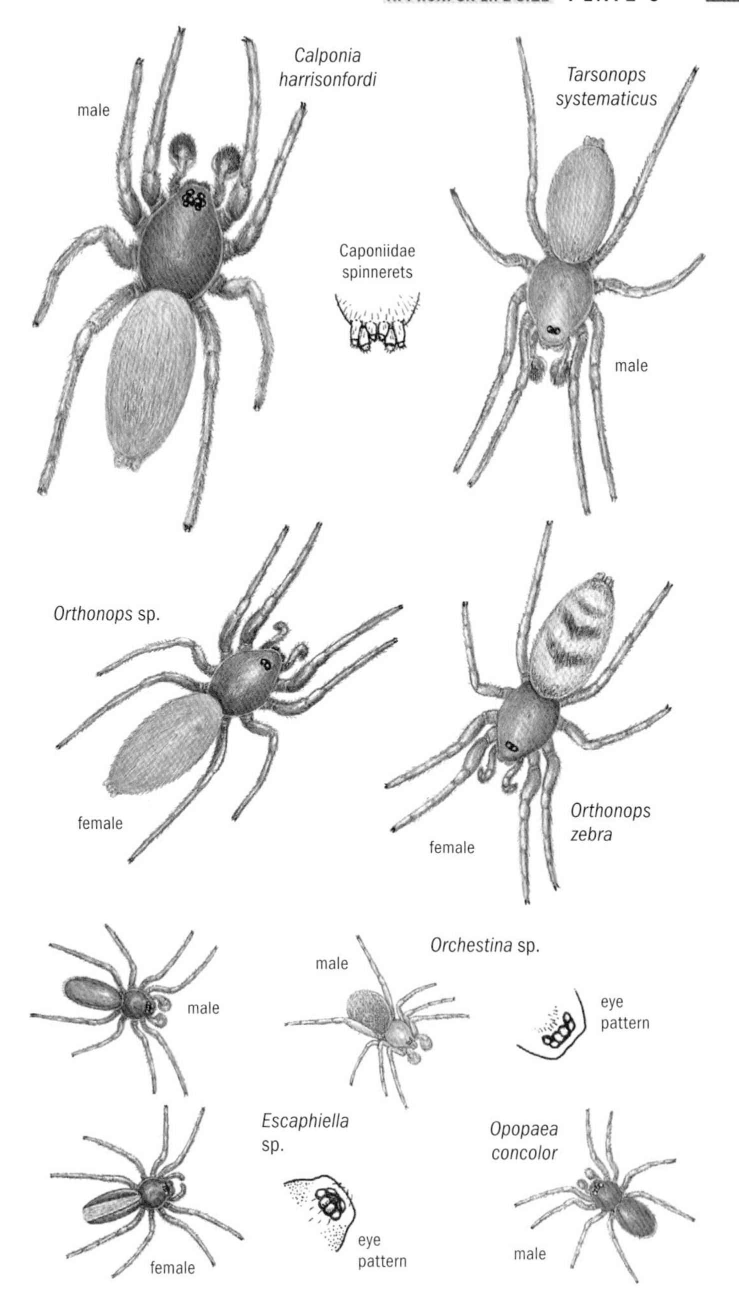

Calponia harrisonfordi
male
Caponiidae spinnerets
Tarsonops systematicus
male
Orthonops sp.
female
Orthonops zebra
female
male
Orchestina sp.
male
eye pattern
Escaphiella sp.
female
eye pattern
Opopaea concolor
male

Family Dysderidae

Wood Louse Hunters PAGE 61

One genus, one regional species. California, Oregon, Washington.

IDENTIFICATION: Distinctively colored terrestrial spiders with massive chelicerae. Their six eyes are arranged in an unusual procurved pattern.

Dysdera PAGE 61

Native to the Mediterranean area, the Wood Louse Hunter, *Dysdera crocata,* is found around lawns and gardens across the Pacific coast. Its coloration and eye arrangement are unique.

Family Trogloraptoridae PAGE 63

One genus, one described and one unidentified regional species. California, Oregon.

IDENTIFICATION: Highly secretive, six-eyed, haplogyne spiders with unique raptorial tarsi. Their anterior and posterior lateral eyes are connected, while their posterior medians are clearly separated.

Trogloraptor PAGE 63

One described species. Southern Oregon.

Trogloraptor marchingtoni is a large cave dweller endemic to a small region of southwestern Oregon. Juveniles of an undescribed second species have been found in a stand of old-growth redwood forest in the extreme northern part of California.

Family Scytodidae

Spitting Spiders PAGE 64

One genus, two identified and several unidentified regional species. California.

IDENTIFICATION: Easily recognized by their massive, dome-shaped carapaces. Their six eyes are arranged in a recurved arc.

Scytodes PAGE 64

Two identified species. California.

The known Pacific coast species are all introduced and synanthropic and range in color from pale yellow with black bands to dark brown.

(continued)

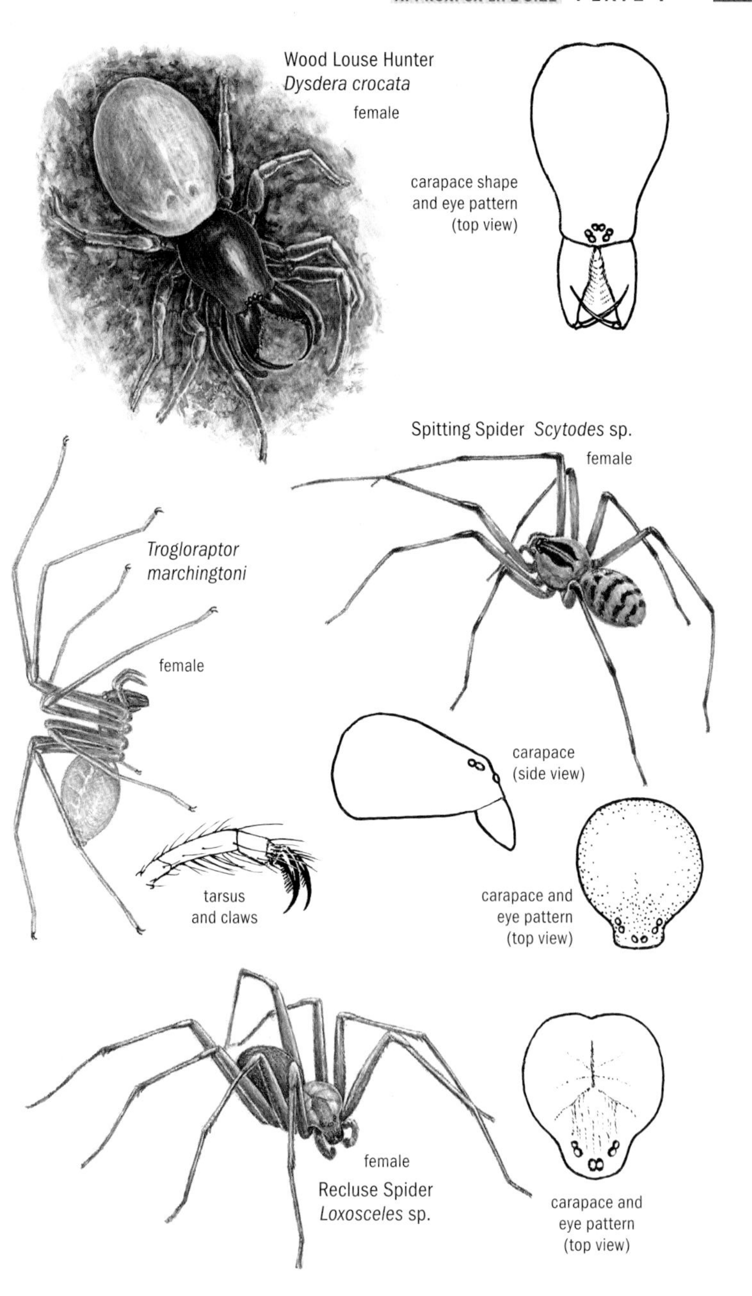
Wood Louse Hunter
Dysdera crocata
female
carapace shape
and eye pattern
(top view)
Spitting Spider Scytodes sp.
female
Trogloraptor
marchingtoni
female
tarsus
and claws
carapace
(side view)
carapace and
eye pattern
(top view)
female
Recluse Spider
Loxosceles sp.
carapace and
eye pattern
(top view)

Family Sicariidae

Recluse Spiders

PAGE 65

One genus, seven regional species. California, Washington (unestablished).

IDENTIFICATION: Spiders with six eyes set in a recurved arc with a flat carapace. They are small and vary from beige to reddish brown in color, and the "violin" mark on the carapace can be obvious or lacking.

Loxosceles

PAGE 66

Seven species. Southern and Central California, Washington.

Four native species live in Southern and Central California. Isolated populations and the occasional individual of one of the introduced species have been found in California and Washington. Despite popular misconceptions, there are no populations of the Brown Recluse, *Loxosceles reclusa,* established anywhere along the Pacific coast.

Plates continue on the following page.

Family Diguetidae

Desertshrub Spiders

PAGE 68

One genus, five regional species. California.

IDENTIFICATION: Six eyes arranged in three dyads, three tarsal claws, and chelicerae that are fused at their bases. A desertshrub spider web is a complex, distinctive structure consisting of an inverted cone of silk and debris resting on a platform and suspended within a large tangle.

Diguetia

PAGE 69

Five species. Southern and Central California.

Desertshrub spiders are chaparral and desert specialists. The region's species vary from boldly patterned (*Diguetia canities*) to chalky white (*D. mojavea*).

Family Plectreuridae

PAGE 70

Two genera, 12 regional species. California, Washington.

IDENTIFICATION: Haplogyne spiders whose eight eyes are aligned in two broad rows. They also have relatively thick legs and chelicerae that are fused at their bases. These spiders build loose tangle webs near the ground.

Plectreurys

PAGE 71

Nine species. California, central Washington.

Plectreurys are dark colored with thick, stocky legs. Femur I has a distinct curve, is thicker than femur II, and lacks a spread of dorsal spines. Adult males have large spurs at the ventrodistal end of tibia I. Most species are limited to Southern and Central California, but *Plectreurys tristis,* a widespread Great Basin species, has been found in Washington.

Kibramoa

PAGE 72

Three species. Southern and Central California.

Dark spiders with contrasting chestnut to reddish carapaces. Femur I is darker than the other leg segments, thin, straight, and decorated dorsally with numerous spines. Adult male *Kibramoa* lack ventral spurs on tibiae I.

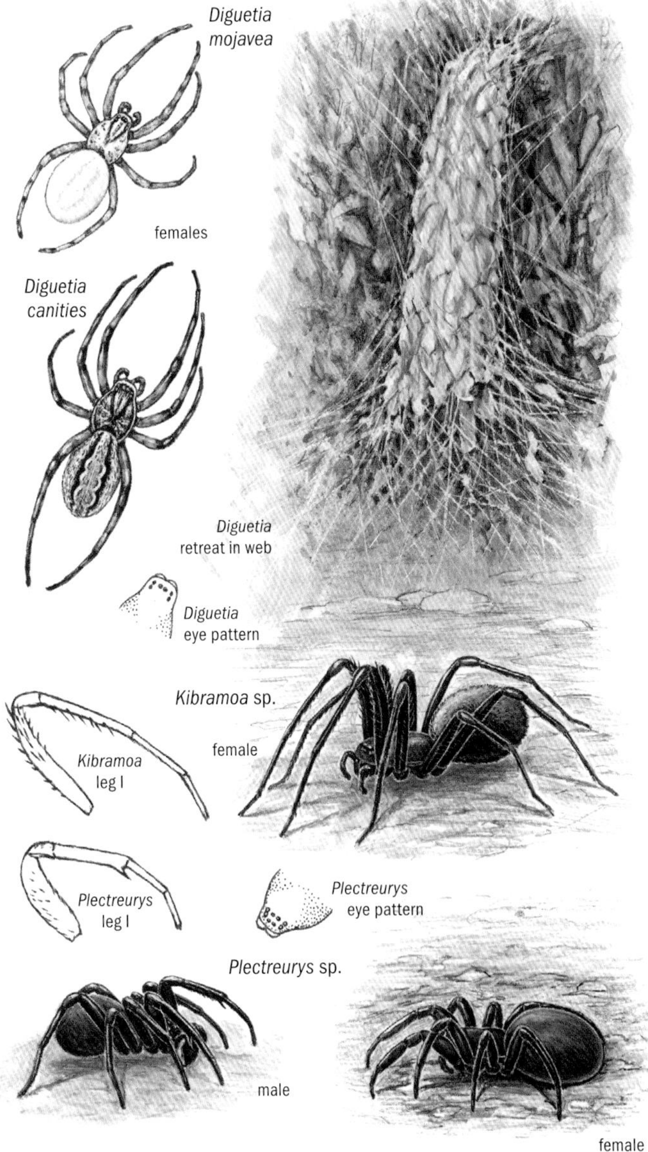
Diguetia mojavea
females
Diguetia canities
Diguetia retreat in web
Diguetia eye pattern
Kibramoa sp.
female
Kibramoa leg I
Plectreurys leg I
Plectreurys eye pattern
Plectreurys sp.
male
female

Family Pholcidae (in part)

Cellar Spiders

PAGE 73

Eight genera, 20 regional species. California, Oregon, Washington.

IDENTIFICATION: Exceptionally long, thin legs with pseudosegmented tarsi. Most species have eight eyes arranged into paired triads on either side of their minute anterior median eyes.

Pholcus

PAGE 73

One species. California through Washington.

The introduced Long-bodied Cellar Spider, *Pholcus phalangioides,* is extremely common in homes throughout the region, where its tangle webs are frequently found in cupboards and in the upper corners. It is beige with dusky markings on its cephalothorax and abdomen.

Holocnemus

PAGE 73

One species. Southern California through southern Oregon.

The introduced Marbled Cellar Spider, *Holocnemus pluchei,* has a diagnostic wide black band across the underside of both its abdomen and cephalothorax and a marbled dorsum. It is generally found in warmer, more exposed areas than Long-bodied Cellar Spiders. The female Marbled Cellar Spider builds a distinctive, spherical nursery web for her newly hatched young.

Crossopriza

PAGE 74

One species. Southern California.

The Tailed Cellar Spider, *Crossopriza lyoni,* has a greenish-gray, marbled abdomen that is accentuated by a triangular tubercle off its dorsoposterior margin. Small populations of this introduced spider have been found in southeastern California.

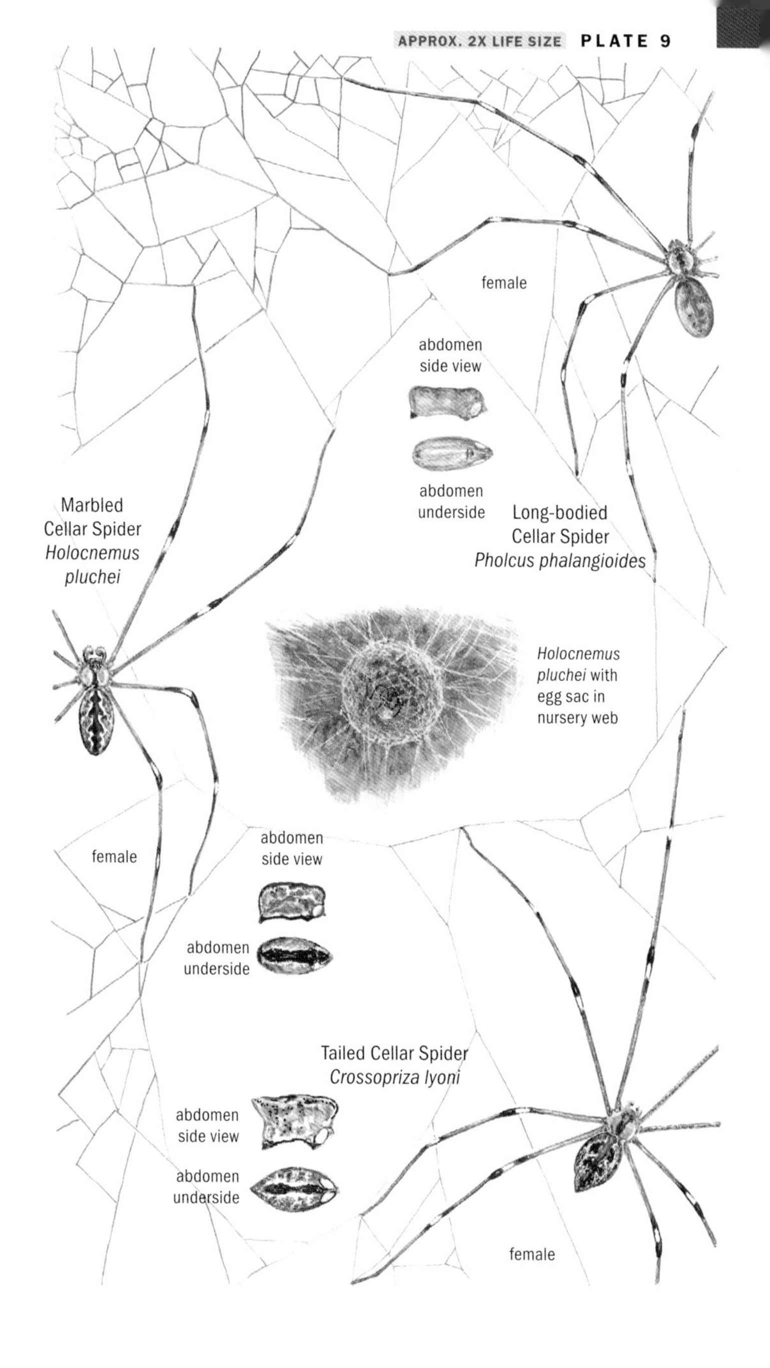
female
abdomen
side view
abdomen
underside
Long-bodied
Cellar Spider
Pholcus phalangioides
Marbled
Cellar Spider
Holocnemus
pluchei
Holocnemus
pluchei with
egg sac in
nursery web
female
abdomen
side view
abdomen
underside
Tailed Cellar Spider
Crossopriza lyoni
abdomen
side view
abdomen
underside
female

Family Pholcidae (cont.)

PAGE 73

Artema

PAGE 74

One species. Southern California.

Artema Atlanta is one of the world's largest cellar spiders. Its formidable size, globular abdomen, and spotted pattern and the massive black growths on the male's chelicerae render it nearly unmistakable. It was introduced to North America. Small populations have been found near the Salton Sea.

Spermophora

PAGE 74

One species. California, Oregon.

Introduced from the Asia-Pacific region, the tiny, six-eyed, *Spermophora senoculata* has been found sporadically along the Pacific coast. Because of its small size and synanthropic habits, it is easily overlooked and could turn up in buildings almost anywhere.

Psilochorus

PAGE 75

12 species. California through Washington.

Also known as round-bodied cellar spiders, *Psilochorus* are small, native cellar spiders (less than 4.0 mm long) that build unobtrusive tangle webs in protected corners and under debris. Adult male *Psilochorus* have narrow palpal femora armed with spurs on their ventral margins.

Physocyclus

PAGE 75

Two species. California.

Physocyclus are generally greater than 4.0 mm long, and adult males have exceptionally large palpal femora without ventral spurs. *Physocyclus* are uncommon to very rare along the Pacific coast.

Pholcophora

PAGE 75

One species. California through Washington.

The Short-legged Cellar Spider, *Pholcophora americana*, is found in montane forests. While still long and thin, its legs are stockier in comparison to other cellar spiders.

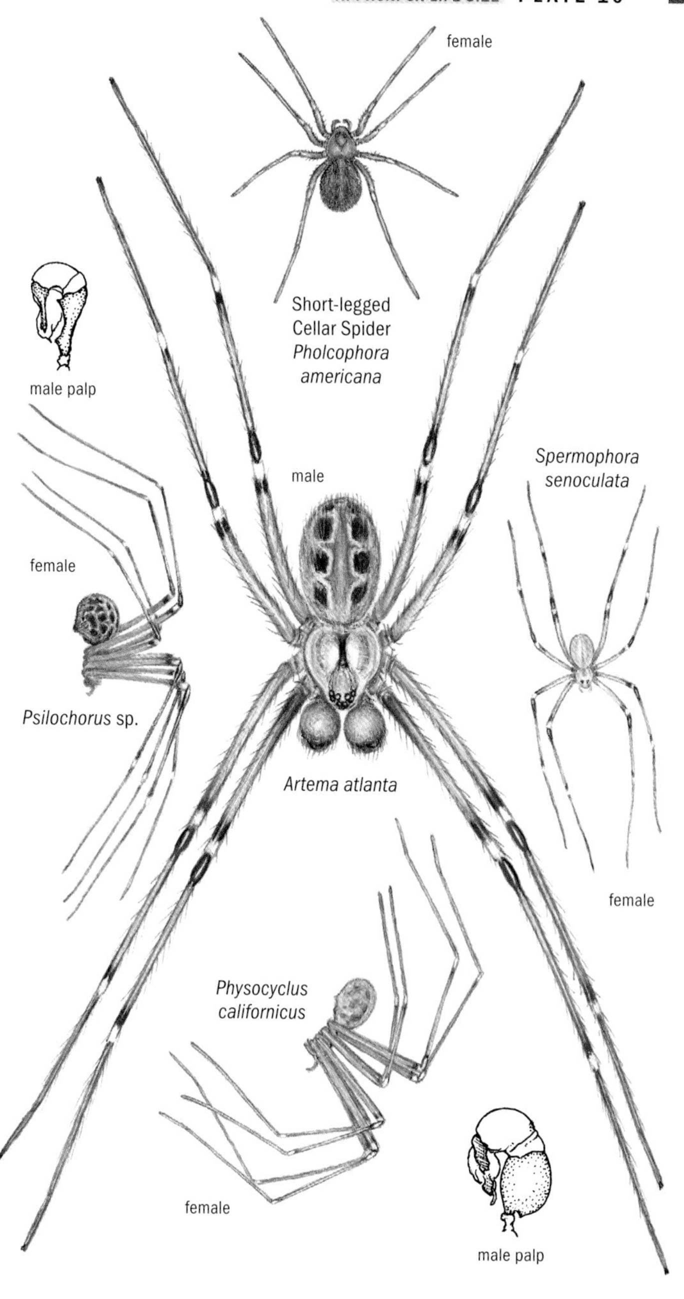
female
Short-legged
Cellar Spider
Pholcophora
americana
male palp
male
Spermophora
senoculata
female
Psilochorus sp.
Artema atlanta
female
Physocyclus
californicus
female
male palp

Family Leptonetidae

PAGE 77

Two genera, 11 regional species. California, Oregon.

IDENTIFICATION: Tiny, cryptic spiders whose six eyes are arranged in one of two distinctive patterns. On living spiders, the legs are uniquely iridescent. Leptonetidae live in protected areas, including caves, hollow trees, and inside rotting logs.

Calileptoneta

PAGE 77

Nine species. Southern California to southern Oregon.

Calileptoneta can be recognized by the placement of their posterior median eyes well behind the main ocular cluster.

Archoleptoneta

PAGE 78

Two species. California.

All of the spiders in the genus have eyes arranged in a compact ocular cluster.

Family Telemidae

PAGE 79

One genus, three described and several undescribed regional species. California, Oregon, Washington.

IDENTIFICATION: Minute spiders with a distinctive W-shaped sclerotized abdominal ridge just above the pedicel. Each of the described species has six eyes arranged in three pairs, although undescribed eyeless individuals have been collected in caves in the Sierra Nevada.

Usofila

PAGE 80

Three species. Central California through northern Washington. Along the Pacific coast, *Usofila* have been collected from caves and thick leaf litter. They are round bodied and make inconspicuous sheet webs.

Family Mysmenidae

PAGE 81

One genus, one regional species. California, Oregon, Washington.

IDENTIFICATION: Tiny, cryptic, eight-eyed spiders that can be recognized by the presence of a sclerotized spot on the undersides of the first femora and spine-like clasping spurs on the adult males' first metatarsi.

(continued)

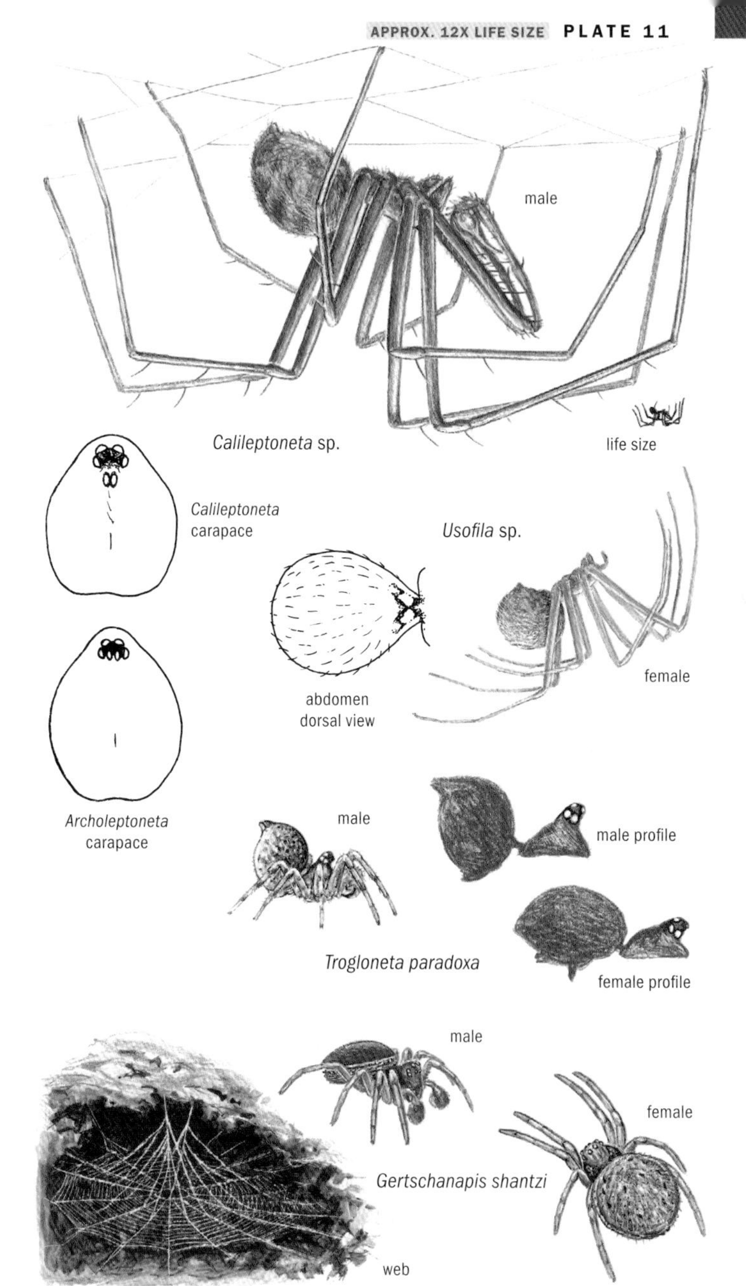
male
life size
Calileptoneta sp.
Calileptoneta carapace
Usofila sp.
abdomen dorsal view
female
Archoleptoneta carapace
male
male profile
Trogloneta paradoxa
female profile
male
female
Gertschanapis shantzi
web

(continued)

Trogloneta PAGE 81

Trogloneta paradoxa is a rare member of the region's spider fauna and has been found in dark, humid places. Both sexes have ocular turrets and elongate clypei, though they are more pronounced on the male, giving the spider's "face" a concave look when seen in profile.

Family Anapidae PAGE 82

One genus, one regional species. California, Oregon.

IDENTIFICATION: Tiny, eight-eyed spiders that build unique, horizontal orbs whose radial threads extend above the web's main body. On females, the palps are reduced to minute nubs. Males have a large abdominal scute, and both sexes have tubercles below femora I and II.

Gertschanapis PAGE 82

Gertschanapis shantzi is a rare spider that has been found in cool, humid areas along the coast and in the Sierra Nevada.

Plates continue on the following page.

Family Uloboridae

Hackled Band Orb Weavers

PAGE 83

Two genera, three regional species. California, Oregon, Washington.

IDENTIFICATION: Small, cribellate spiders whose genera build distinctive webs. The spiders have eight eyes, exceptionally long calamistra, dorsally concave fourth metatarsi, and nondivided cribella.

Hyptiotes

PAGE 83

Two species. California through Washington.

Also known as triangle-web spiders, they are small and squat, with tiny anterior lateral eyes and posterior lateral eyes on tubercles. They make unique triangle-shaped webs.

Uloborus

PAGE 84

One species. California through Washington.

Uloborus diversus, the region's only feather-legged spider, has a roughly triangular abdomen and extremely long fore legs. At the ends of tibiae I are conspicuous brushes of setae. Its orb web is often horizontal and always has a stabilimentum across its hub.

Family Oecobiidae

Wall Spiders

PAGE 85

One genus, three regional species. California, Oregon, Washington.

IDENTIFICATION: Small, fairly flat spiders with distinctive, enlarged anal tubercles. Their eight eyes are arranged in a tight cluster (fig. 9).

Oecobius

PAGE 86

Three species. Southern California to central Washington.

The introduced and synanthropic *Oecobius navus* is widespread across the Pacific coast. Two additional species, the introduced *O. putus* and the native *O. isolatus,* are found in Southern California.

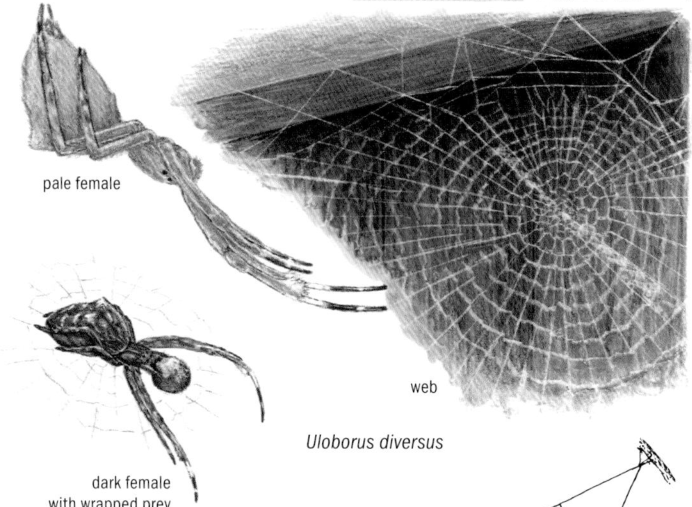

pale female

web

Uloborus diversus

dark female
with wrapped prey

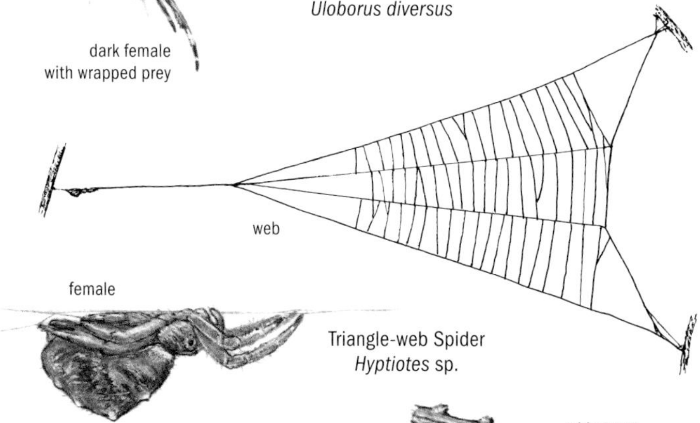

web

female

Triangle-web Spider
Hyptiotes sp.

egg sac on
pine branch

Wall Spider
Oecobius navus

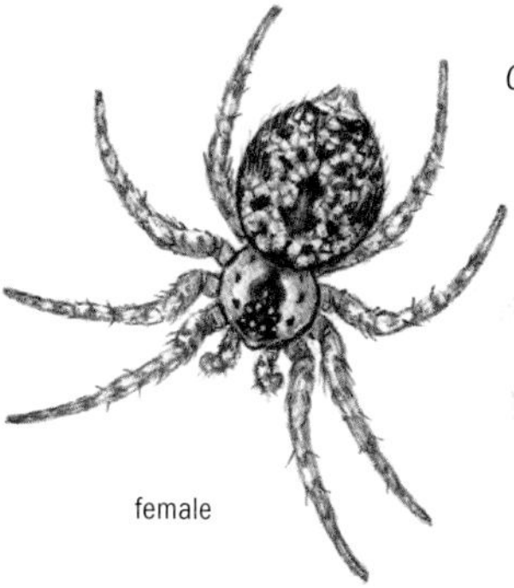

under web on wall

female

Family Mimetidae

Pirate Spiders PAGE 88

Three genera, four described regional species. California, Oregon, Washington.

IDENTIFICATION: Spiders that specialize in hunting other spiders and have unique rows of stiff spines on their anterior tarsi and metatarsi.

Mimetus PAGE 88

Two species. Southern California to central Washington.

Reo (not illustrated) PAGE 89

One species. Southern and Central California.

Mimetus are generally mottled yellow in color and have short clypei. Their long chelicerae are adorned, each with a spine on its anterior face. Several undescribed or unidentified *Mimetus* are also known from California. *Reo eutypus* is structurally similar to *Mimetus*, but it is darker in color and differs in the fine details of its reproductive structures. *Mimetus* have highly unusual pendulous egg sacs, which they wrap in a distinctive coating of thick, coppery threads.

Ero PAGE 89

One identified species. California (rare), Washington.

Ero have higher clypei than either *Mimetus* or *Reo* and short chelicerae that lack anterior spines. *Ero canionis* is a dark spider with a pale abdominal patch. While nearly all Pacific coast records are from Washington, at least one unidentified *Ero* has been found in Northern California.

Family Nesticidae

Cave Cobweb Spiders PAGE 90

Two genera, four regional species. California, Oregon, Washington.

IDENTIFICATION: Small spiders that make distinctive "split-foot" webs in caves, under rotting logs, and in other cool, damp places. Cave cobweb weavers have thinly rebordered labia, and adult males have large paracymbia. Nesticids generally have eight eyes, although some cave-dwelling species are eyeless.

(continued)

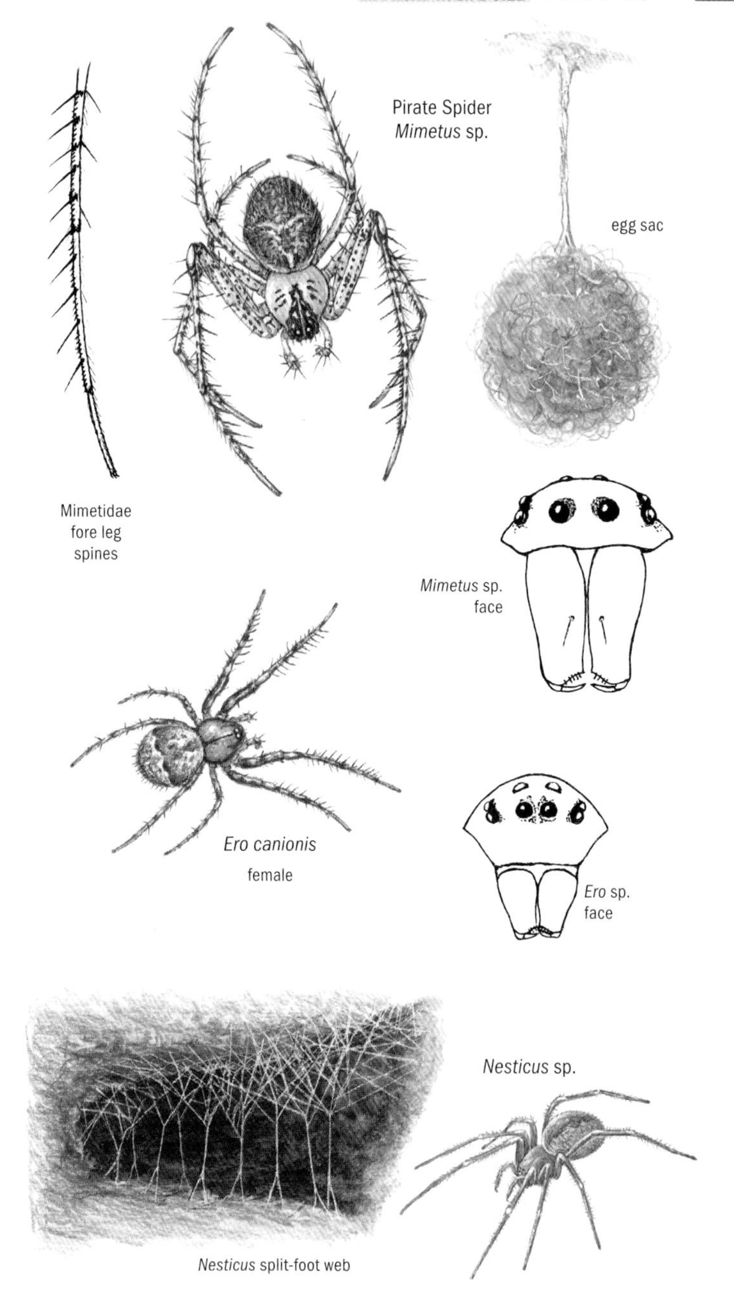
Pirate Spider
Mimetus sp.
egg sac
Mimetidae
fore leg
spines
Mimetus sp.
face
Ero canionis
female
Ero sp.
face
Nesticus sp.
Nesticus split-foot web

(continued)

Nesticus PAGE 91

Three species. Central California through northern Washington.

Eidmannella (not illustrated) PAGE 92

One species. California, Oregon.

Both of these genera range from orangish to pale yellow, becoming paler and longer-legged as they move deeper into caves. *Nesticus* includes both eyed and eyeless species. *Eidmannella pallida* is a regionally rare species that has been found in Southern California and at least once in Oregon. Differentiating the two genera requires examining their reproductive structures.

Plates continue on the following page.

Family Theridiidae (in part)

Cobweb Weavers, Comb-Footed Spiders PAGE 93

29 genera, 93 regional species. California, Oregon, Washington.

IDENTIFICATION: A row of curved, serrated setae on the ventral surfaces of the fourth tarsi (which may be lost on some genera). They have very few or no leg spines and a labium without a swollen anterior edge. For the most part, they build chaotic-looking tangle webs. Several of the region's smallest and rarest genera are not illustrated but are discussed in the main text.

Latrodectus PAGE 95

Two species. Southern California to northern Washington.

The Pacific coast's species include the native Western Black Widow, *Latrodectus hesperus,* and the recently introduced Brown Widow, *L. geometricus.* Both are easily recognized by the presence of a red or orange hourglass-shaped mark on the underside of the abdomen. The Western Black Widow is found widely across the Pacific coast, and its egg sacs are round or ovoid with a papery outer coat. The Brown Widow is common in Southern California, and it constructs unique, spiked egg sacs. Widow spiders are the only regional spiders of regular and significant medical concern.

Steatoda PAGE 96

15 species. Southern California to northern Washington.

Steatoda vary from white to brick red in color, although most are dark purple, and many have a white line around the abdomen's anterior margin. The native species are fairly small, and the sexes are similar in shape. The introduced False Black Widow, *Steatoda grossa,* is common in homes throughout the region. It is highly sexually dimorphic, and its egg sacs are wrapped in a thick layer of fluffy white silk. *S. nobilis* (fig. 10) is a large spider recently introduced to Southern California.

Parasteatoda PAGE 98

One species. California through Washington.

The introduced and synanthropic Common House Spider, *Parasteatoda tepidariorum,* is especially common in Southern California. The female is variable in color but normally has a series of gray chevrons along her abdomen's posterior margin, and her spinnerets are pointed ventrally. The male is distinctly smaller with a narrow, oblong abdomen and is often found around females' webs. The egg sacs are round with a tan, papery outer layer.

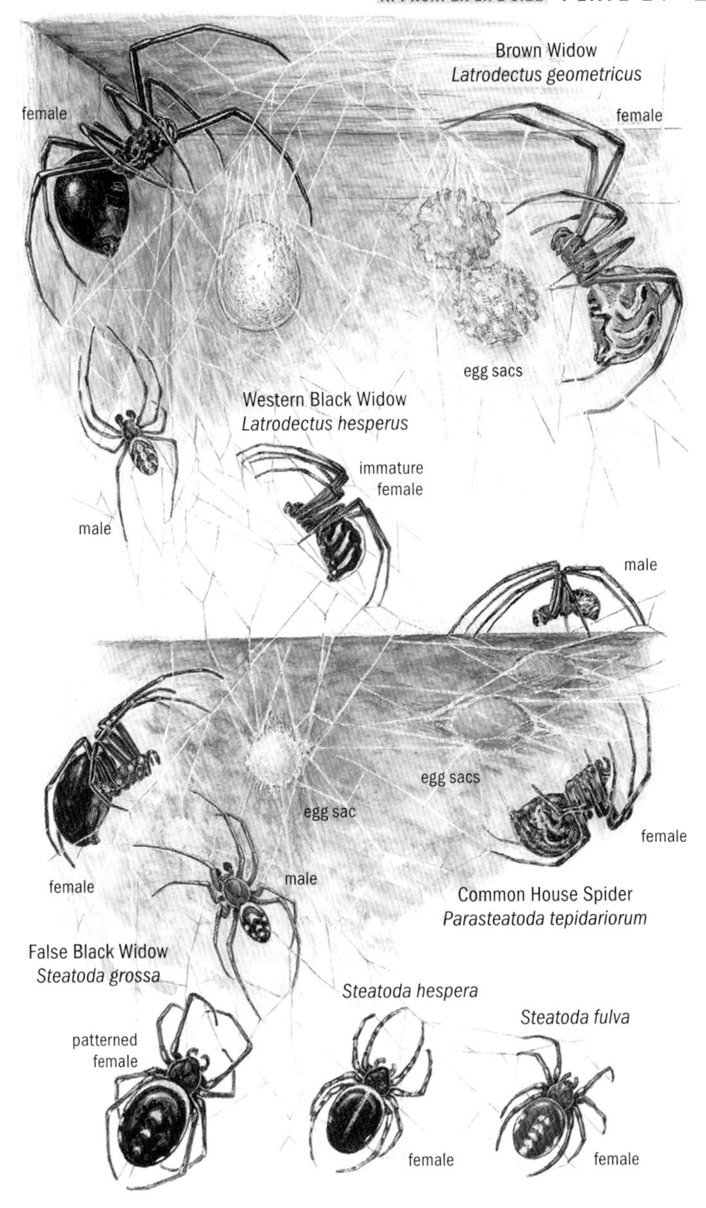

Brown Widow
Latrodectus geometricus
female
female
egg sacs
Western Black Widow
Latrodectus hesperus
immature
female
male
male
egg sacs
egg sac
female
female
male
Common House Spider
Parasteatoda tepidariorum
False Black Widow
Steatoda grossa
Steatoda hespera
Steatoda fulva
patterned
female
female
female

Family Theridiidae (cont.) PAGE 93

Euryopis PAGE 103

Four species. Southern California to central Washington.

Dipoena PAGE 104

11 species. Southern California to northern Washington.

Euryopis and *Dipoena* have flat fangs and toothless chelicerae, and rather than hunting with webs, they specialize in stalking and ambushing ants. *Euryopis* are small spiders with teardrop-shaped abdomens, and they vary from pale with a single dark abdominal spot *(Euryopis californica)* to predominately dark *(E. formosa)*. *Dipoena* have globular abdomens, and the males have unusual elevated carapaces that are often etched with deep, radiating grooves. *Dipoena* can be uniformly dark like *Dipoena nigra* or distinctly patterned like *D. malkini*.

Crustulina PAGE 98

One species. Southern California to northern Washington.

Crustulina sticta is a small, forest-dwelling, purplish-brown spider similar in appearance to some *Steatoda*. It can be recognized by its distinctive, heavily granulated carapace and sternum.

Tidarren PAGE 98

Two species. Southern and Central California.

These unusual spiders exhibit extreme sexual dimorphism. The females are small to fairly large, round-bodied spiders with mottled abdomens. Males are tiny, each with a single, massive palp (one is removed just before reaching maturity). *Tidarren* build multilevel sheet webs surrounded by tangle webs.

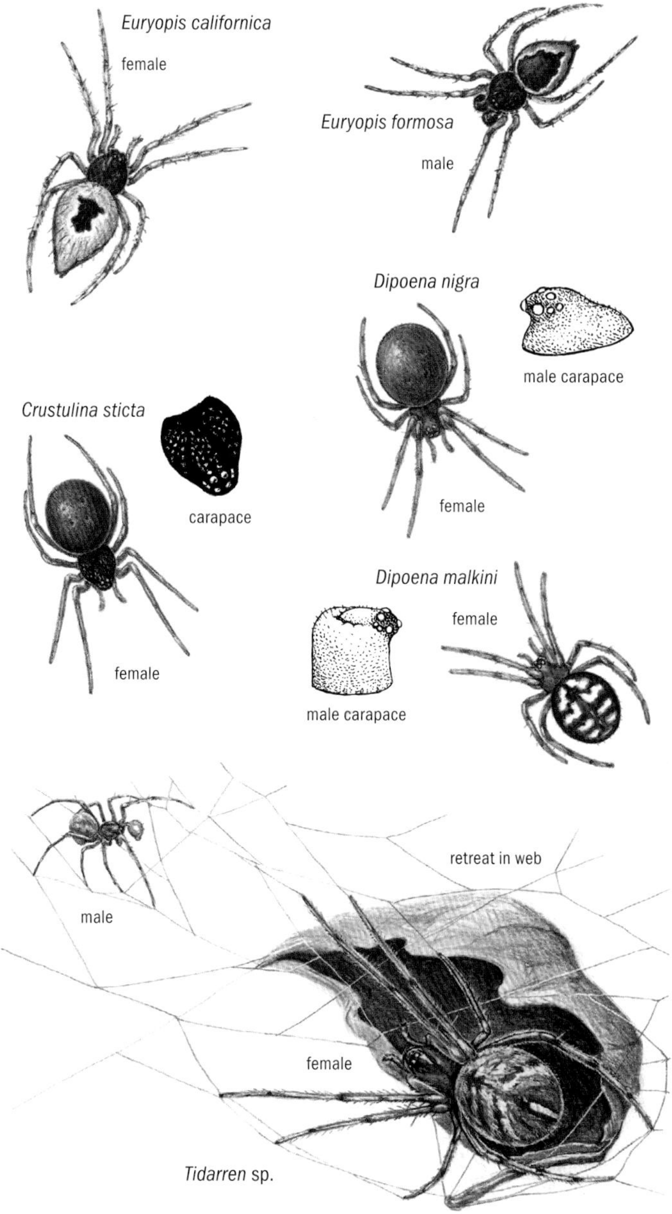

Euryopis californica

female

Euryopis formosa

male

Dipoena nigra

male carapace

female

Crustulina sticta

carapace

female

Dipoena malkini

female

male carapace

retreat in web

male

female

Tidarren sp.

Family Theridiidae (cont.)

PAGE 93

Rhomphaea

PAGE 100

One species. Central California to northern Washington.

The *Rhomphaea fictilium* female has an extremely long, slender abdomen. While the male's abdomen is shorter, it is still distinctive. *R. fictilium* is an araneophage that lives in cool, coastal forests.

Argyrodes

PAGE 99

One species. Southern California.

Neospintharus

PAGE 99

One species. Southern California.

Each of these genera is regionally represented by a single kleptoparasitic species. The Dewdrop Spider, *Argyrodes elevatus,* has a rounded triangular abdomen with an extensive silvery cast. *Neospintharus furcatus* has a more pointed triangular abdomen with only sporadic silver spotting. These spiders live in the webs of larger spiders and feed on the remains of their meals and on insects too small to garner the hosts' attention.

Cryptachaea

PAGE 101

Four species. Southern California to Washington.

Cryptachaea are small theridiids that build their webs in protected areas, including under peeling bark and rock piles. Their globular abdomens are higher than they are long, are highly variable in color, and often show a mottled pattern.

Enoplognatha

PAGE 104

Seven species. Southern California to northern Washington.

Enoplognatha can be pale-bodied foliage dwellers or dark-bodied and live near the ground. All have fairly ovoid abdomens and conspicuous coluli. The most common of the native species, the Marbled Cobweb Spider, *Enoplognatha marmorata,* is dark, distinctly patterned, and found from Southern California to northern Washington. Also widespread is the introduced *E. ovata*. Females can exhibit one of three different sex-linked abdominal patterns known as the *lineata, redimata,* and *ovata* forms. Males are limited to the *lineata* abdominal pattern.

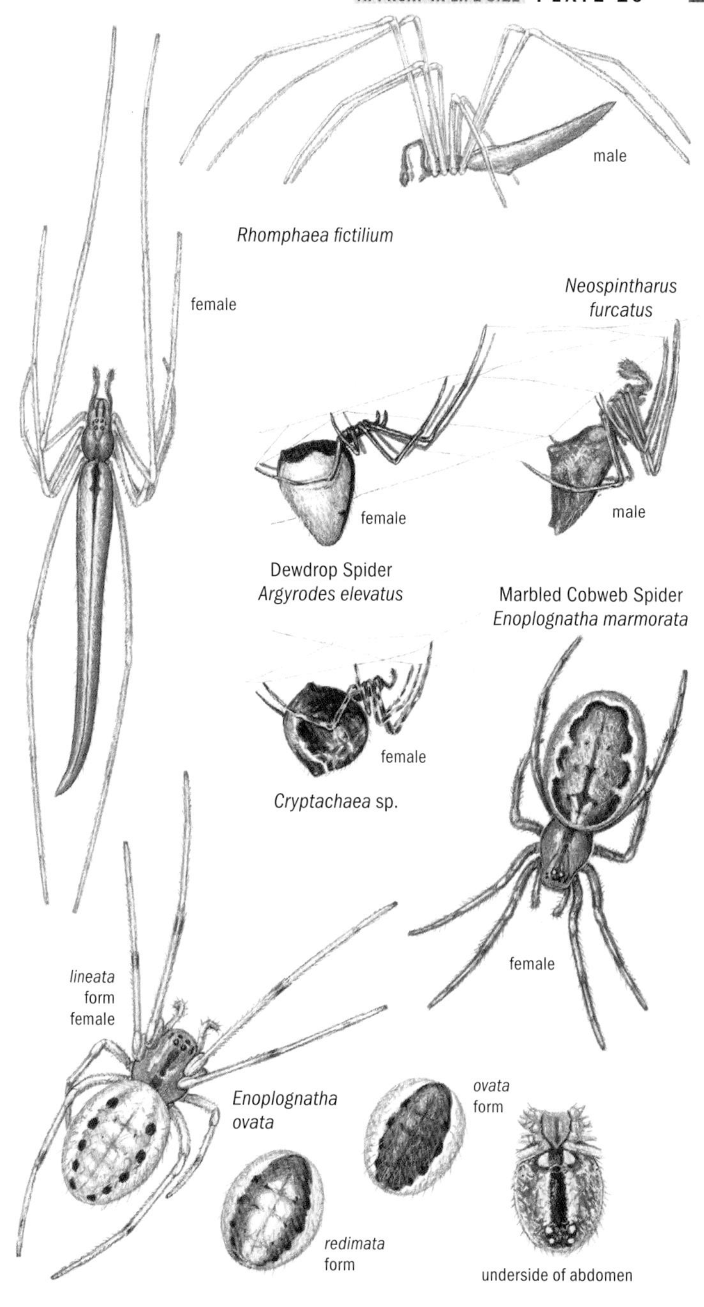
male
Rhomphaea fictilium
female
Neospintharus
furcatus
female
male
Dewdrop Spider
Argyrodes elevatus
Marbled Cobweb Spider
Enoplognatha marmorata
female
Cryptachaea sp.
female
lineata
form
female
Enoplognatha
ovata
ovata
form
redimata
form
underside of abdomen

Family Theridiidae (cont.) PAGE 93

Theridion PAGE 101

23 species. Southern California through northern Washington.
Theridion is the most diverse of the region's Theridiidae genera. *Theridion* are small, vary dramatically in color, and lack coluli. The females generally have a spherical abdomen decorated with a colorful, irregular central stripe.

Rugathodes PAGE 102

One species. Southern California to northern Washington.

Neottiura PAGE 102

One species. Washington.

Nesticodes PAGE 102

One species. Sporadic in California, Washington.

Wamba PAGE 102

One species. Central California to southern Oregon.
The spiders housed in these genera were all once considered members of the genus *Theridion*. *Rugathodes sexpunctatus* is a small spider whose abdomen is decorated with six variably shaped brown and white spots. *Neottiura bimaculata* is a holarctic species with numerous records from Washington. The female is dark reddish brown and black with a large yellow abdominal patch. The male (not illustrated) lacks an abdominal patch and is particularly long legged and ant-like. The introduced Red House Spider, *Nesticodes rufipes,* is found sporadically across the Pacific coast. Predominately a spider of the tropics, it is often imported along with shipments of fresh fruit and has occasionally built up small, isolated regional populations. *Wamba crispulus* is a small spider with an unusual white-and-black abdomen that, when viewed from above, is wider than it is long.

Anelosimus PAGE 105

One species. Southern California.
Anelosimus analyticus is a small spider with a highly reduced colulus and an ovoid abdomen marked with a broad, sinuous, gray band along its midline.

Robertus PAGE 105

One species. Central California through northern Washington.
Robertus vigerens is a small, dark spider with whitish eyes, an exceptionally large colulus, and a slightly ovoid, dorsoventrally compressed abdomen. It lives in leaf litter and under woodland debris.

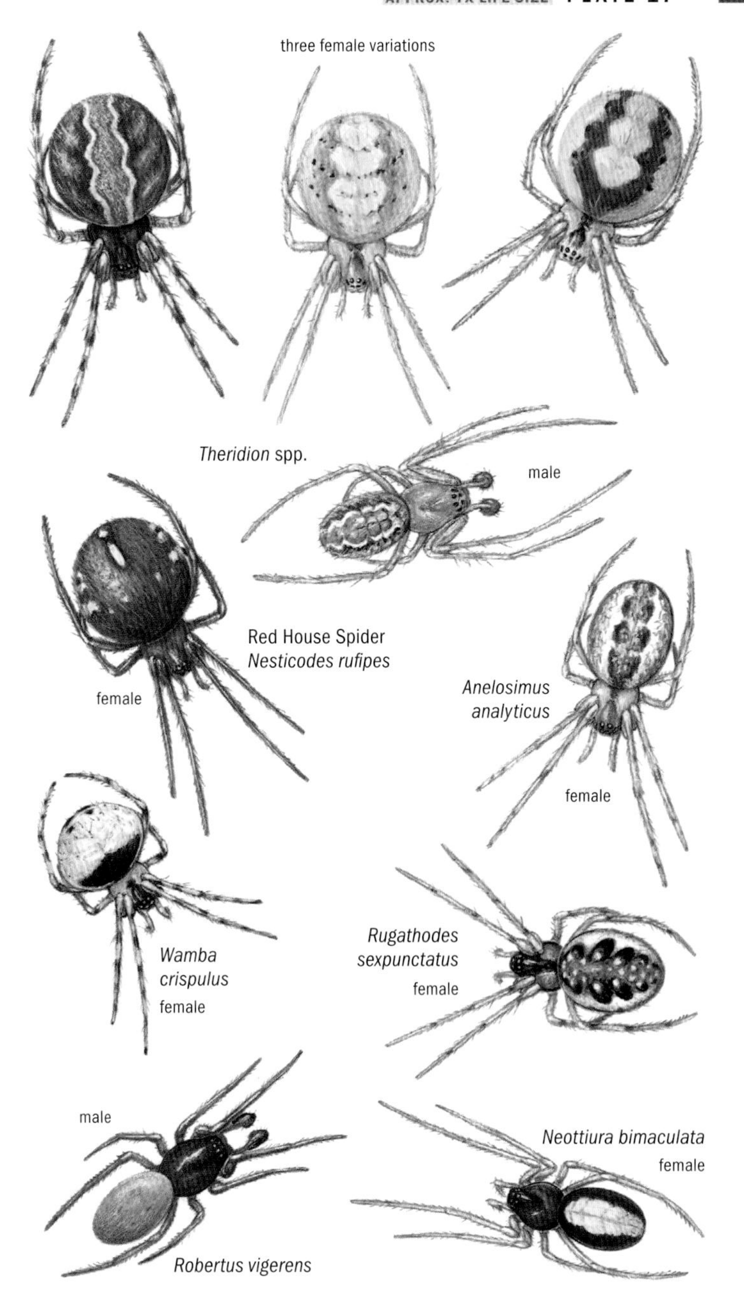
three female variations
Theridion spp.
male
Red House Spider
Nesticodes rufipes
female
Anelosimus analyticus
female
Wamba crispulus
female
Rugathodes sexpunctatus
female
male
Neottiura bimaculata
female
Robertus vigerens

Family Araneidae (in part)

Orb Web Weavers

PAGE 110

17 genera, 55 regional species. California, Oregon, Washington.
IDENTIFICATION: Vertically inclined orb webs (for most) and heavily spined legs and squarish endites. As a family, orb web weavers vary dramatically in size, shape, and color. With a few exceptions, female Araneidae have ornate epigyna and males have short palpal tibiae. While the anatomical features that unite the family may be subtle, many genera and species are fairly distinctive and, with some experience, are easily recognized.

Argiope

PAGE 111

Four species. Southern California to northern Washington.
Female garden orb weavers are large and distinctively patterned, while the males are smaller and less ornate. They generally build their webs in open, sunny areas and often include either a transverse bar or an X-shaped stabilimentum through the hub. The Banded Garden Spider, *Argiope trifasciata,* is found from Southern California to northern Washington. The Black and Yellow Garden Spider, *A. aurantia,* is more coastally inclined and lives from Southern California to northern Oregon. The Silver Garden Spider, *A. argentata,* is found in Southern and Central California, often in dry, rocky areas. The Mexican Garden Spider, *A. blanda* (not illustrated), is very similar to the Silver Garden Spider but is regionally known only from Southern California.

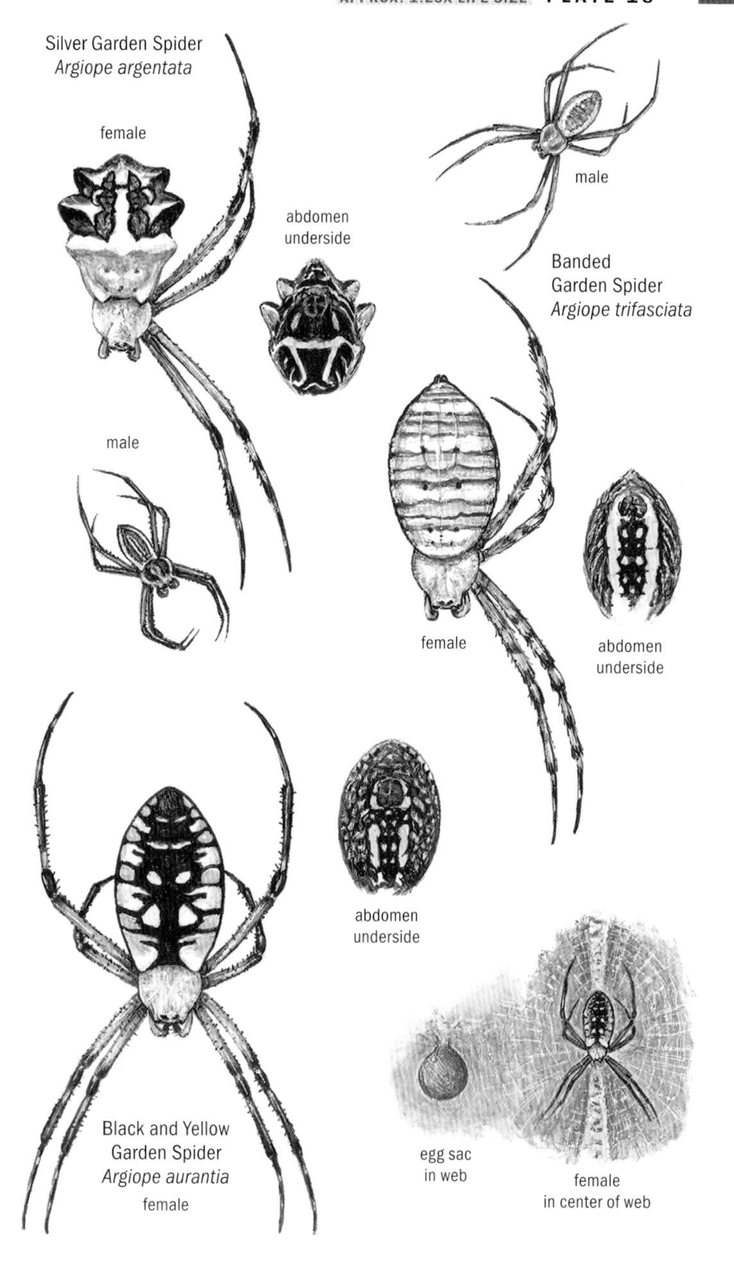
Silver Garden Spider
Argiope argentata
female
abdomen
underside
male
male
Banded
Garden Spider
Argiope trifasciata
female
abdomen
underside
abdomen
underside
Black and Yellow
Garden Spider
Argiope aurantia
female
egg sac
in web
female
in center of web

Family Araneidae (cont.)

PAGE 110

Araneus

PAGE 113

16 species. Southern California to northern Washington.
Araneus is the most widespread and diverse of the region's orb web weaver genera and can vary dramatically in color and size, even within species. Some *Araneus* species, known as round-shouldered orb weavers, have smoothly spherical abdomens, while others, referred to as angulate orb weavers, have distinctive "humps" across the fronts of their abdomens. Larger species are more common in bushes and grassy fields, while the smaller species are more common in the tree canopies. The examples shown include some of the more abundant, widespread, or easily recognized Pacific coast species.

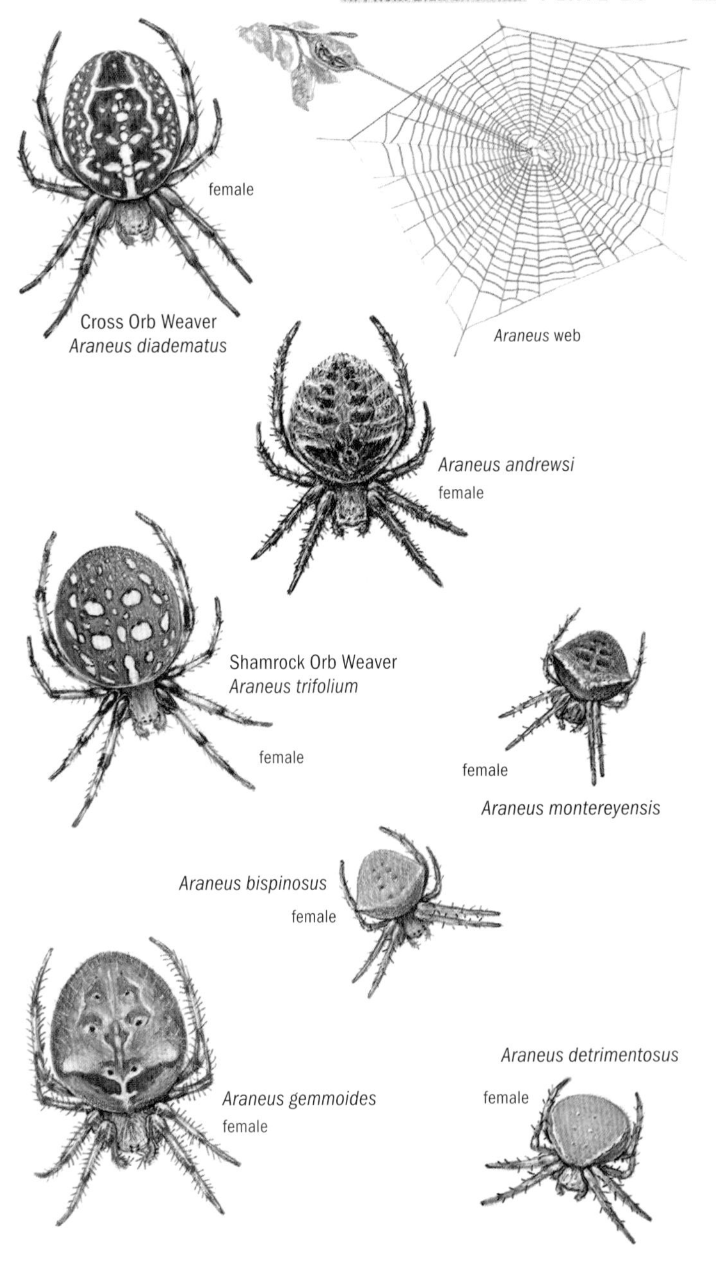
female
Cross Orb Weaver
Araneus diadematus
Araneus web
Araneus andrewsi
female
Shamrock Orb Weaver
Araneus trifolium
female
female
Araneus montereyensis
Araneus bispinosus
female
Araneus detrimentosus
female
Araneus gemmoides
female

Family Araneidae (cont.)

PAGE 110

Neoscona

PAGE 113

Three species. California to Washington.
Neoscona are also known as spotted orb weavers. Unlike *Araneus* they have a longitudinal thoracic groove. On the underside of the abdomen, there are black patches bordered by four light spots, although this should be used as a field mark only with caution because this design can occur on several other genera. The Western Spotted Orb Weaver, *Neoscona oaxacensis,* and the Arabesque Orb Weaver, *N. arabesca,* are found in all three of the contiguous Pacific coast states. *N. crucifera* is regionally limited to Southern California.

Larinioides

PAGE 114

Three species. Central California through northern Washington.
Larinioides have dorsoventrally flattened, oval abdomens decorated dorsally with a folium and ventrally by a black patch and two large, pale dashes. *Larinioides* are common near fresh water, especially on docks and bridges. *Larinioides patagiatus* is fairly common throughout the region. *L. cornutus* is widespread but rare, and the European Bridge Orb Weaver, *L. sclopetarius* (not illustrated), was recently introduced to northern Washington.

Larinia

PAGE 118

Three species. California, Washington (very rare).
Larinia are elongate, striped orb weavers whose abdomens are marked ventrally by black patches divided by a thin white line. Along the Pacific coast, only *Larinia directa* occurs regularly, with numerous records from coastal Southern California.

Aculepeira

PAGE 118

One species. Southern California to northern Washington.
The female *Aculepeira packardi* is fairly large with a folium on her dorsal abdominal surface and a series of three white stripes on her ventral surface. An uncommon species, *A. packardi* prefers cool mountain woodlands and fields.

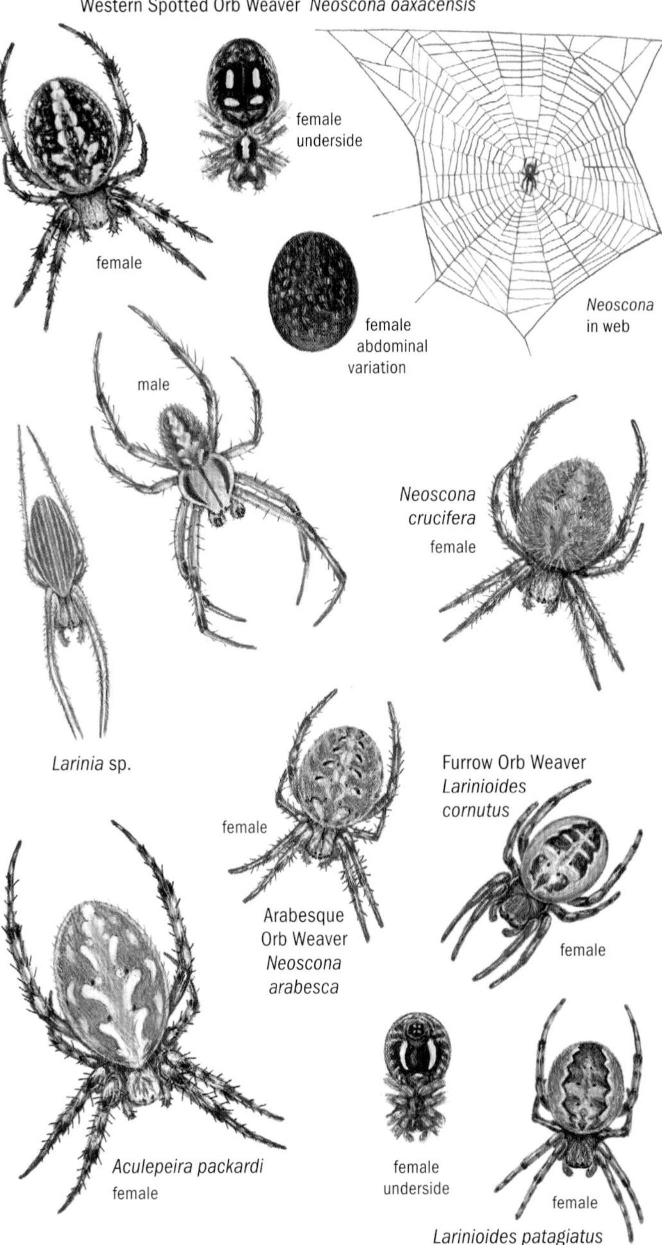
Western Spotted Orb Weaver *Neoscona oaxacensis*
female
underside
female
female
abdominal
variation
Neoscona
in web
male
Neoscona
crucifera
female
Larinia sp.
Furrow Orb Weaver
Larinioides
cornutus
female
Arabesque
Orb Weaver
Neoscona
arabesca
female
Aculepeira packardi
female
female
underside
female
Larinioides patagiatus

Family Araneidae (cont.)

PAGE 110

Gea

PAGE 112

One species. Southern California.
Gea heptagon is a small spider with a roughly heptagonal abdomen that is often decorated with a large, dark, triangular patch. When disturbed, *G. heptagon* falls out of its web and is quickly able to change the white portions of its body brown.

Cyclosa

PAGE 115 (FIG. 11)

Three species. Southern California to northern Washington.
Also known as trash line orb weavers, *Cyclosa* have small webs that are easily recognized by the row of refuse strung through the midline. A spider rests in the empty space at the hub, camouflaged amid the debris, its outline further broken up by the distinctive pattern of tubercles on its abdomen (fig. 11).

Eriophora

PAGE 120

One species. Southern and Central California.
Eriophora edax is a large, nocturnal spider. The female has a broad, triangular abdomen, often with a white stripe down the dorsal midline. Ventrally, she has a diagnostic black trapezoidal abdominal patch surrounded by a yellowish border. She also has an exceptionally long, swordlike scape, although it is often broken during mating.

Eustala

PAGE 116

Four species. California, Oregon, Washington (rare).
Eustala vary considerably in color and form, both within and between species. They have triangular abdomens with rounded anterior corners and many have tubercles along the abdomen's posterior midline. The females also have anteriorly pointed scapes, a feature unique among the region's orb web weavers. *Eustala rosae* is the region's most common species.

Metepeira

PAGE 117

Seven species. Southern California to northern Washington.
Metepeira build a distinctive orb web surrounded by a tangle with a cuplike retreat of leaves suspended in the middle. *Metepeira* are fairly large spiders with a folium pattern on the dorsum of the abdomen and a black patch bisected by a white line on the ventral surface. One unusual coastal species, *Metepeira spinipes,* regularly gathers together in large colonies, and while the colony's tangle webs are interconnected, each spider is still territorial over its individual orb.

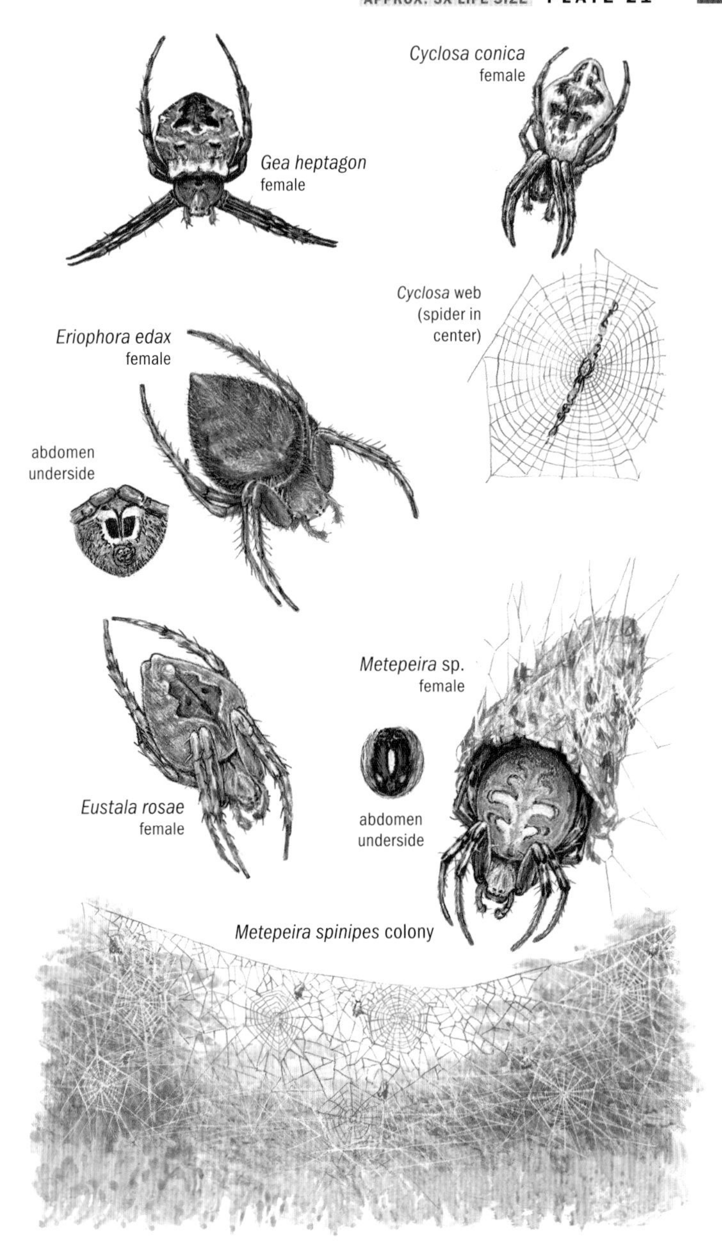
Gea heptagon
female
Cyclosa conica
female
Cyclosa web
(spider in
center)
Eriophora edax
female
abdomen
underside
Metepeira sp.
female
abdomen
underside
Eustala rosae
female
Metepeira spinipes colony

Family Araneidae (cont.) PAGE 110

Zygiella PAGE 119

Two species. California to Washington.

Parazygiella PAGE 119

Two species. California to Washington

These two genera are very similar and for many years were lumped together within *Zygiella*. They share features of their web design and aspects of their reproductive structures that are unusual among the Araneidae. *Zygiella* is regionally represented by two introduced species, including the widespread *Zygiella x-notata*. They are grayish yellow with dark foliate marks on their abdomens. Their distinctive orb webs always have a piece missing from the upper half and a thread leading to the spider's retreat. The region's two *Parazygiella* are native to the Pacific coast. *Parazygiella dispar* builds webs similar to those of *Zygiella* and is found along the coast from Central California to northern Washington. The second species, *P. carpenteri,* weaves a complete orb, often across the deeply creviced trunk of a large conifer. *Parazygiella* are similar in appearance to *Zygiella,* and separating them may require a detailed examination of their reproductive structures.

Araniella PAGE 121

One species. Southern California to northern Washington.

The Six-spotted Orb Weaver, *Araniella displicata,* is a small spider that, despite being highly variable in color, nearly always has six (sometimes eight) small black spots on its abdomen. A foliage-dweller, it often builds its web across a single large leaf.

Hypsosinga PAGE 120

Two species. California, Washington.

Hypsosinga are small, boldly patterned spiders. *Hypsosinga pygmaea* is a boldly striped, black-and-white species known sporadically from Washington. *H. funebris* varies in color from orange to black and regularly sports a folium and lateral stripes on its abdomen. It has been found in Southern and Central California.

Gasteracantha PAGE 122

One species. Southern California.

With its distinctive shape and abdominal spines, the female Spiny-backed Orb Weaver, *Gasteracantha cancriformis,* is unmistakable. The male (not illustrated) is small and beetle-like and lacks abdominal spines.

(continued)

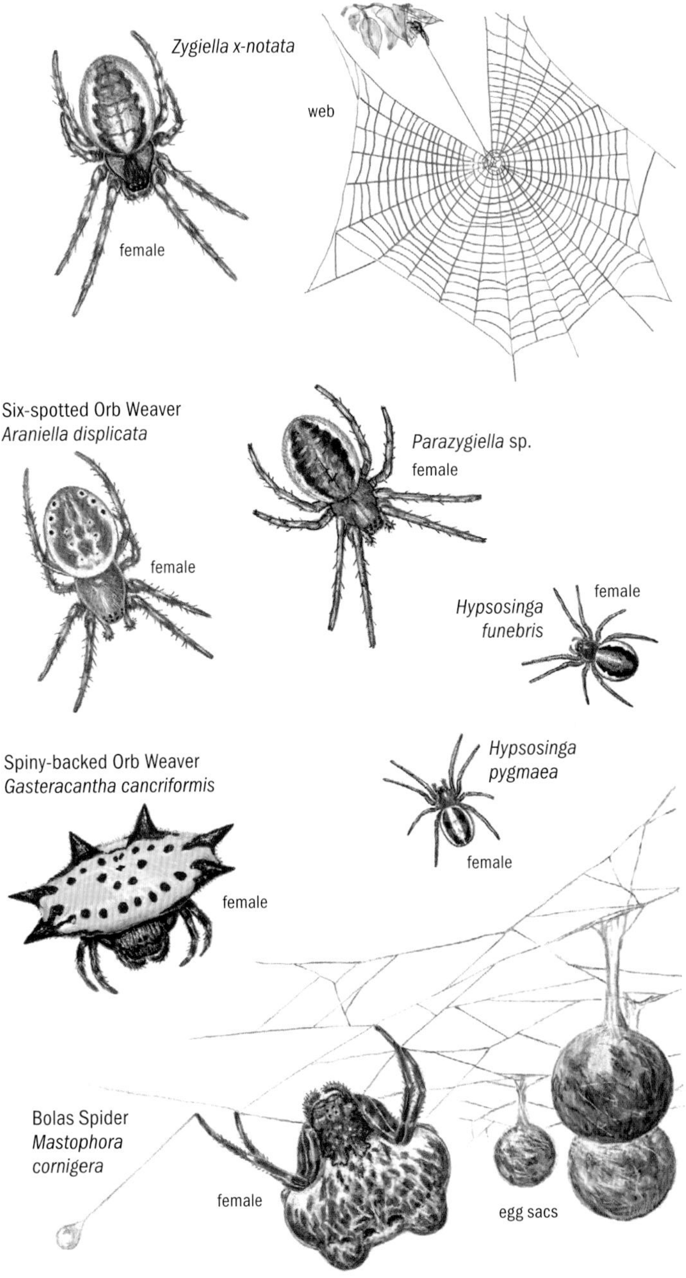

Zygiella x-notata

web

female

Six-spotted Orb Weaver
Araniella displicata

female

Parazygiella sp.
female

Hypsosinga funebris
female

Hypsosinga pygmaea
female

Spiny-backed Orb Weaver
Gasteracantha cancriformis

female

Bolas Spider
Mastophora cornigera

female

egg sacs

(continued)

Mastophora

PAGE 122

One species. Southern and Central California.

Mastophora cornigera is a distinctive spider. The female has a bulbous abdomen and coral-like tubercles on her carapace. She also constructs clusters of easily recognized, pendulous egg sacs. The male is tiny and is often found around the female's web. Unlike other spiders, a bolas spider catches its prey using a glob of glue swung about on a silk thread.

Plates continue on the following page.

Family Tetragnathidae

Long-Jawed Orb Weavers

PAGE 126

Seven genera, 20 regional species. California, Oregon, Washington.

IDENTIFICATION: A diverse family whose genera are often easier to recognize than the family as a whole. Tetragnathids have eight eyes and rectangular endites and, with one rare exception, build horizontal or inclined orb webs with open hubs. In many species, the males have exceptionally large, strongly toothed chelicerae. Except in the genus *Meta,* the males' palpal tibiae are elongate and conical. Female long-jawed orb weavers have very simple epigyna.

Tetragnatha

PAGE 127

10 species. Southern California to northern Washington.

Among the most distinctive spiders in North America are the long-jawed orb weavers, *Tetragnatha.* They have narrow bodies and long, thin legs that are often held flat against their bodies, either in the hub of the web or against a nearby blade of grass. Male *Tetragnatha* have especially long, robust chelicerae and fangs. They are most common around water.

Metellina

PAGE 127

Three species. Southern California to northern Washington.

Metellina are fairly small spiders with egg-shaped or broadly triangular abdomens that are decorated ventrally by a series of black and white bars. Unlike related spiders, *Metellina* have three large teeth on their cheliceral promargins. They are found predominantly in damp, shaded areas.

Leucauge

PAGE 128

One species. Southern California.

The Orchard Spider, *Leucauge venusta,* is a beautifully patterned but uncommon spider that builds its tightly woven orb web in a tree or bush.

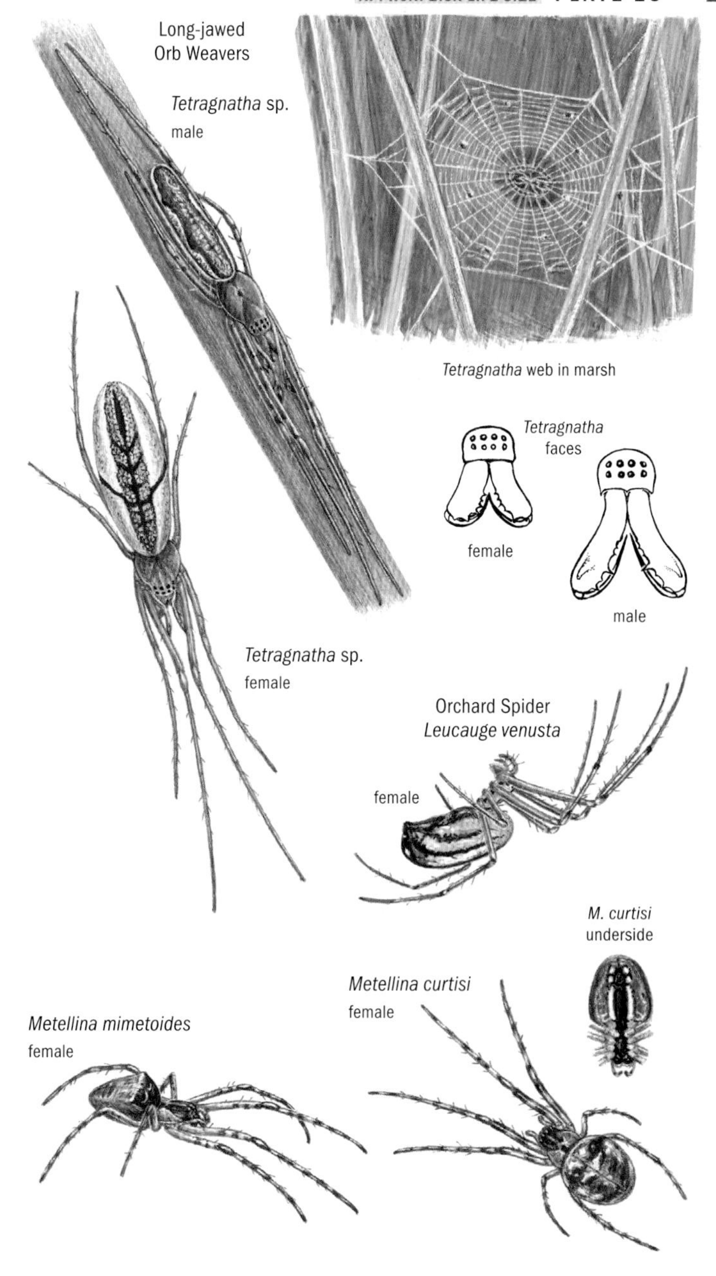
Long-jawed
Orb Weavers
Tetragnatha sp.
male
Tetragnatha web in marsh
Tetragnatha
faces
female
male
Tetragnatha sp.
female
Orchard Spider
Leucauge venusta
female
M. curtisi
underside
Metellina curtisi
female
Metellina mimetoides
female

Family Tetragnathidae (cont.) PAGE 126

Pachygnatha PAGE 128

Three species. Oregon, Washington.
Unlike other tetragnathids, *Pachygnatha* have abandoned capture webs and have become cursorial hunters. They are small spiders with large jaws, fairly high clypei, and a folium pattern on the abdomen. They are rare denizens of damp forests and bogs.

Glenognatha PAGE 129

One species. Southern California.
Glenognatha foxi is the smallest regional species of the long-jawed orb weaver. It is tiny and round bodied and, unlike other tetragnathids, is found in arid regions. While the female has a rather variable abdominal pattern, the male is normally unmarked and orangish red.

Metleucauge PAGE 129

One species. California.
Metleucauge eldorado is a large spider most frequently found along riverbanks in the shadows of boulders and fallen logs. It has a silvery folium pattern on its back and sinuous white bands bordering a black patch on its venter.

Meta PAGE 129

One species. Central California.
The Dolloff Cave Spider, *Meta dolloff,* is one of the rarest spiders in North America. It is a large brown spider with a globular, vertically aligned abdomen and is endemic to several small cave systems along the Central California coast.

Family Pimoidae PAGE 131

Two genera, 14 regional species. California, Oregon, Washington.
IDENTIFICATION: With one extremely rare exception, medium to fairly large spiders with long, setae-covered legs. They build messy sheet webs. They have patterned, oval abdomens and deep, pit-like thoracic furrows, and when grabbed, their legs break at the patella-tibia joint. *Nanoa enana* (not illustrated) is a tiny, rarely encountered, leaf litter–dwelling spider found in a small portion of Northern California and southern Oregon.

(continued)

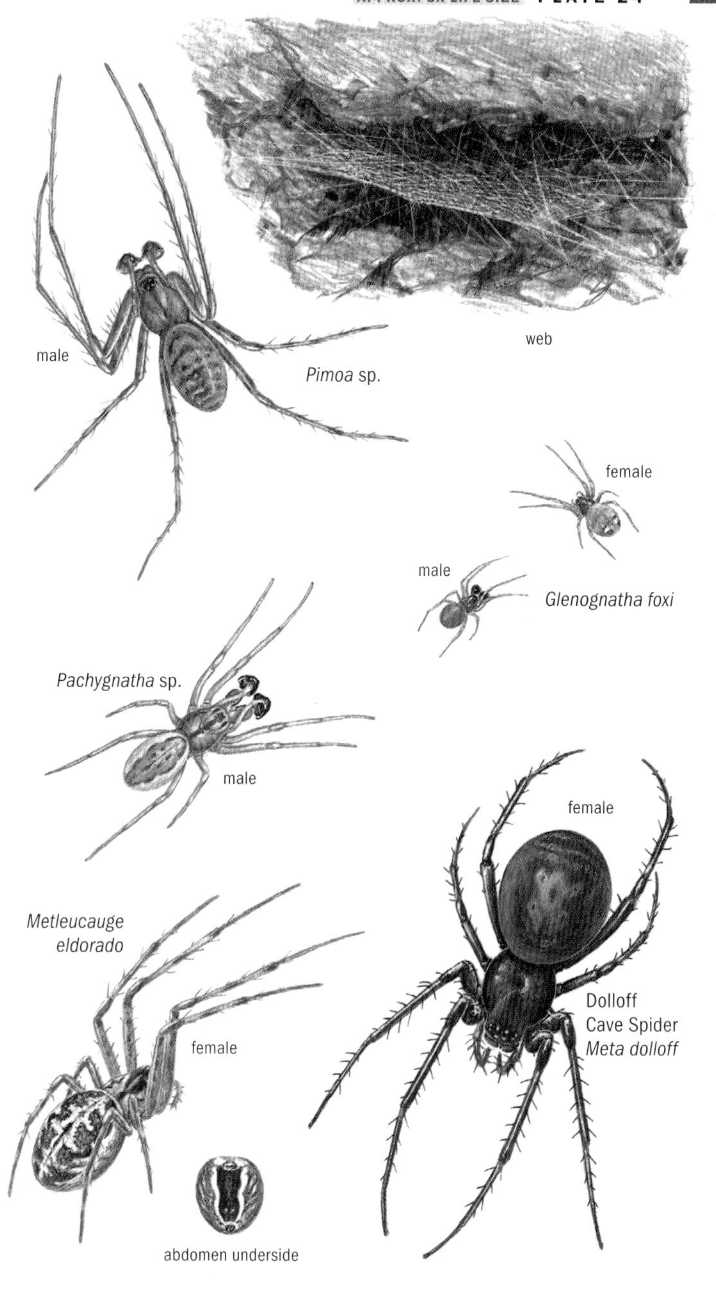
web
male
Pimoa sp.
female
male
Glenognatha foxi
Pachygnatha sp.
male
female
Metleucauge eldorado
Dolloff Cave Spider *Meta dolloff*
female
abdomen underside

(continued)

Pimoa PAGE 132

13 species. Central California to northern Washington.
Pimoa are long-legged spiders that live in damp, shaded forests. Their messy sheet webs are normally built in sheltered areas and are up to a meter across. Their bodies vary from reddish tan to dark mahogany brown.

Plates continue on the following page.

Family Linyphiidae (in part)

Sheet Web Weavers, Dwarf Spiders, Money Spiders PAGE 133

71 genera, approximately 300 species. California, Oregon, Washington.

IDENTIFICATION: A diverse group of small to minute spiders that build single to multilayer sheet webs, some of which are pulled into dome or bowl shapes. Their legs are lightly spined and easily break at the patella-tibia joint (autospasy). Linyphiidae is broadly divided into two subfamilies: the larger, more ornately patterned Linyphiinae, and the tiny, generally uniformly colored Erigoninae. Although Erigoninae contains the greatest diversity of regional genera, because of their small size and enigmatic nature, only few individuals are shown to illustrate the depth of the subfamily's variation.

Frontinella PAGE 135

One species. California to Washington.

The Bowl and Doily Spider, *Frontinella communis,* has a boxy, brown-and-white striped abdomen and builds a distinctive bowl-shaped web above a flat sheet. It is found in tall grasses and shrubs and, while common in California, becomes progressively rarer to the north.

Neriene PAGE 135

Three species. Southern California through northern Washington.

The region's *Neriene* are uniquely patterned and build characteristic webs. The Filmy Dome Spider, *Neriene radiata,* and the Sierra Dome Spider, *N. litigiosa,* both have brown-and-white striped, oblong bodies and build shallow to deep dome webs. Found widely across the Pacific coast, the Filmy Dome Spider is more common in Washington, while the Sierra Dome Spider is found in wooded areas throughout much of the region. Less ornate but still recognizable is *N. digna.* Abundant along the coast from Central California to northern Washington, it builds a small flat sheet web.

Microlinyphia PAGE 137

Four species. Southern California to northern Washington.

Only two of the region's four species are common. *Microlinyphia dana* is the area's only larger sheet web weaver with an entirely dark-gray venter, while *M. mandibulata* has pale wavy bands along either side of its abdomen. Both species build flat sheet webs in damp forested areas and gardens across the region.

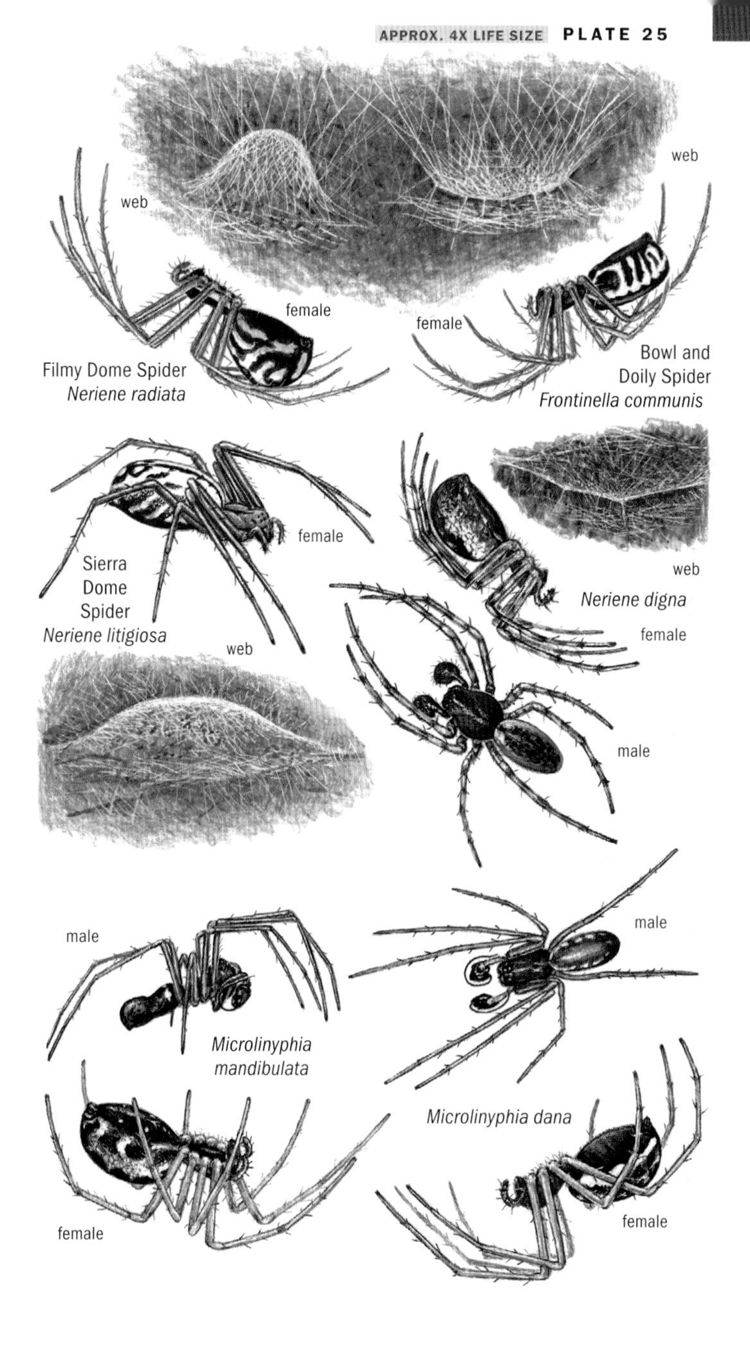
web
web
female
female
Filmy Dome Spider
Neriene radiata
Bowl and
Doily Spider
Frontinella communis
female
Sierra
Dome
Spider
Neriene litigiosa
web
web
Neriene digna
female
male
male
male
Microlinyphia
mandibulata
Microlinyphia dana
female
female

Family Linyphiidae (cont.) PAGE 133

Pityohyphantes PAGE 136

Eight species. California through Washington.
Also known as hammock spiders, *Pityohyphantes* are best recognized by the tuning fork–shaped mark on the carapace and their serrated dorsal abdominal stripe. They are fairly large for linyphiids and, unlike other sheet web weavers, are more prevalent in trees, especially conifers, than near the forest floor.

Bathyphantes, Lepthyphantes, Linyphantes (not illustrated) PAGE 138

Over 36 combined species. Southern California to northern Washington.
These three very similar, but poorly studied, genera represent the most diverse of the Pacific coast's Linyphiinae sheet web weavers. Each genus contains a dozen or more regional species, all of which are exceptionally small and normally require a detailed examination of their leg spination and reproductive structures to differentiate. Examples of abdominal patterning are shown to illustrate the variation within this group.

Wubana PAGE 137

Six species. Northern California through northern Washington.
Male *Wubana* are easily recognized by their distinctive carapace "horn," which is often adorned with a tuft of setae. They have grayish-yellow to orange carapaces and gray abdomens and prefer damp areas.

Erigoninae PAGE 138

Several hundred species. Southern California to northern Washington.
Due to their extremely small size and great diversity, only a few genera are shown, mainly to illustrate the unusual carapace sculpting found on some males. The size of these spiders is demonstrated by the male *Erigone dentosa* crawling across a penny.

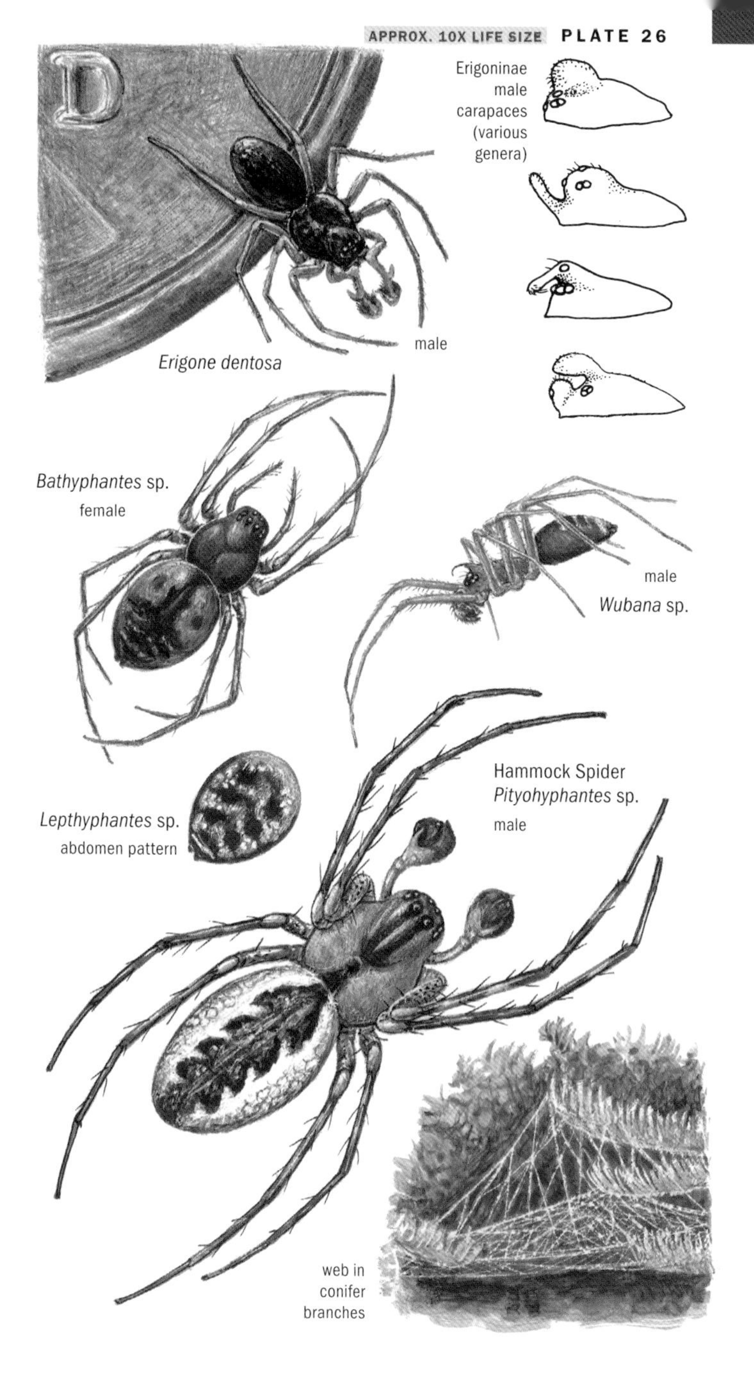
Erigoninae
male
carapaces
(various
genera)
male
Erigone dentosa
Bathyphantes sp.
female
male
Wubana sp.
Hammock Spider
Pityohyphantes sp.
male
Lepthyphantes sp.
abdomen pattern
web in
conifer
branches

Family Anyphaenidae

Ghost Spiders

PAGE 140

Three genera, eight regional species. California, Oregon, Washington.

IDENTIFICATION: Wandering hunters with unique, lamelliform setae in their claw tufts and tracheal spiracles that are located either near the middle of the abdomen or close to the epigastric furrow.

Anyphaena

PAGE 141

Six species. Southern California to northern Washington.

Hibana

PAGE 141

One species. Southern and Central California.

These two genera are fairly similar, differing most conspicuously in the placement of the tracheal spiracle. On *Hibana incursa* it is located near the epigastric furrow, while on *Anyphaena* it is located near the middle of the abdomen. *H. incursa* is generally a pale spider with brown chelicerae and diffuse, dark carapace bands. *Anyphaena* range in color from pale yellow to dark orange and frequently have pale to dark paramedian carapace bands. Both genera commonly hunt in foliage.

Lupettiana

PAGE 142

One species. Southern and Central California.

Lupettiana mordax is a distinctive spider with a dark purplish-brown carapace, grayish-green abdomen decorated with a chevron, and long, forward-pointing chelicerae.

Family Miturgidae

Prowling Spiders

PAGE 143

Two genera, four described regional species. California, Oregon, Washington.

IDENTIFICATION: Long-legged, wandering hunters whose anterior lateral spinnerets are conical and nearly joined at their bases. The tips of their posterior lateral spinnerets are fairly elongate.

Cheiracanthium

PAGE 143

Two species. California to Washington.

(continued)

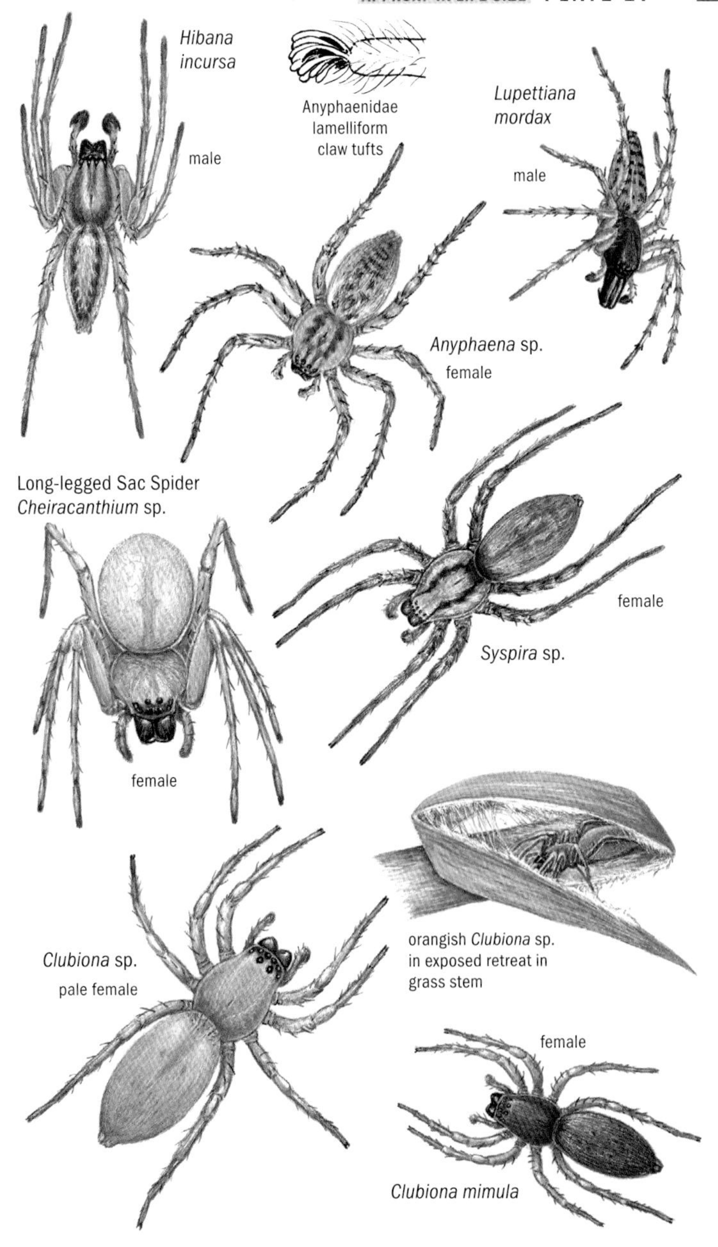
Hibana incursa
male
Anyphaenidae lamelliform claw tufts
Lupettiana mordax
male
Anyphaena sp.
female
Long-legged Sac Spider
Cheiracanthium sp.
female
female
Syspira sp.
orangish Clubiona sp. in exposed retreat in grass stem
Clubiona sp.
pale female
female
Clubiona mimula

(continued)

The introduced long-legged sac spiders, *Cheiracanthium,* are essentially unmarked, pale yellowish green, and especially common around homes and orchards in California, becoming progressively rarer farther north. Their bites have repeatedly been blamed for causing necrotic wounds, although cases of confirmed *Cheiracanthium* bites have found little evidence of necrosis.

Syspira PAGE 144

Two described species. Southern and Central California.
Syspira are distinctively patterned, long-legged, cursorial spiders found in deserts and grasslands.

Family Clubionidae

Sac Spiders PAGE 145

One genus, 16 regional species. California, Oregon, Washington.
IDENTIFICATION: Subtle distinguishing features including conical, nearly conjoined anterior lateral spinnerets; posterior lateral spinnerets with short, rounded tips; and the presence of precoxal triangles. Although they vary considerably in size and color, with experience their generally consistent body shape and habits are easy to recognize.

Clubiona PAGE 146

16 species. Southern California to northern Washington.
Clubiona are nocturnal wandering hunters that are found in a wide range of habitats where they build saclike retreats in loose foliage and other protected spots. Their abdomens can either be undecorated or have a heart mark.

Family Corinnidae

Ant-Mimic Sac Spiders

PAGE 146

Nine genera, 32 described regional species. California, Oregon, Washington.

IDENTIFICATION: An exceptionally diverse group of wandering hunters, many of which, through color and behavior, mimic ants or flightless wasps. Excepting the rare genus *Drassinella* (not illustrated), they have conical, conjoined, or nearly touching anterior lateral spinnerets.

Scotinella

PAGE 149

Six species. Southern California to northern Washington.

Phrurotimpus

PAGE 150

Six species. Central California to Northern Washington.

Piabuna

PAGE 150

At least one unidentified species. Southern California.

These three genera represent small spiders with precoxal triangles and four or more spine pairs beneath their anterior tibiae. *Phrurotimpus* have a spine on the dorsal surface of each femur and often show dark paramedian carapace bands. *Scotinella* lack dorsal femoral spines and are generally less conspicuously marked. *Piabuna* are minute spiders without dorsal femoral spines whose anterior median eyes are large, dark, and arranged dorsally on the carapace.

Castianeira

PAGE 148

Seven species. Southern California to northern Washington.

With the comparatively large sizes, bold patterns, and diurnal habits of its members, *Castianeira* is the most recognizable of the Pacific coast's Corinnidae genera. *Castianeira thalia* and *C. occidens* are illustrated as regional representatives of this diverse genus.

Meriola

PAGE 148

Three species. Southern California to northern Washington.

Trachelas

PAGE 148

One species. California.

Instead of spines, these closely related genera have spur-like cusps peppering their legs. *Trachelas pacificus* has a comparatively narrow carapace and a recurved posterior eye row. The two native and one introduced species of *Meriola* have rounded carapaces and straight posterior eye rows.

(continued)

(continued)

Creugas PAGE 149

One species. Southern California.

Septentrinna PAGE 149

One species. Southern California.

These two genera of medium-size ant-mimic sac spiders have precoxal triangles and between five and eight pairs of spines on the ventral faces of tibiae I. *Septentrinna steckleri* is a yellowish-brown spider that lives in colonies of desert ants. *Creugas bajulus* is a fairly large corinnid with a grayish abdomen and reddish-brownish carapace and legs. Little is known about its natural history.

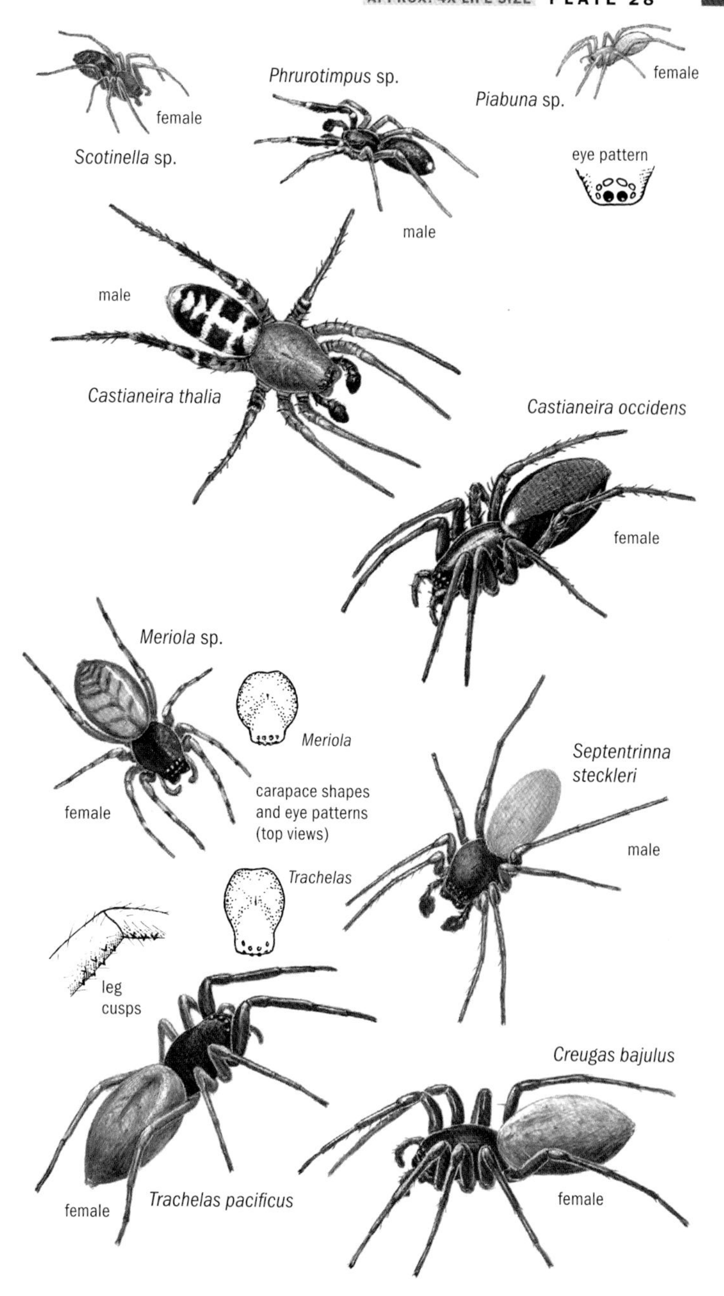

Scotinella sp.
female
Phrurotimpus sp.
male
Piabuna sp.
female
eye pattern
Castianeira thalia
male
Castianeira occidens
female
Meriola sp.
female
Meriola
carapace shapes
and eye patterns
(top views)
Trachelas
Septentrinna steckleri
male
leg
cusps
Trachelas pacificus
female
Creugas bajulus
female

Family Liocranidae

PAGE 152

Four genera, seven regional species. California, Oregon, Washington.

IDENTIFICATION: Two-clawed spiders without claw tufts and with closely conjoined, conical anterior lateral spinnerets. Most genera lack precoxal triangles. Because of their exceptional morphological diversity, it is often easier to identify a spider to its genus than place it as a member of the family as a whole.

Neoanagraphis

PAGE 154

Two species. Southern California.

The region's two *Neoanagraphis* are similar in appearance and have uniquely elongate, smooth-edged tarsal claws on their hind legs. They are found in arid areas.

Agroeca

PAGE 154

Three species. California, Washington.

Agroeca are small spiders that live in thick leaf litter. As a genus, *Agroeca* lacks any conspicuous diagnostic features. Instead it must be identified through a careful examination of the structures defining the family as a whole and then by elimination of other, similar genera. *Agroeca* lack precoxal triangles and have either two or three pairs of spines beneath their anterior tibiae.

Apostenus

PAGE 153

One species. Southern California.

Apostenus californicus is a montane spider and differs from other members of the region's Liocranidae by having five pairs of spines beneath each anterior tibia, more than any similar genus.

Hesperocranum

PAGE 153

One species. Central California to central Oregon.

Hesperocranum rothi is a small, montane spider easily recognized by the two distinctive rows of bristles on the ventral tibial surfaces of legs I through III.

Family Zoridae

PAGE 197

One genus, one regional species. California, Oregon, Washington.

IDENTIFICATION: Small spiders with recurved posterior eye rows and between six and eight pairs of spines beneath tibiae I and II.

(continued)

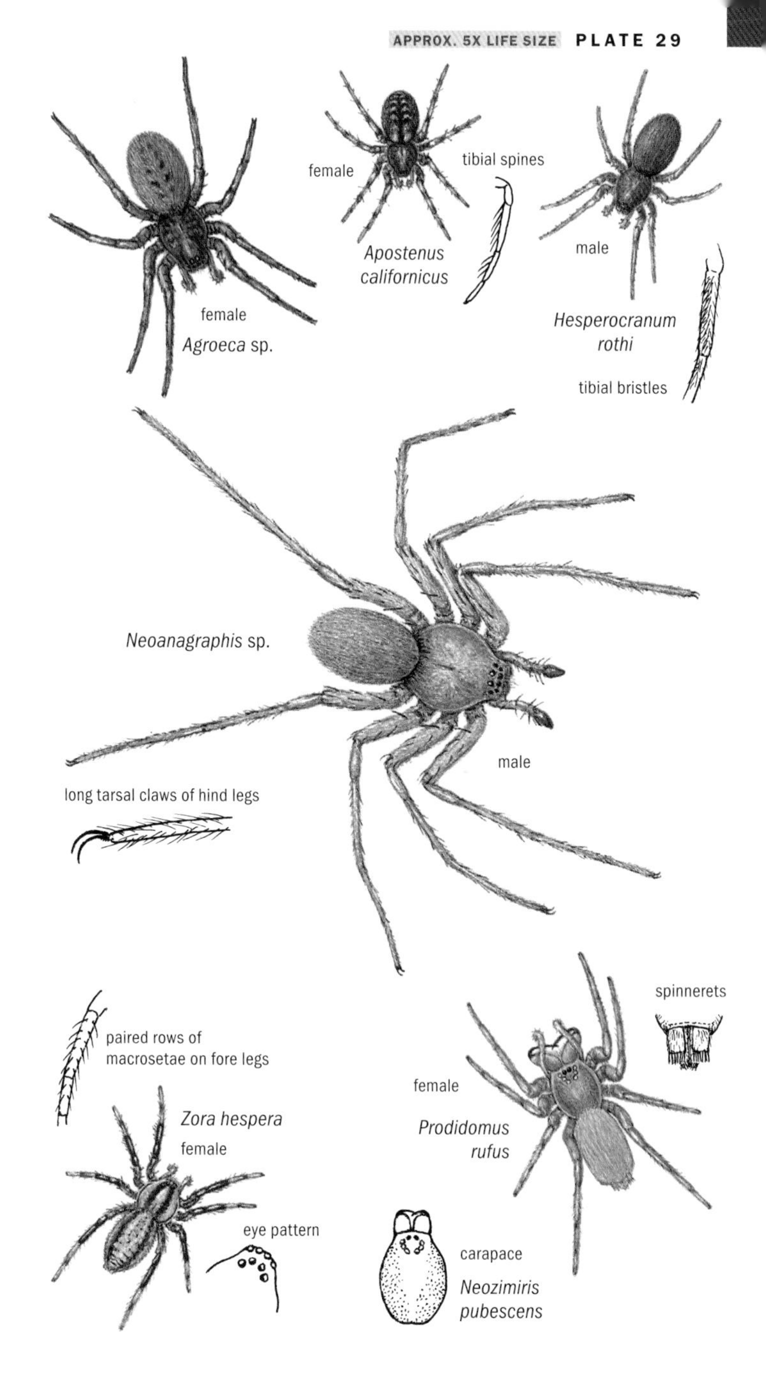
female
Agroeca sp.
female
Apostenus californicus
tibial spines
male
Hesperocranum rothi
tibial bristles
Neoanagraphis sp.
male
long tarsal claws of hind legs
paired rows of macrosetae on fore legs
Zora hespera
female
eye pattern
spinnerets
female
Prodidomus rufus
carapace
Neozimiris pubescens

(continued)

Zora
PAGE 197

Zora hespera is a uniquely patterned spider that has been collected in a wide variety of habitats from Southern California to northern Washington.

Family Prodidomidae
PAGE 155

Two genera, two regional species. California.

IDENTIFICATION: Small, very rare spiders with enlarged, flatted anterior lateral spinnerets and deeply procurved posterior eye rows.

Prodidomus
PAGE 155

One species. Southern California.

Neozimiris
PAGE 155

One species. Southern California.

Both of these genera have been found in arid portions of Southern California. *Prodidomus rufus* has enlarged chelicerae that diverge significantly away from one another. On *Neozimiris pubescens* the posterior eye row is so procurved that it forms a nearly circular arrangement. Its chelicerae have a more traditional, parallel alignment.

Plates continue on the following page.

Family Gnaphosidae (in part)

Stealthy Ground Spiders

PAGE 156

21 genera, 97 regional species. California, Oregon, Washington.
IDENTIFICATION: Widely separated, cylindrical, elongate anterior lateral spinnerets. Although some genera are boldly patterned, most are nondescript and require a close examination of their leg spination, cheliceral structure, eyes, and reproductive structures to confirm an identification. Several of the region's least common genera are not illustrated but are discussed in the family account.

Micaria

PAGE 157

21 species. Southern California to northern Washington.
Micaria are thin-legged ant mimics whose bodies are often banded and peppered with iridescent scales. Unlike most other stealthy ground spiders, they are diurnal and their anterior lateral spinnerets are comparatively small.

Cesonia (fig. 12)

PAGE 157

Five species. Southern and Central California.
Each of the different species of *Cesonia* is uniquely patterned with black and white stripes crossing both the abdomen and carapace. Additionally, the posterior median eyes are closer to the posterior laterals than they are to each other.

Sergiolus (fig. 13)

PAGE 158

Five species. Southern California through northern Washington.
Sergiolus often have species-specific patterns of white banding across their abdomens, although several species demonstrate an exceptional degree of intraspecific variation. Either their posterior eyes are equidistant or their posterior median eyes are closer to each other than they are to the posterior laterals.

Herpyllus

PAGE 160

Six species. California to Washington.
While most *Herpyllus* are unpatterned, the Western Parson Spider, *Herpyllus propinquus,* is very distinctive. Male *Herpyllus* have abdominal scuta (a feature also found on *Cesonia* and *Sergiolus*). Positively confirming their identification often requires a detailed look at their reproductive structures. Most *Herpyllus* are regionally limited to California and Oregon. Only one species, *H. hesperolus,* is known to regularly occur in Washington.

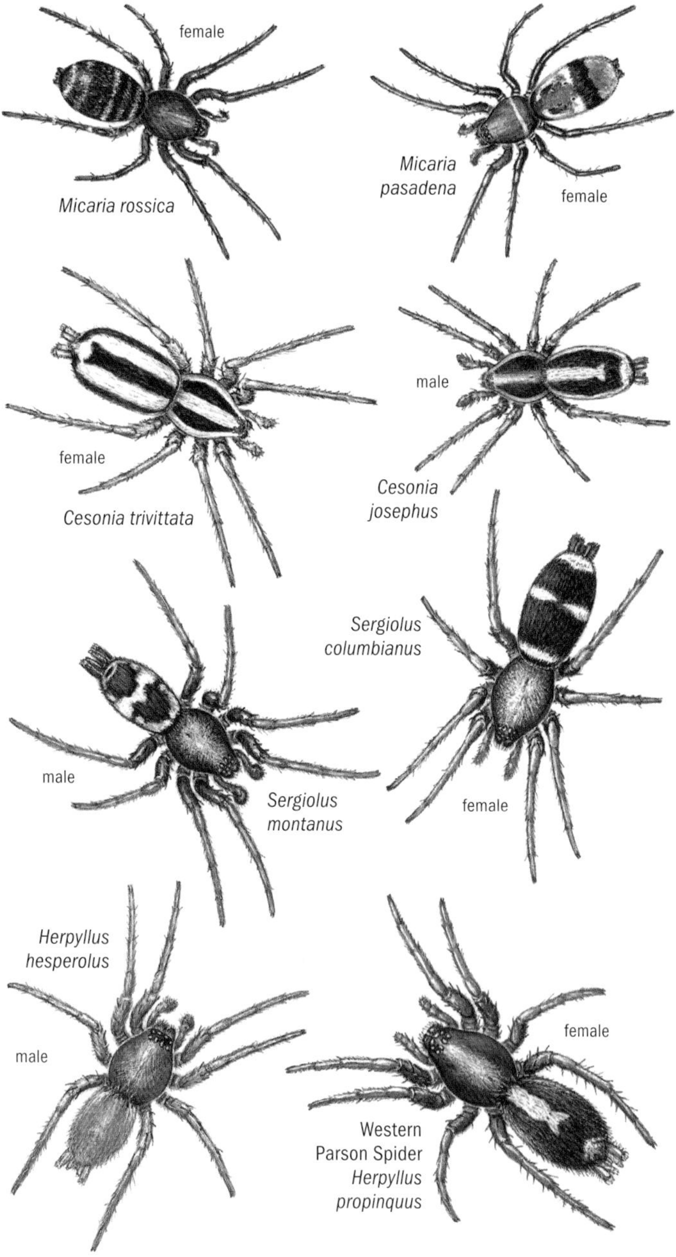

female

Micaria rossica

Micaria pasadena

female

female

Cesonia trivittata

male

Cesonia josephus

Sergiolus columbianus

male

Sergiolus montanus

female

Herpyllus hesperolus

male

female

Western Parson Spider *Herpyllus propinquus*

Family Gnaphosidae (cont.) PAGE 156

Scopoides PAGE 161

Four species. Southern California.

Scotophaeus PAGE 161

One species. Southern California to northern Washington.

Synaphosus PAGE 162

One species. Southern California.

Nodocion PAGE 162

Five species. Southern California to central Washington.

The males of these genera have small abdominal scuta and vary in color from pinkish gray to tawny brown. Among these unpatterned gnaphosids, the four regional *Scopoides* are easily recognized by their distinctive procurved posterior eye row and elongated anterior lateral spinnerets. The Mouse Spider, *Scotophaeus blackwalli,* is an introduced, reddish-brown, synanthropic species whose abdomen is covered with a silky gray pubescence. It is especially common in homes near the coast. *Synaphosus syntheticus* is a small, introduced, synanthropic spider with preening brushes on its third metatarsi and is regionally established only in Southern California. *Nodocion* can be distinguished from similar genera by the presence of shallow notches on the ventrodistal margins of the trochanters and posterior median eyes that are slightly larger than the posterior laterals.

Zelotes PAGE 163

35 species. Southern California through northern Washington.

Trachyzelotes PAGE 164

Three species. California, Oregon.

Drassyllus PAGE 164

16 species. Southern California through northern Washington.

Urozelotes PAGE 164

One species. California, southern Oregon.

All four of these genera have preening combs on the undersides of metatarsi III and IV. While their identification often depends on the details of their reproductive structures, their eye patterns can also be highly informative. By far the most common genus in this group is *Zelotes.* They are generally dark greenish-black spiders whose posterior median eyes are round and roughly the

(continued)

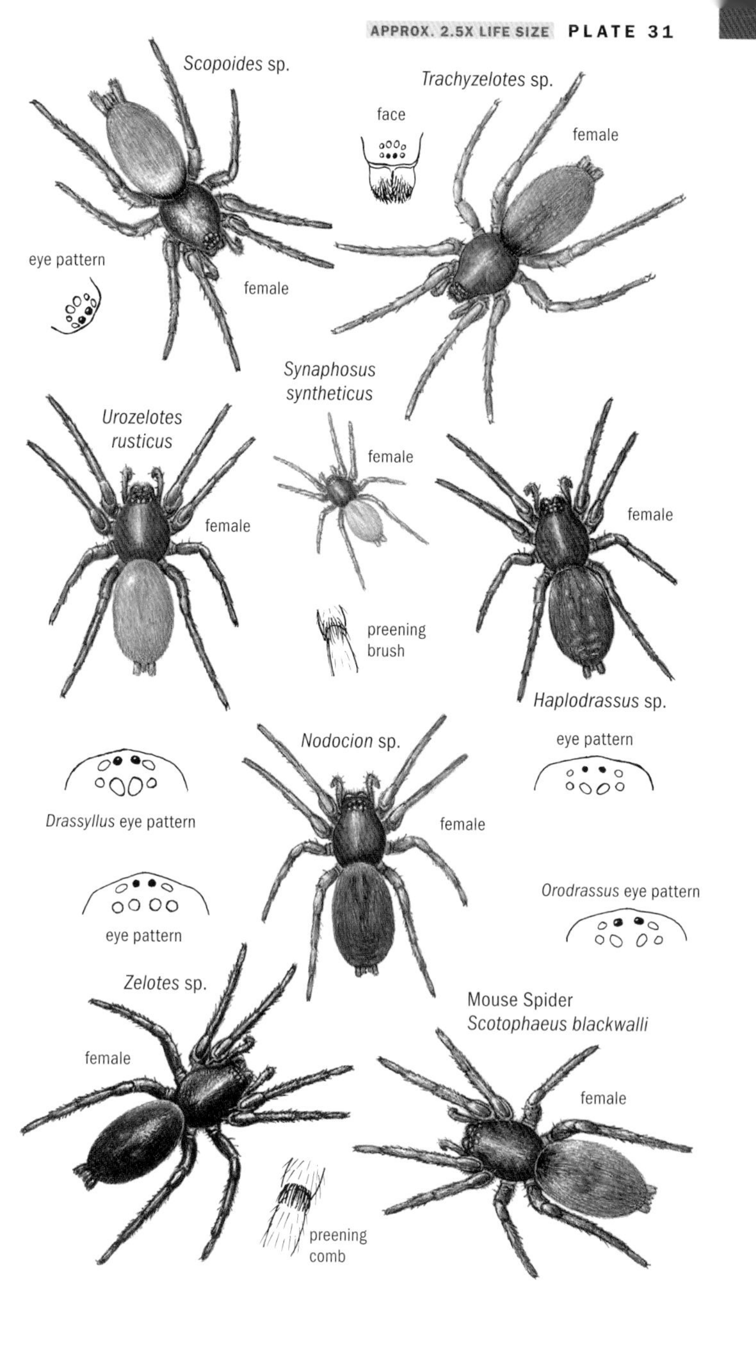

Scopoides sp.
eye pattern
female
Trachyzelotes sp.
face
female
Synaphosus syntheticus
female
preening brush
Urozelotes rusticus
female
female
Haplodrassus sp.
eye pattern
Nodocion sp.
female
Drassyllus eye pattern
eye pattern
Orodrassus eye pattern
Zelotes sp.
female
preening comb
Mouse Spider
Scotophaeus blackwalli
female

(continued)

same size as the posterior laterals. Introduced to North America, *Trachyzelotes* are easily recognized by the thick layer of setae coating the fronts of their chelicerae. Both *Drassyllus* and *Urozelotes* have posterior median eyes that are oval in shape, larger than the posterior laterals, and close together. *Drassyllus* range in color from greenish brown to nearly black, while *Urozelotes rusticus,* an introduced, synanthropic spider, has a rusty-orange carapace and tan abdomen.

Haplodrassus PAGE 162

Six species. Southern California to northern Washington.

Orodrassus PAGE 163

Three species. Southern California to northern Washington.

These two genera lack both preening combs and abdominal scuta and are generally orangish brown with gray abdomens. *Haplodrassus* have enlarged, oval posterior median eyes that are closer to one another than they are to the posterior laterals (much like *Drassyllus*). Compared with those of *Haplodrassus,* the posterior median eyes on *Orodrassus* are smaller and farther apart.

Plates continue on the following page.

Family Gnaphosidae (cont.)

PAGE 156

Gnaphosa

PAGE 165

11 species. Southern California to northern Washington.
Gnaphosa are found in a wide variety of habitats, and while diverse in color and size, they are united by the presence of a serrated cheliceral keel.

Callilepis

PAGE 165

Three species. Southern California to northern Washington.

Eilica

PAGE 166

One species. Southern California.
These two closely related genera possess translucent lobes on their cheliceral retromargins. *Callilepis* are found in dry sandy or rocky areas. They all have a single cheliceral lobe, and some species, especially *Callilepis gosoga,* are richly coated in silvery setae. *Eilica bicolor* is a desert-dwelling spider with two to three lobes on its cheliceral retromargins. *Eilica* are closely associated with ant colonies, occasionally even found living within them.

Drassodes

PAGE 164

Three species. Southern California to northern Washington.
Drassodes are generally buffy to gray spiders that can be easily recognized by the presence of a deep notch along the distoventral edge of each trochanter.

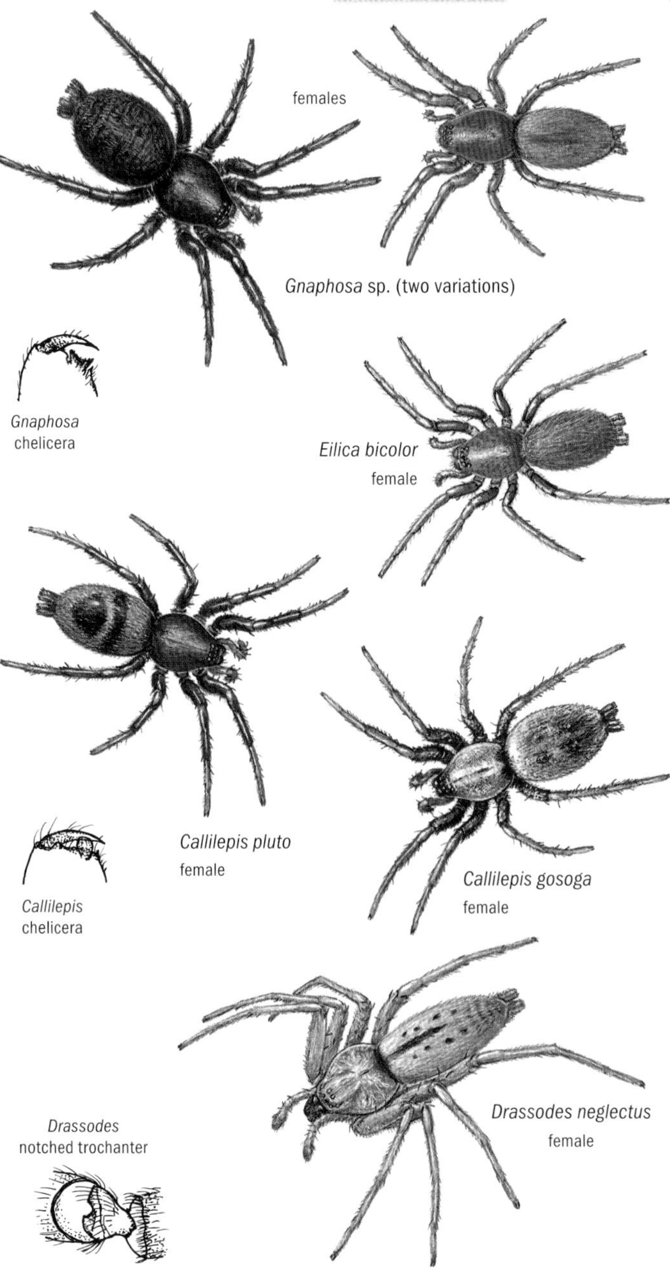
females
Gnaphosa sp. (two variations)
Gnaphosa
chelicera
Eilica bicolor
female
Callilepis pluto
female
Callilepis
chelicera
Callilepis gosoga
female
Drassodes neglectus
female
Drassodes
notched trochanter

Family Salticidae

Jumping Spiders

PAGE 167

33 genera, 119 described regional species. California, Oregon, Washington.

IDENTIFICATION: Exceptionally large anterior median eyes and stocky, often colorful bodies and powerful legs. Salticids are among the most distinctive spiders in North America. They are excellent jumpers that regularly move in a hesitant, jerky manner. In many species, the males are more ornately decorated with ornamentation on their legs and chelicera. The females are frequently more drably attired.

Phidippus

PAGE 167

19 species. Southern California to northern Washington.

Phidippus contains the largest and some of the most conspicuous jumping spiders in North America. Found in a wide range of habitats, they vary dramatically in color and pattern. Their carapaces are curved along their sides, and most species have iridescent scales on their chelicerae. Additionally, female *Phidippus* sport tufts of setae behind their posterior median eyes. The *Phidippus* illustrated represent some of the region's most common and distinctive species.

Thiodina

PAGE 176

One species. California.

Thiodina hespera is found in neighborhoods throughout much of the state. While the defining characteristic of the genus emphasizes the structure and alignment of several setae on their tibiae I, the color pattern on both sexes is unique and easily recognized.

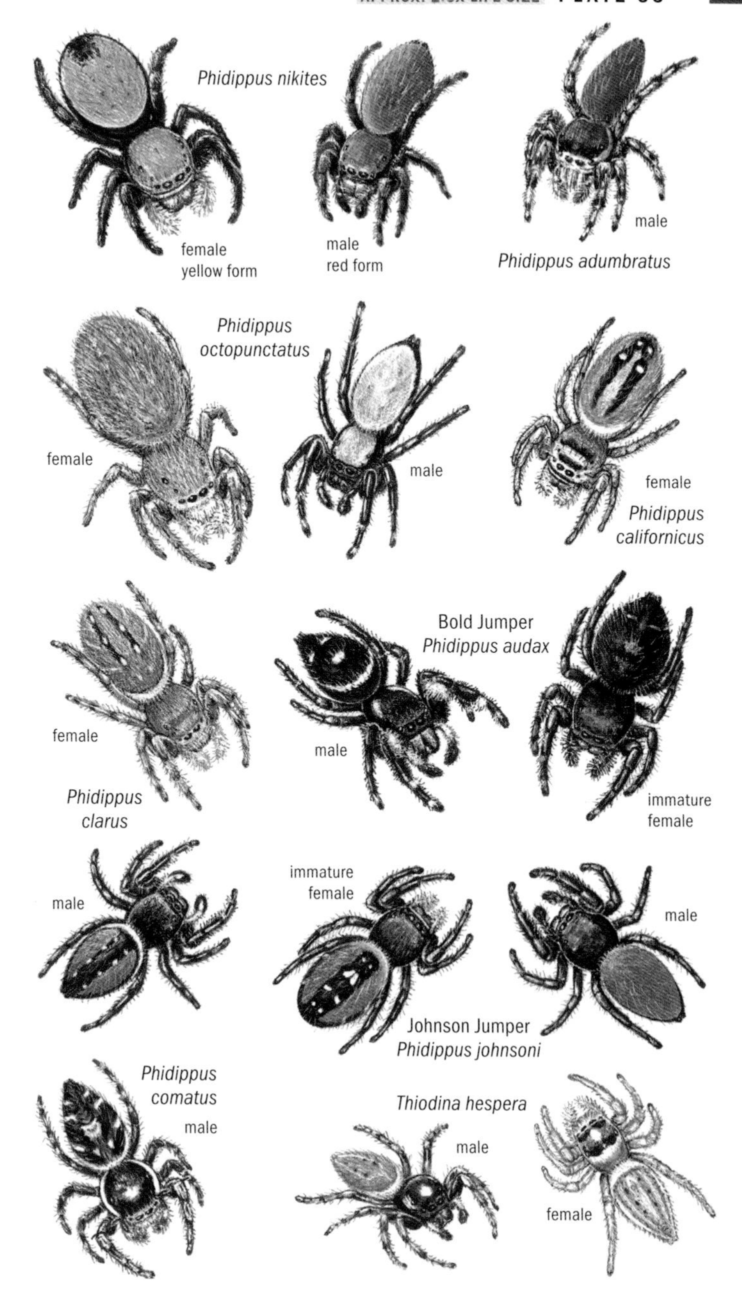
Phidippus nikites
female
yellow form
male
red form
male
Phidippus adumbratus
Phidippus
octopunctatus
female
male
female
Phidippus
californicus
Bold Jumper
Phidippus audax
female
male
Phidippus
clarus
immature
female
male
immature
female
male
Johnson Jumper
Phidippus johnsoni
Phidippus
comatus
male
Thiodina hespera
male
female

Family Salticidae (cont.) PAGE 167

Eris PAGE 168

One species. California to Washington.
Also known as the Bronze Jumper, the male *Eris militaris* has distinctive "brassy" setae on his body, while the female is gray with black and white abdominal patches. Like *Phidippus*, its posterior median eyes are closer to the anterior laterals than they are to the posterior laterals.

Metaphidippus PAGE 169

Six species. California to northern Washington.

Pelegrina PAGE 169

Six species. California to Washington.
The females of these two widespread genera are nearly identical, and differentiating the males can be challenging. Male *Metaphidippus* generally have distinctive patches of setae on their chelicerae and lack a white band connecting their anterior median eyes. *Metaphidippus* are more common in California, with only one species, *M. mannii*, ranging into northern Washington. *Pelegrina* are more diverse north of California. Males lack bold cheliceral patches, and with the exception of *P. aeneola*, they have a white stripe connecting their anterior median eyes.

Phanias PAGE 169

Six species. California to Washington.
Male *Phanias* are fairly elongate and yellowish brown with white lateral stripes on their abdomens and carapaces. Females are similarly elongate but are pale gray with brownish mottling. Only one species, *Phanias albeolus*, is regularly found as far north as Washington.

Evarcha PAGE 175

One species. Southern California through northern Washington.
The male *Evarcha proszynskii* has a broad white band that almost completely encircles his carapace and has several pale spots on the posterior portion of his white-bordered abdomen. The female (not shown) is dark without distinctive markings.

Marchena PAGE 173

One species. Central California through northern Washington.
The *Marchena minuta* male is uniquely patterned with a short, white central stripe that runs across both his carapace and his

(continued)

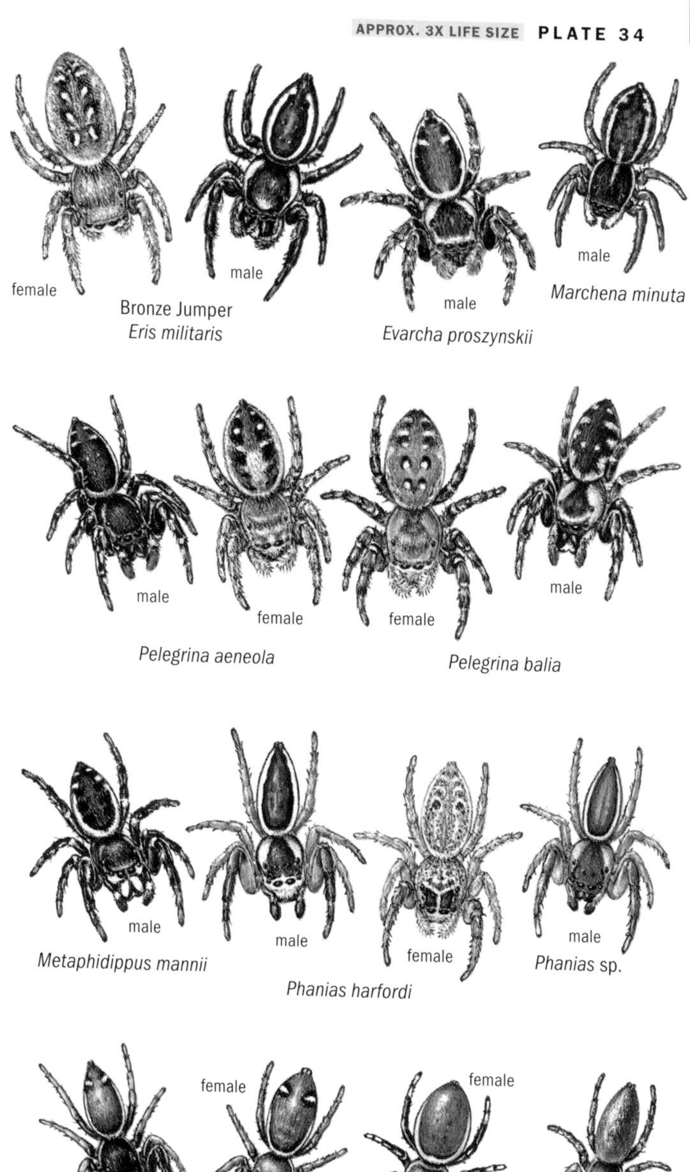

Bronze Jumper
Eris militaris

Evarcha proszynskii

Marchena minuta

Pelegrina aeneola

Pelegrina balia

Metaphidippus mannii

Phanias harfordi

Phanias sp.

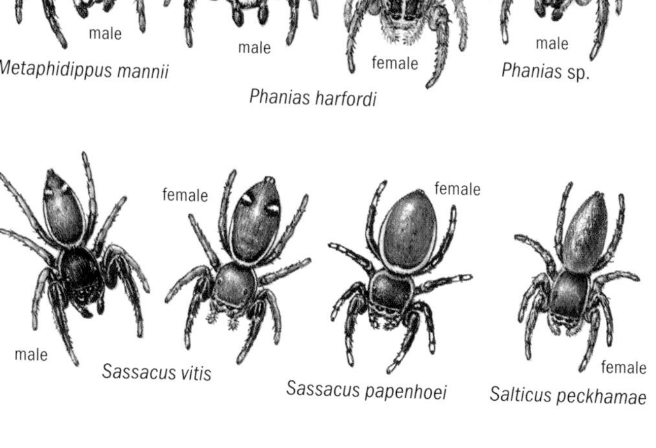

Sassacus vitis

Sassacus papenhoei

Salticus peckhamae

(continued)

abdomen. The female (not shown) is grayish brown with black and white mottling. This spider is a woodland specialist, especially preferring conifer forests. With magnification, distinctive rows of tubercles on the spider's femora and carapace are visible.

PAGE 171

Sassacus

Three species. Southern California through Washington.

Sassacus are small, iridescent, beetle-like jumping spiders found predominately in foliage. While *Sassacus vitis* and *S. papenhoei* are both widespread along the Pacific coast, *S. paiutus* (not shown) is regionally limited to southeastern California.

PAGE 171

Salticus (in part)

Three species. Southern California to northern Washington.

Due to its superficial similarity to members of the genus *Sassacus*, the *Salticus peckhamae* is illustrated here. It is a small spider covered in iridescent scales and is found from Northern California to northern Washington. More information on other regional *Salticus* can be found on plate 36.

Plates continue on the following page.

Family Salticidae (cont.)

PAGE 167

Habronattus

PAGE 170

30 described and several undescribed species. Southern California to northern Washington.

Habronattus is the most species-rich genus of jumping spider along the Pacific coast. *Habronattus* are primarily terrestrial and live in a wide variety of habitats. They are fairly small spiders whose third leg is longer than their fourth, an uncommon condition among jumping spiders. While female *Habronattus* are fairly dull in their overall coloration, males are often brightly decorated with conspicuous ornamentation on their legs, chelicerae, and carapaces. The species chosen for illustration were selected to demonstrate the diversity of color and form found amongst the region's *Habronattus* fauna.

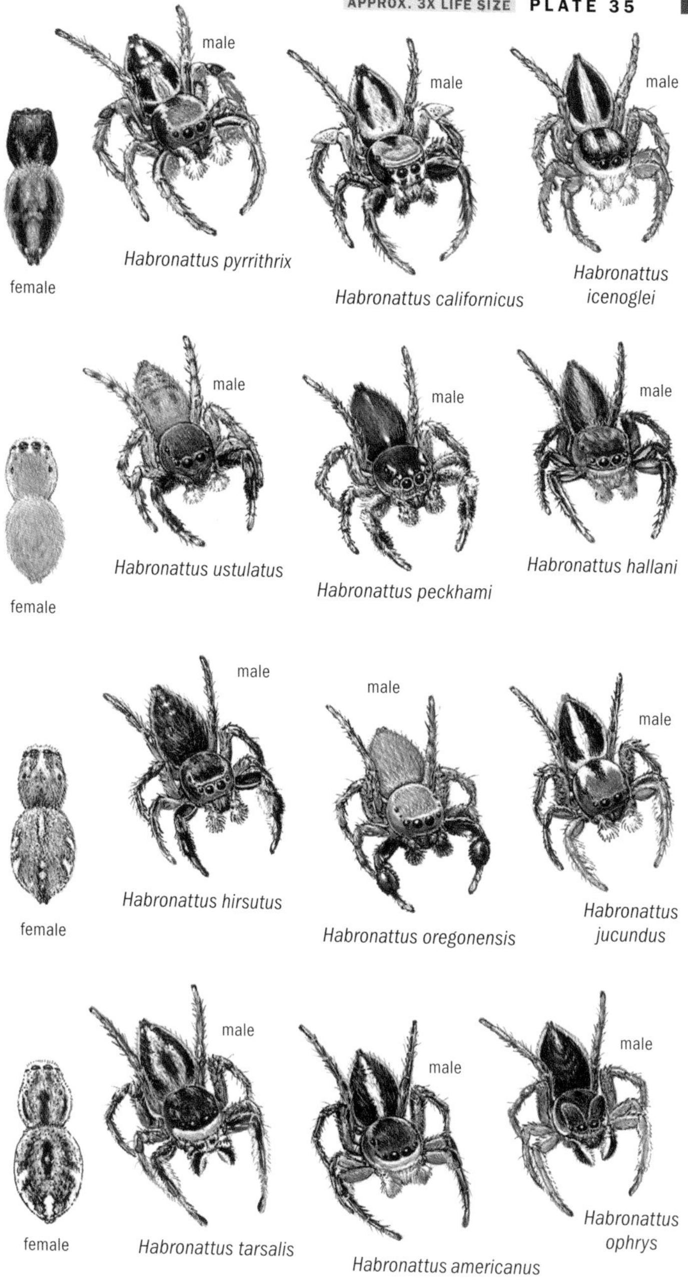

male

female

Habronattus pyrrithrix

male

Habronattus californicus

male

Habronattus icenoglei

male

female

Habronattus ustulatus

male

Habronattus peckhami

male

Habronattus hallani

male

female

Habronattus hirsutus

male

Habronattus oregonensis

male

Habronattus jucundus

male

female

Habronattus tarsalis

male

Habronattus americanus

male

Habronattus ophrys

Family Salticidae (cont.)

PAGE 167

Salticus

PAGE 171

Three species. Southern California to northern Washington.
Salticus are small jumping spiders with conspicuous patterns or iridescent scales. The most widespread regional species is the Zebra Jumper, *Salticus scenicus*. It has distinctively angled abdominal bars. California's *S. palpalis* is similar but has solid, transverse abdominal bands. *S. peckhamae* (pl. 34) is unpatterned but covered in iridescent scales.

Menemerus

PAGE 174

Two species. California.
Both of the region's *Menemerus* species, *Menemerus bivittatus* (also known as the Gray Wall Jumper) and *M. semilimbatus*, are introduced and regionally limited to California. They are fairly large and distinctively patterned, and the females have conspicuous white palps.

Platycryptus

PAGE 173

Two species. Southern California to northern Washington.

Metacyrba

PAGE 173

One species. Southern and Central California.
These two closely related genera are most frequently found on tree bark, old wood fences, and stone walls. *Platycryptus californicus* is found from Southern California to northern Washington, while *P. arizonensis* is regionally limited to Southern California. Both are fairly elongate, grayish spiders. *Metacyrba taeniola* is a small, dark spider with enlarged fore legs that frequently has two thin, broken, white lines running down its abdomen.

Marpissa

PAGE 172

One species. Southern and Central California.

Paramarpissa

PAGE 172

One species. Central and Northern California.
Each of these genera is regionally represented by a single long-bodied species with exceptionally powerful, elongate fore legs. *Marpissa robusta* is sexually dimorphic and uniquely patterned. The male *Paramarpissa griswoldi* has a striking white band running the length of his abdomen, while the female is grayer and more cryptically colored.

(continued)

male
Zebra Jumper
Salticus scenicus
female
Menemerus semilimbatus
Platycryptus
californicus
female
female
male
male
Paramarpissa griswoldi
male
Salticus palpalis
male
Menemerus
bivittatus
female
female
Marpissa
robusta
male
female
Metacyrba taeniola
male
Pantropical Jumper
Plexippus paykulli
female
female
Terralonus
californicus

(continued)

Plexippus

PAGE 174

One species. Southern California.
The introduced Pantropical Jumper, *Plexippus paykulli,* is a large, distinctively patterned spider that has become established in Southern California. The male's carapace pattern gives it three white "facial" stripes, and both sexes have a pair of white spots on either side of the abdominal midline.

Terralonus

PAGE 169

Three species. Southern California to northern Washington.
Unlike many jumping spiders, both sexes of *Terralonus* are mottled and grayish brown. They often live under rocks in barren areas. *Terralonus californicus* is one of the few species of jumping spider that lives almost entirely on beaches.

Family Salticidae (cont.)

PAGE 167

Neon

PAGE 177

Four species. Southern California to northern Washington.

Neon are tiny, dusky-yellow to brown jumping spiders found in thick beds of leaf litter. *Neon ellamae* can be recognized by the white tips to its legs.

Talavera

PAGE 177

One species. Southern California to Washington.

Chalcoscirtus

PAGE 177

One species. Southern California.

Euophrys

PAGE 177

One species. Central California to Washington.

These three genera include very small jumping spiders most frequently found in leaf litter. *Talavera minuta* is tiny and grayish brown with a coat of short, white setae that gives it a "frosted" appearance. *Chalcoscirtus diminutus* is a shiny, dark spider with a band of white setae across the front of its abdomen and on the clypei of the male. The male also has a very shiny scute covering his abdomen. *Euophrys monadnock* is a rare species whose male has bright-orange femora on legs III and IV. The female is grayish brown and lacks conspicuous field marks.

Mexigonus

PAGE 178

Two species. Southern California.

Mexigonus are small Salticidae most common in open rocky areas. They are brown and, depending on the species, are either mottled or coarsely striped.

Phlegra

PAGE 174

One species. Southern California.

Phlegra hentzi was introduced from the eastern United States to Southern California. Both sexes are similarly patterned with two white stripes on the carapace and three on the abdomen. The male has a bright-blue clypeus while on the female it is rusty brown.

Sitticus

PAGE 175

Four species. Central California to northern Washington.

Most *Sitticus* are brown with a "spread wing" pattern on the abdomen and a pale line running up the middle of the carapace. The exception is *Sitticus dorsatus,* whose males are bright orange, a color pattern thought to mimic flightless wasps.

(continued)

(continued)

Peckhamia PAGE 176
Two species. California.

Synageles PAGE 176
Three species. Southern California to northern Washington.

Sarinda PAGE 176
One species. Southern California.
All three of these genera are ant mimics. Their extraordinary similarity to ants is at times enhanced by their behaviors, in which they hold their fore legs above the head in an antenna-like manner. *Peckhamia* and *Synageles* are extremely similar, and differentiating them requires a close examination of their eye arrangement, spination, and reproductive structures. Both of these genera hold their second pair of legs over their heads to mimic antenna. *Sarinda cutleri* is a dark-bodied ant mimic known regionally only from southeastern California. Both sexes have enlarged palps and the males have large, forward-pointing chelicerae. When mimicking antenna, *S. cutleri* holds its first leg pair over its head.

Tutelina PAGE 171
Three species. California to Washington.
Tutelina are uncommon, iridescent jumping spiders with distinctive, longitudinal stripes on their legs. They are small spiders whose long, thin legs give them a particularly ant-like appearance, a look further enhanced by their habit of waving their first pair of legs over their heads in an antenna-like manner. Although three species have been reported along the Pacific coast, only *Tutelina similis* has been found in recent years.

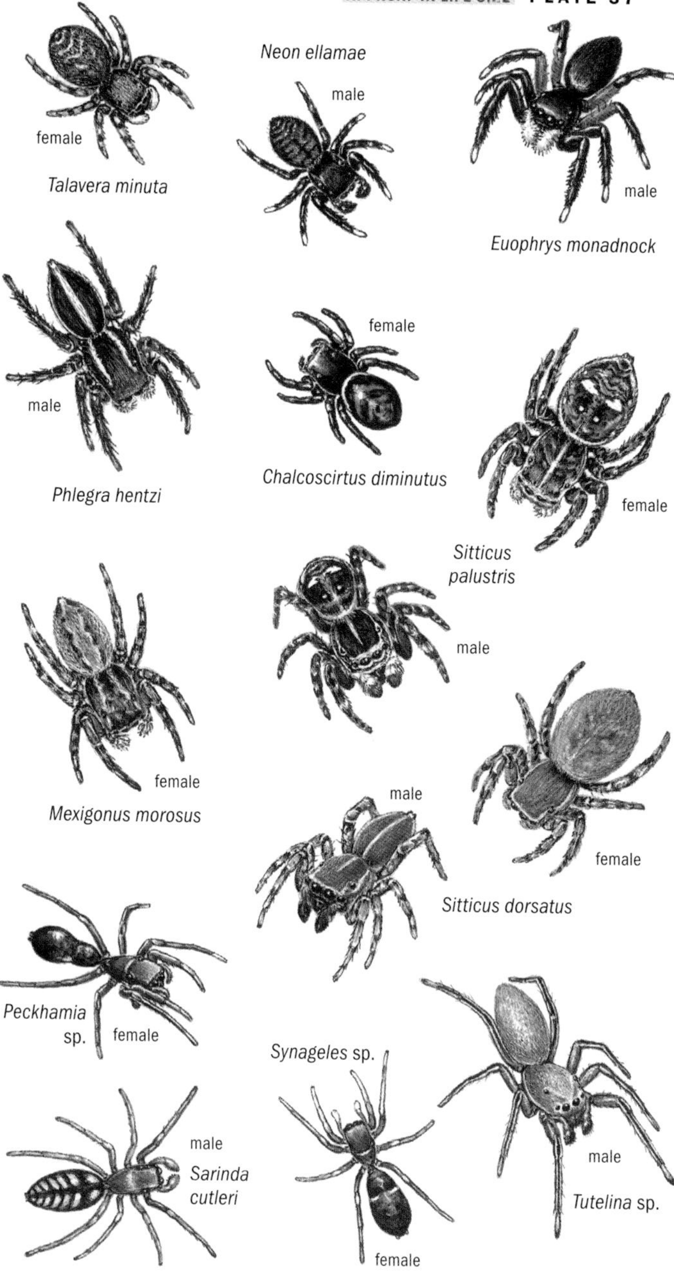
female
Talavera minuta
Neon ellamae
male
male
Euophrys monadnock
male
Phlegra hentzi
female
Chalcoscirtus diminutus
female
Sitticus palustris
male
female
Mexigonus morosus
male
female
Sitticus dorsatus
Peckhamia sp.
female
Synageles sp.
female
male
Sarinda cutleri
male
Tutelina sp.

Family Thomisidae

Crab Spiders PAGE 180

Nine genera, 55 regional species. California, Oregon, Washington.

IDENTIFICATION: Stocky build and laterigrade fore legs. Crab spiders are one of the most recognizable families in North America. They are found in a wide variety of habitats, from tree bark and leaf litter to flower heads.

Diaea PAGE 181

One species. California.

Diaea livens is a unique, translucent green spider with a variably colored abdomen. It is almost entirely associated with coast live oaks *(Quercus agrifolia)* along the California coast.

Misumenoides PAGE 182

One species. Southern and Central California.

Misumena PAGE 182

One species. Central California to Washington.

The White-banded Crab Spider, *Misumenoides formosipes,* and the Goldenrod Crab Spider, *Misumena vatia,* are sexually dimorphic spiders whose anterior eyes are all nearly the same size. Both species also show a conspicuous dearth of body setae. The most reliable way to separate them is by looking at their faces. The White-banded Crab Spider has a distinctive white ridge between its chelicerae and anterior eyes, a feature missing from the Goldenrod Crab Spider (fig. 14). The two species also have only a limited overlap in distribution. Both species are most frequently found hunting among the petals of blooming flowers.

Misumessus PAGE 182

One species. Southern and Central California.

Mecaphesa PAGE 183

16 species. Southern California to northern Washington.

On both of these genera, the anterior median eyes are distinctly smaller than the anterior laterals and their bodies sport a peppering of setae. *Misumessus oblongus* is essentially unpatterned, with only a sparse coating of body setae. The 16 species of *Mecaphesa* have long, plentiful body setae and are often decorated with assorted spots and stripes. Both of these genera are found on blooming flowers and at the tips of budding branches.

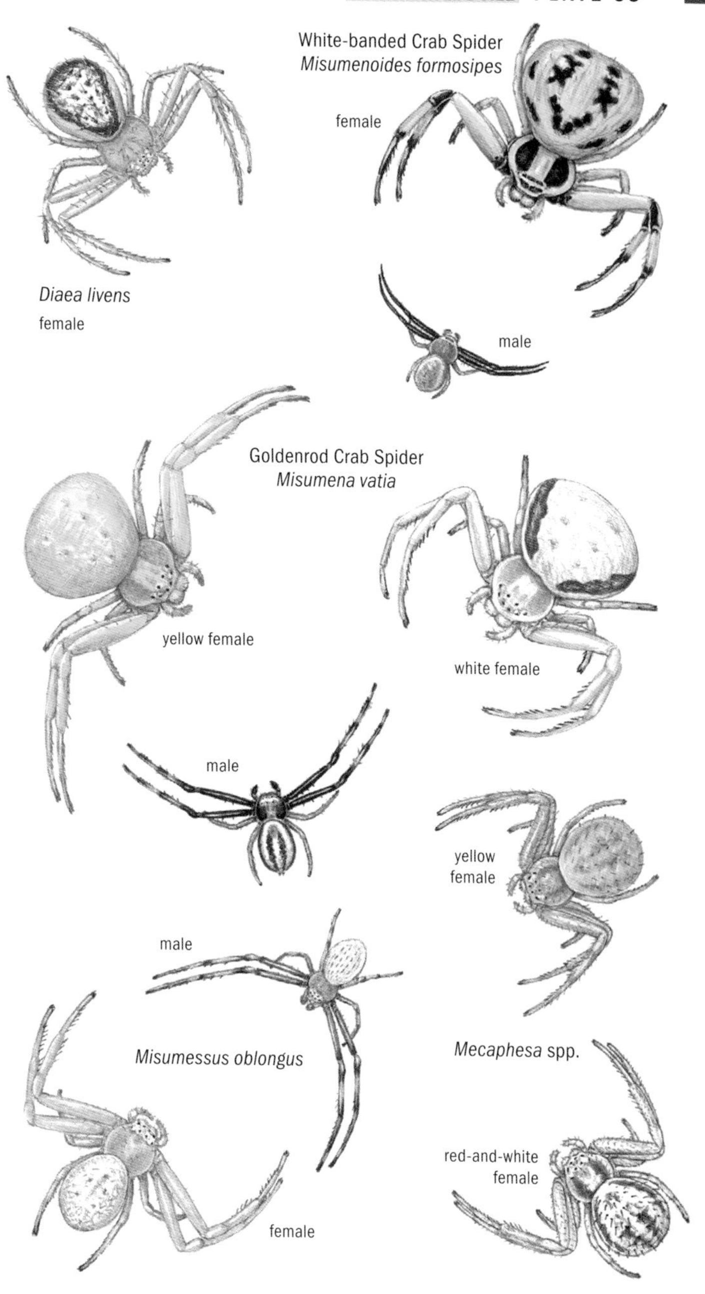
White-banded Crab Spider
Misumenoides formosipes
female
male
Diaea livens
female
Goldenrod Crab Spider
Misumena vatia
yellow female
white female
male
yellow
female
male
Misumessus oblongus
female
Mecaphesa spp.
red-and-white
female

Family Thomisidae (cont.) PAGE 180

Xysticus PAGE 185

26 species. Southern California to northern Washington.

Ozyptila PAGE 185

Six species. Central California to Washington.
These two cryptically colored genera are predominately found in leaf litter and are best differentiated by the shape of their fore legs. *Xysticus* have three or more spines on the ventral surfaces of tibiae I and long, narrow femora. *Ozyptila* have only two pairs of ventral spines on tibiae I, and their femora are distinctly shorter and thicker. Both regularly have a broad, pale stripe down the midline of the carapace.

Tmarus PAGE 181

Two species. California to Washington.
Tmarus are easily recognized by their unusual abdominal shape, elongate clypei, and especially long fore legs. They are most frequently found on small branches where they look like leaf buds or broken twigs. *Tmarus salai* is known only from Southern California, while *T. angulatus* is found throughout much of California, becoming less common in Oregon and Washington.

Coriarachne PAGE 184

Two species. California to Washington.
Coriarachne have exceptionally flat carapaces and are most frequently found in tight crevices and under tree bark. Both of the regional species are found from California through Washington. *Coriarachne utahensis* has a slight roundness to its carapace, and it is decorated with broad, round-tipped setae. *C. brunneipes* has an especially flat carapace, and its setae are thin and pointed.

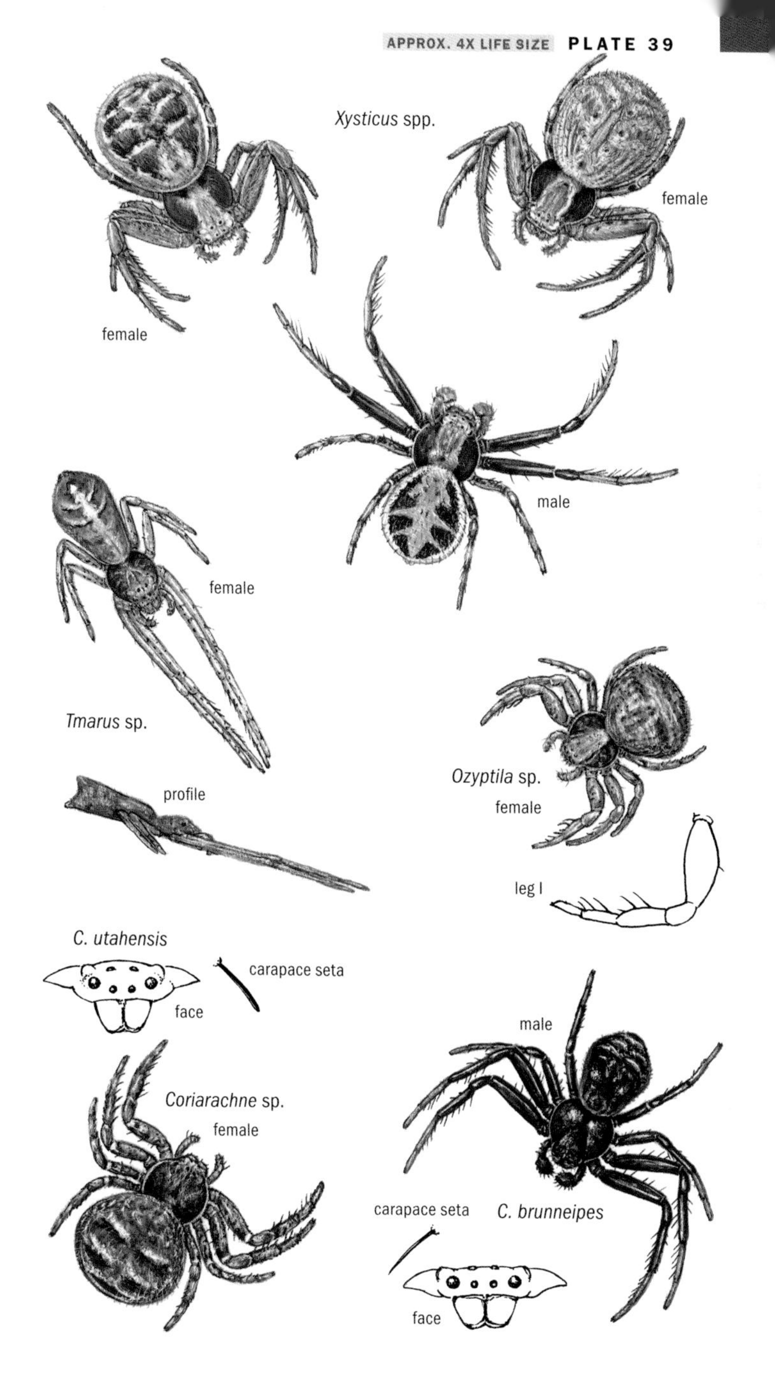

Xysticus spp.
female
female
male
female
Tmarus sp.
profile
Ozyptila sp.
female
leg I
C. utahensis
carapace seta
face
Coriarachne sp.
female
male
carapace seta
C. brunneipes
face

Family Philodromidae

Running Crab Spiders PAGE 187

Six genera, 60 regional species. California, Oregon, Washington.
IDENTIFICATION: Spiders with leg II longer than leg I. Despite this family's exceptional diversity, all of the genera share this one unifying feature. In some genera this difference is conspicuous while in others it is more subtle.

Titanebo PAGE 187

10 species. Southern California to central Washington.

Ebo PAGE 187

Four species. California to Washington.
In these very similar genera, leg II is nearly twice as long as leg I. *Titanebo* are most common in brushy foliage, have lateral spines on their tibiae and metatarsi, and have fairly high clypei. *Ebo* are much less common, lack lateral spines on their tibiae and metatarsi, have short clypei, and are most common in leaf litter, caves, and under debris.

Philodromus PAGE 188

33 species. Southern California to northern Washington.
Philodromus is the most diverse of the running crab spider genera. *Philodromus* are found in a wide range of habitats, and while they vary considerably in size, color, and pattern, they are joined together by the arrangement of their eyes. On *Philodromus*, the posterior medians are closer to the posterior laterals than they are to one another.

Thanatus PAGE 189

Five species. Southern California to northern Washington.

Apollophanes PAGE 189

Two species. Southern California to Washington.
In these genera, the second leg is only a little longer than the first, and the posterior median and posterior lateral eyes are roughly equidistant from each other. The differences between these two cursorial genera are subtle, but *Apollophanes* tend to be flatter bodied and longer legged than *Thanatus*, and their abdominal markings are normally less sharply defined.

(continued)

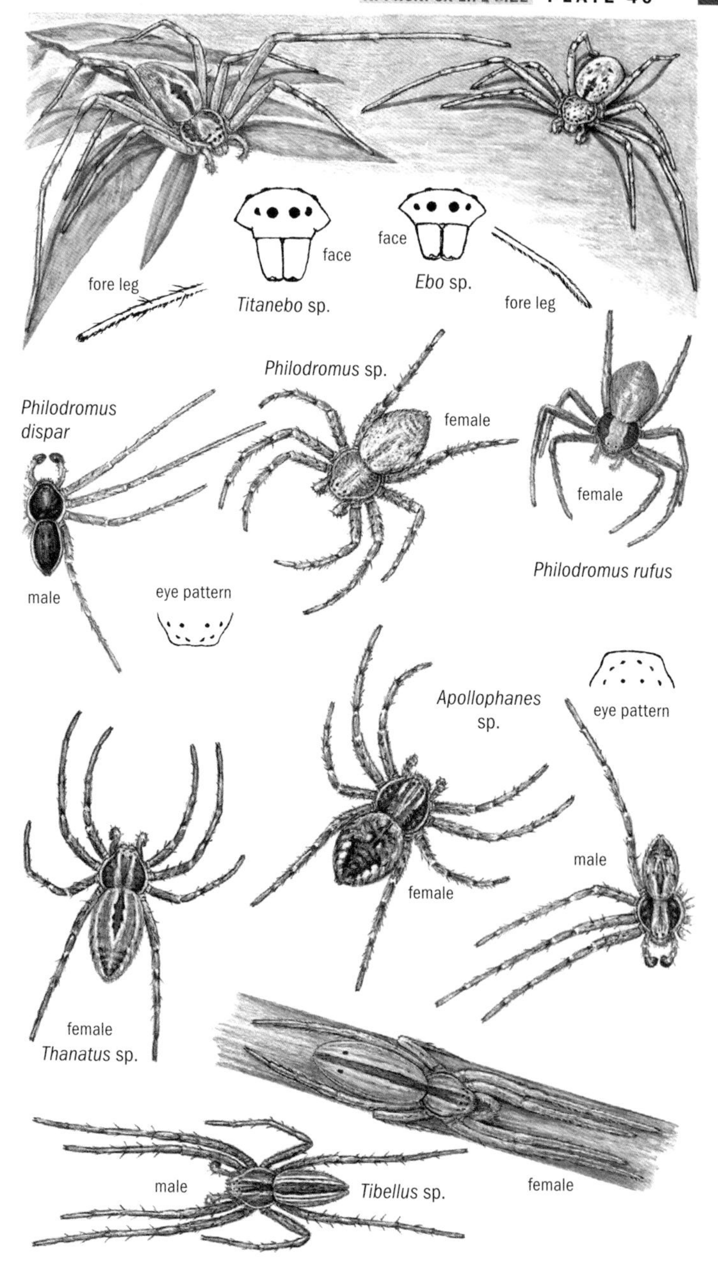
face
Titanebo sp.
face
Ebo sp.
fore leg
fore leg
Philodromus sp.
female
Philodromus dispar
male
eye pattern
female
Philodromus rufus
Apollophanes sp.
eye pattern
male
female
female
Thanatus sp.
male
Tibellus sp.
female

(continued)

Tibellus PAGE 188

Six species. Southern California to northern Washington. *Tibellus* are elongate spiders with longitudinal stripes on both the carapace and abdomen. Also known as slender crab spiders, they are especially common in grassy fields and meadows where their patterning provides excellent camouflage.

Family Selenopidae

Flatties PAGE 191

One genus, one described species. California.

IDENTIFICATION: Extremely flat-bodied spiders with six eyes in their anterior eye row, two in their posterior, and long, laterigrade legs.

Selenops PAGE 191

Selenops actophilus is a cryptically colored spider regionally endemic to dry, rocky areas of Southern California.

Family Sparassidae

Giant Crab Spiders, Huntsman Spiders PAGE 192

Three genera, five species. California.

IDENTIFICATION: Dorsoventrally flattened, long-legged spiders with a unique trilobed extension on the metatarsal-tarsal joints.

Olios PAGE 193

Three species. California.

Macrinus (not illustrated) PAGE 193

One species. Southern California.

Olios vary dramatically in size and range from pale tan to chestnut brown in color and often show a thin Y-shaped mark on the abdomen. *Macrinus mohavensis* is the only regional representative of this otherwise tropical genus and is essentially identical in appearance to *Olios*. It differs in the structure of its reproductive organs and is known only from San Bernardino County, California.

Heteropoda PAGE 194

One species. Southern California.

The introduced Huntsman Spider, *Heteropoda venatoria*, is not established along the Pacific coast, but it has been collected numerous times around commercial ports. This uniquely patterned spider is exceptionally large, with leg spans up to 12 cm across.

Family Homalonychidae PAGE 195

One genus, two regional species. California.

IDENTIFICATION: Desert-dwelling, cursorial spiders whose females and juveniles camouflage their bodies with a coating of fine sand.

(continued)

(continued)

Adult males can be recognized by their carapace pattern, abdominal shape, and strongly recurved posterior eye row.

Homalonychus

PAGE 195

Two species. Southern California.
Because of their coating of sand, juvenile and adult female *Homalonychus* are often exceedingly well camouflaged. When alarmed, they pull their legs together into a distinctive, stiff X-shaped position.

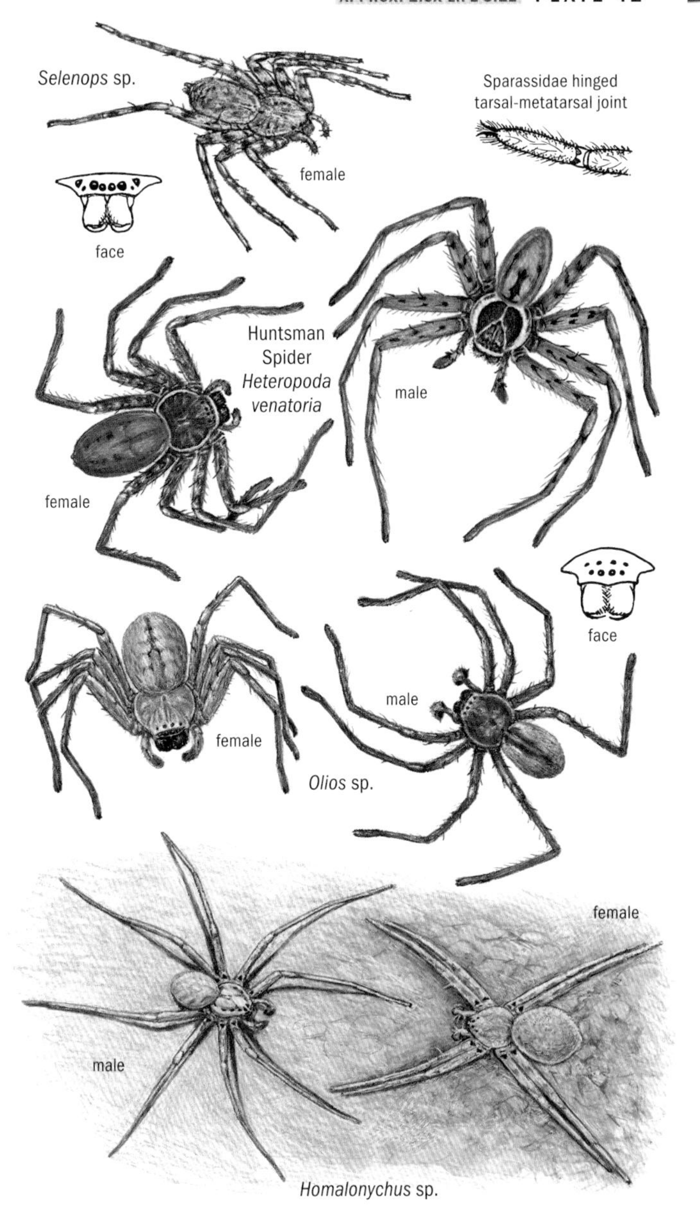
Selenops sp.
Sparassidae hinged
tarsal-metatarsal joint
female
face
Huntsman
Spider
Heteropoda
venatoria
male
female
face
male
female
Olios sp.
female
male
Homalonychus sp.

Family Dictynidae

Mesh Web Weavers (Cribellate Species) PAGE 198

15 genera, 107 described regional species. California, Oregon, Washington.

IDENTIFICATION: A diverse family whose anterior lateral spinnerets are separated and either shorter than or equal in length to their posterior lateral spinnerets. The cribellate species are generally under 4.0 mm in length and lack leg spines, and they often build distinctive mesh-like webs at the tips of flowers and twigs. The ecribellate species are generally larger, up to 8.0 mm in length, have heavily spined legs, and are ground-dwelling hunters. Due to the rarity, physical similarities, and taxonomic inconsistencies of some genera, not all are illustrated, but they are discussed in the main text.

Dictyna PAGE 202

21 species. California to Washington.

Emblyna (not illustrated) PAGE 202

33 species. California to Washington.

Phantyna (not illustrated) PAGE 202

Three species. California to Washington.

Tivyna (not illustrated) PAGE 202

One species. California.

This assortment of small, cribellate spiders makes up the majority of the Pacific coast's Dictynidae. They all have high clypei and, even within species, often present a great deal of variation in color and pattern. In most cases, identifying the genus of the males requires looking at the fine details of their reproductive structures, while the females are frequently inseparable.

Tricholathys PAGE 203

Eight species. California to Washington.

Although similar in appearance and behavior, *Tricholathys* have significantly lower clypei than the *Dictyna*-type mesh web weavers.

Mallos PAGE 201

Four species. Southern California to northern Washington.

(continued)

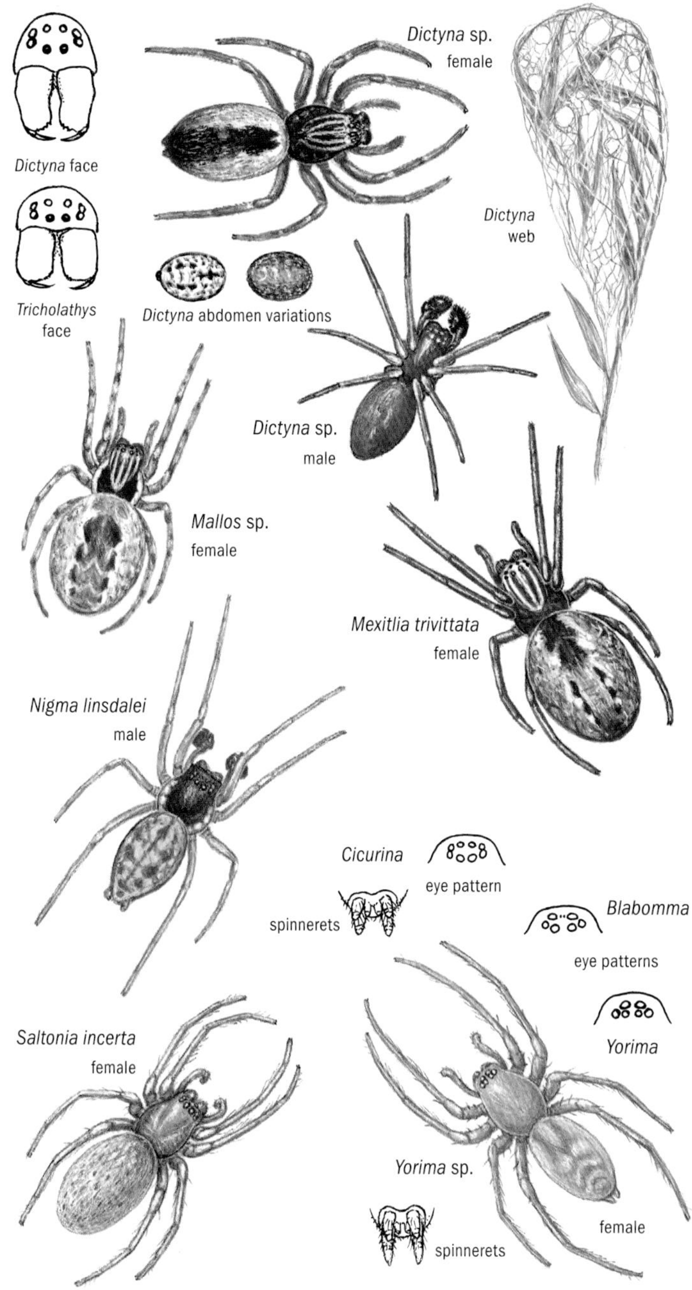

Dictyna face

Dictyna sp.
female

Dictyna
web

Tricholathys
face

Dictyna abdomen variations

Dictyna sp.
male

Mallos sp.
female

Mexitlia trivittata
female

Nigma linsdalei
male

Cicurina
eye pattern

spinnerets

Blabomma

eye patterns

Yorima

Saltonia incerta
female

Yorima sp.

female

spinnerets

(continued)

Mexitlia PAGE 201

One species. Southern California.

Both of these genera are cribellate mesh web weavers. *Mallos* have distinctive, wide, white bands around the thoracic portion of the carapace. *Mexitlia trivittata* is a large spider, up to 8.0 mm in length. While similar to *Mallos,* it lacks white carapace bands. It is a communal species that can form aggregations several hundred strong.

Nigma PAGE 202

One species. California.

Nigma linsdalei is a distinctive, translucent green spider known predominately from oak trees within California's Coast Ranges.

Saltonia PAGE 199

One species. California.

Saltonia incerta lives almost entirely under desert salt crusts in Southern California. Unique within its habitat, it can be further recognized by its heavily spined legs and broad colulus.

Yorima PAGE 200

Five species. California.

Blabomma PAGE 200

10 species. Southern California to northern Washington.

Cicurina PAGE 200

14 species. Southern California to northern Washington.

These three genera of ecribellate Dictynidae are extremely similar in appearance and best separated by the arrangement of their eyes and by the structure of their spinnerets, as illustrated and described in the text.

Plates continue on the following page.

Family Cybaeidae PAGE 205

Four genera, 42 currently recognized species (see text). California, Oregon, Washington.

IDENTIFICATION: Most common in forest leaf litter and caves. Cybaeidae is a diverse and morphologically challenging family. While most species have eight eyes, others have six or zero. Their anterior lateral spinnerets are either touching at their bases or extremely close together. They are also thicker and often longer than the posterior lateral spinnerets.

Cybaeina PAGE 206

Four species. Northern California to central Washington.

Cybaeota (not illustrated) PAGE 206

Three species. Southern California to northern Washington.

Both of these genera contain similar, small cybaeids with three or four pairs of spines under tibiae I. They are best distinguished by the fine details of their dentition as described in the text.

Cybaeus PAGE 207

34 described and many undescribed species. Southern California to northern Washington.

Cybaeozyga (not illustrated) PAGE 207

One species. Southern Oregon.

These genera have either two or three pairs of spines beneath tibiae I. While *Cybaeus* is a widespread and diverse genus, *Cybaeozyga heterops* is quite small and known only from southern Oregon. The two genera are best separated by the details of their cheliceral dentition as outlined in the text.

Family Hahniidae PAGE 209

Seven genera, 40 species. California, Oregon, Washington.

IDENTIFICATION: Spiders with rows of trichobothria that increase in length as they progress down their metatarsi and tarsi. This is an extremely varied and difficult to define family. Members of the subfamily Hahniinae have a unique transverse spinneret pattern. The anterior lateral spinnerets on spiders in the subfamily Cryphoecinae are clearly separated and are shorter than the posterior lateral spinnerets.

Hahnia (fig. 15) PAGE 211

Four species. California to Washington.

(continued)

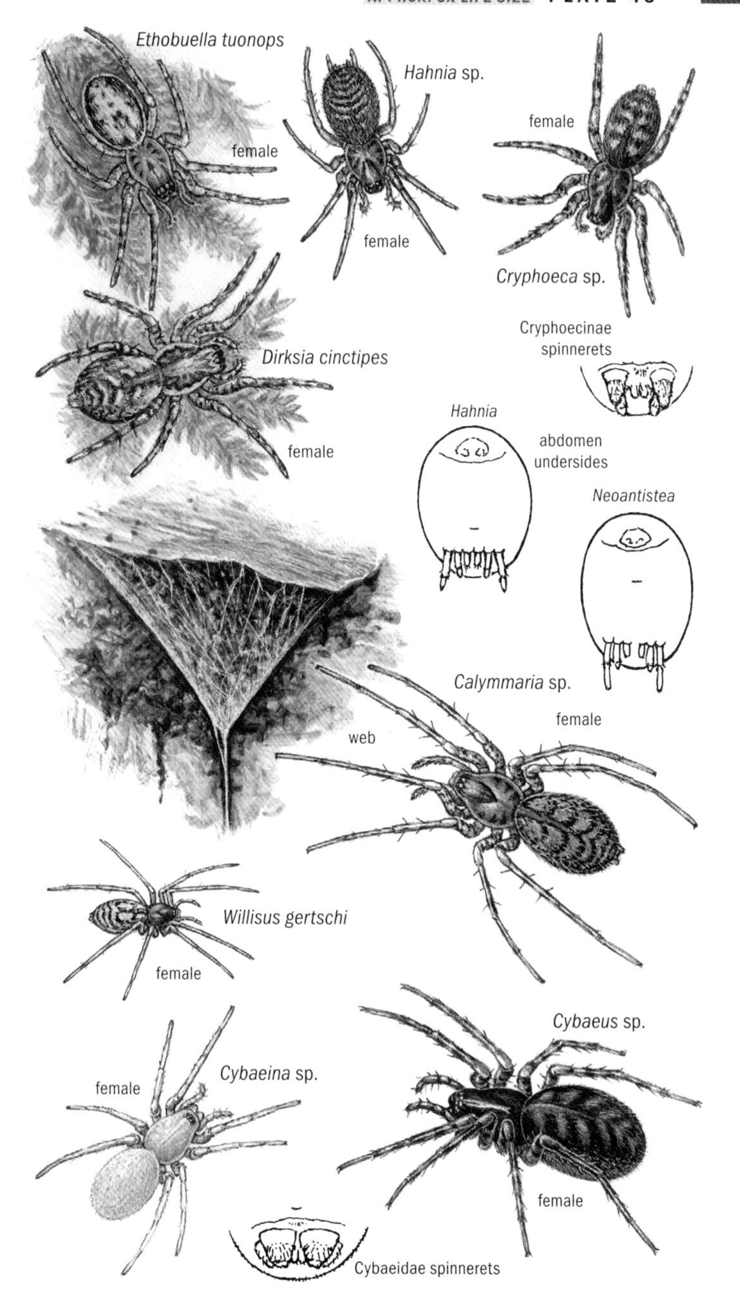
Ethobuella tuonops
female
Hahnia sp.
female
female
Cryphoeca sp.
Cryphoecinae spinnerets
Dirksia cinctipes
female
Hahnia
abdomen undersides
Neoantistea
Calymmaria sp.
female
web
Willisus gertschi
female
Cybaeus sp.
Cybaeina sp.
female
female
Cybaeidae spinnerets

(continued)

Neoantistea (fig. 15) PAGE 211

Four species. California to Washington.
These genera share a distinctive transverse spinneret arrangement. They can be distinguished from each other by the structure of their lateral spinnerets and eye arrangement (fig. 15).

Ethobuella PAGE 211

Two species. Central California, Washington.

Dirksia PAGE 211

One species. Northern California to northern Washington.
These genera of small, foliage-dwelling spiders in the subfamily Cryphoecinae have either four or five pairs of spines beneath tibiae I. Species of *Ethobuella* can be separated from *Dirksia cinctipes* by the details of their spination and eye arrangement, as described in the text.

Calymmaria PAGE 212

27 species. Southern California to northern Washington.
Calymmaria are small to medium-size long-legged spiders whose distinctive cone-shaped webs are frequently found in shaded, protected nooks, especially near water. Away from their webs, they can be identified by their spination and dentition pattern, as discussed in the text.

Willisus PAGE 213

One species. Southern California.
Willisus gertschi is a small spider found under rocks in Southern California's San Bernardino Mountains. It has a shiny carapace and a patterned abdomen and lacks ventral spines on tibiae I.

Cryphoeca PAGE 213

One described and at least one unidentified species (see text). Central California to northern Washington.
Cryphoeca are small, leaf litter–dwelling spiders with two pairs of ventral spines on tibiae I and three to five teeth on their cheliceral retromargins.

Family Pisauridae

Nursery Web Spiders, Fishing Spiders
PAGE 217

Two genera, two regional species. California, Oregon, Washington.

IDENTIFICATION: Long-legged spiders with a moderately recurved posterior eye row. They are found near ponds, lakes, and slow-moving streams where they often rest with their legs stretched across the water's surface. They suspend their egg sacs in distinctive "nursery webs."

Dolomedes
PAGE 218

One species. Northern California to northern Washington.
The Six-spotted Fishing Spider, *Dolomedes triton,* is a fairly large spider with white stripes on each side of its body and two rows of pale spots on its abdomen.

Tinus
PAGE 218

One species. Southern California.
Tinus peregrinus is a fairly small, tan-and-white fishing spider regionally limited to desert ponds and protected backwaters along the Colorado River.

Family Tengellidae
PAGE 215

Three genera, 34 regional species. California.

IDENTIFICATION: Nocturnal, wandering hunters whose distinguishing features are subtle. Often multiple characteristics must be used in conjunction to confirm an individual's familial identity. Tengellidae have five or more pairs of spines on the undersides of the anterior tibiae, have claw tufts, and are medium to large in size. Their eyes are in two straight rows, their trochanters are notched, and their anterior spinnerets are conical and nearly contiguous across their bases.

Titiotus
PAGE 216

16 species. Central and Northern California.

Socalchemmis
PAGE 216

14 species. Southern and Central California.

Anachemmis
PAGE 217

Four species. California.
While habitat and range can provide important clues, confidently differentiating these three very similar genera often requires exam-

(continued)

(continued)

ining their reproductive structures. *Titiotus* is the most diverse of the region's genera and contains the largest species in the family. While it is most common in the rocky grasslands and forests, several *Titiotus* species are found in caves in the Sierra Nevada. Most *Socalchemmis* are limited to Southern California's chaparral and desert scrub, with two species ranging into Central California. Among the four *Anachemmis* species, two have widespread, but non-overlapping, distributions across California, while the other two are endemic to desert caves.

Family Zoropsidae

False Wolf Spider

PAGE 220

One genus, one regional species. California.

IDENTIFICATION: A family with one local representative, *Zoropsis spinimana,* a large, six-eyed, cribellate spider with a strongly recurved posterior eye row and a distinctively marked carapace and abdomen.

Zoropsis

PAGE 220

Native to the Mediterranean region, the False Wolf Spider, *Zoropsis spinimana,* has become a common resident of suburban areas around California's San Francisco Bay.

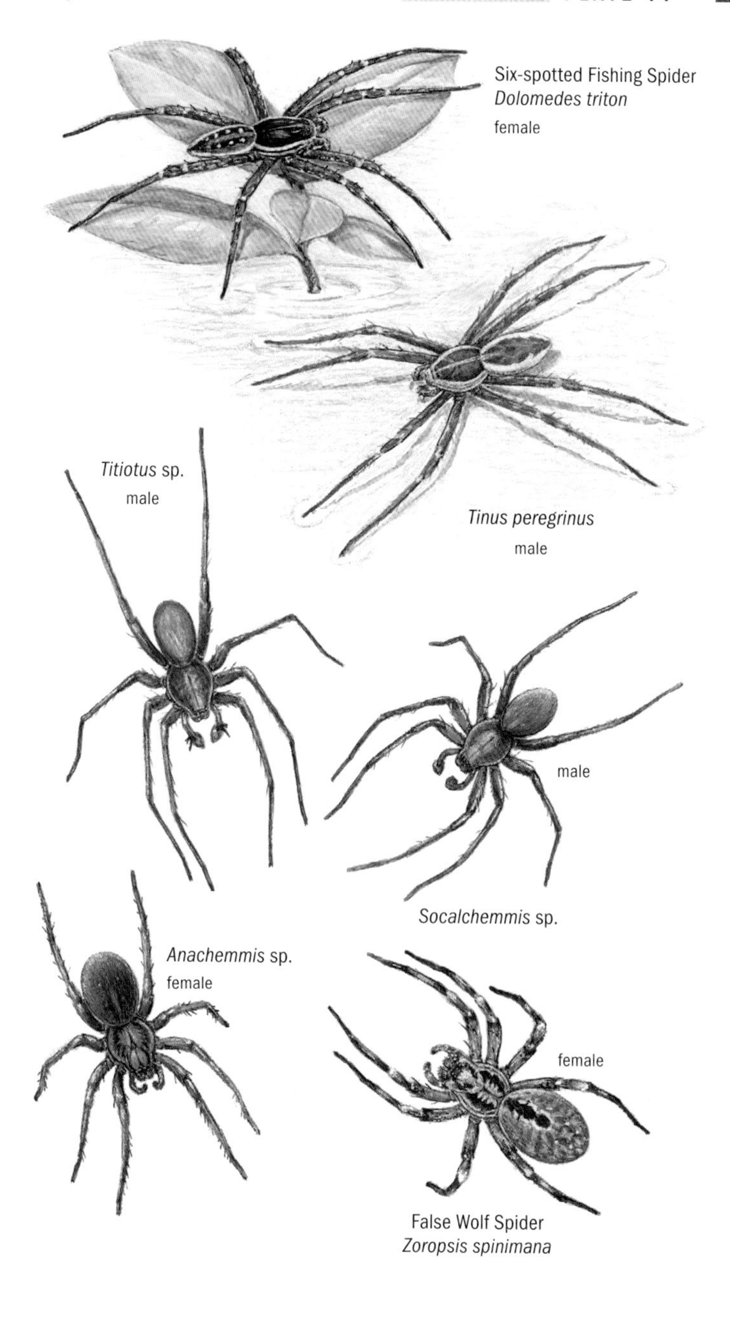
Six-spotted Fishing Spider
Dolomedes triton
female
Titiotus sp.
male
Tinus peregrinus
male
male
Socalchemmis sp.
Anachemmis sp.
female
female
False Wolf Spider
Zoropsis spinimana

Family Oxyopidae

Lynx Spiders PAGE 220

Three genera, six regional species. California, Oregon, Washington.

IDENTIFICATION: Generally foliage dwellers that vary dramatically in size from small to fairly large. All share a distinctive hexagonal eye arrangement and have tapered abdomens and long, thin, heavily spined legs.

Oxyopes PAGE 221

Three species. Southern California to northern Washington.

Oxyopes is the most diverse and widespread of the Pacific coast Oxyopidae genera. While there can be variation in *Oxyopes* coloration, the overall pattern of each sex and species is generally distinctive. The Striped Lynx Spider, *Oxyopes salticus,* is regionally limited to California and western Oregon and can be recognized by the thin, sharply defined lines on the undersides of the femora on its fore legs. The Western Lynx Spider, *O. scalaris,* is found in foliage from Southern California to northern Washington. The uniquely patterned Trident Lynx Spider, *O. tridens,* is terrestrial and regionally limited to arid portions of Southern California.

Peucetia PAGE 221

Two species. California.

The more common of the region's two species, *Peucetia viridans,* can vary in color from pale green with a few white bars to heavily mottled in black, ivory, and red. *P. longipalpis* is less diverse in coloration and has proportionately shorter legs. *Peucetia* are most often found in tall grasses and on woody shrubs.

Hamataliwa PAGE 222

One species. Southern California.

Hamataliwa grisea is found on trees and woody shrubs where its gray body mimics the color and texture of the bark. The female is much hairier than the male, with long fringes of setae on her legs.

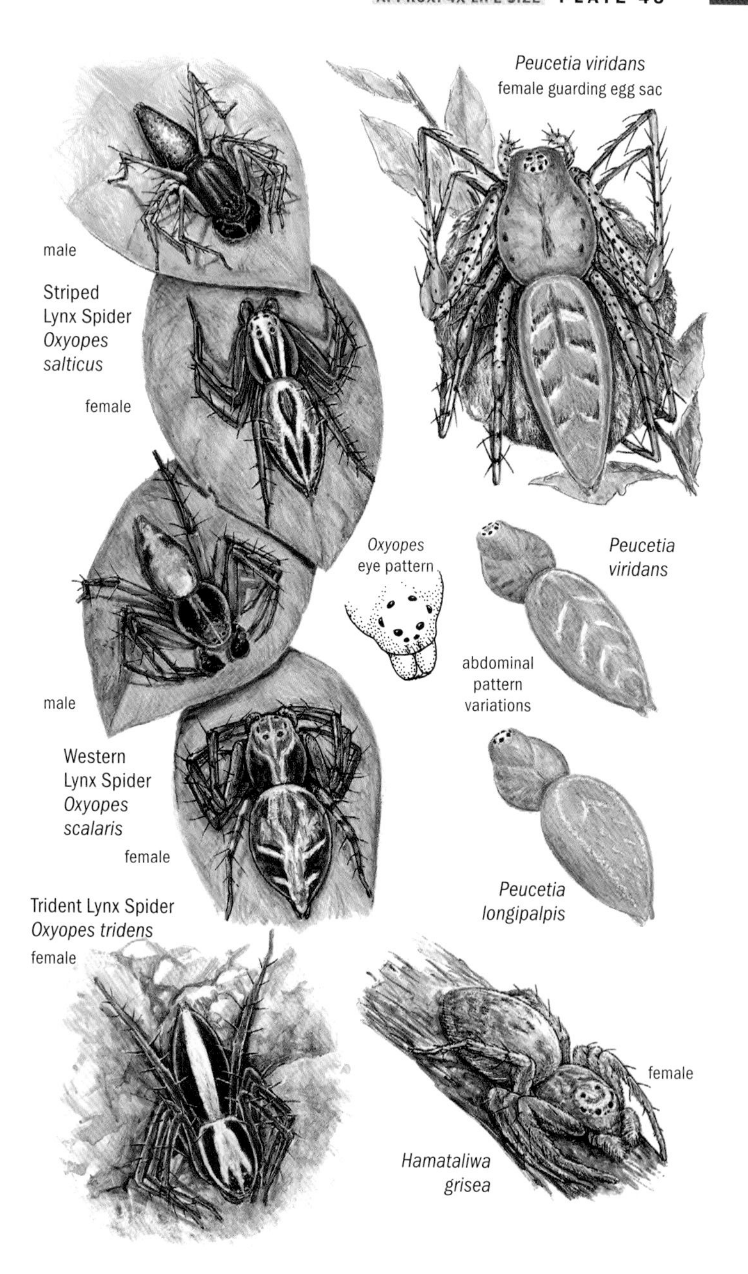
Peucetia viridans
female guarding egg sac
male
Striped
Lynx Spider
Oxyopes
salticus
female
Oxyopes
eye pattern
Peucetia
viridans
abdominal
pattern
variations
male
Western
Lynx Spider
Oxyopes
scalaris
female
Peucetia
longipalpis
Trident Lynx Spider
Oxyopes tridens
female
female
Hamataliwa
grisea

Family Lycosidae

Wolf Spiders PAGE 223

12 genera, 64 described regional species. California, Oregon, Washington.

IDENTIFICATION: Cursorial hunters, with one uncommon exception, easily recognized by their three-row eye arrangement. Females are frequently seen carrying their egg sacs behind them by their spinnerets or with a mass of newly hatched young clinging to their backs. They are among the most abundant and frequently encountered spiders in North America.

Pardosa PAGE 224

40 species. Southern California to northern Washington.

Pardosa is the most diverse and abundant of the Pacific coast's wolf spider genera. Also known as thin-legged wolf spiders, they show considerable variation in color and pattern, from pale gray and brown to nearly black and from uniformly colored to boldly patterned. *Pardosa* are small wolf spiders with long, thin legs, and when seen head on, the anterior sides of their carapaces are straight and nearly vertical, not sloped or bowed like those of other, similar genera.

Allocosa PAGE 228

One species. Southern California to northern Oregon.

Allocosa subparva is most frequently found along beaches and riverbanks. Unlike many other wolf spiders, it has a shiny, dark carapace without a conspicuous coat of setae.

Pirata PAGE 225

Three species. Southern California to northern Washington.

Pirata are best recognized by the V- or tuning fork–shaped pattern on their shiny carapaces and are most abundant near slow-moving streams and areas of standing water.

Varacosa PAGE 228

One species. Southern California.

Trochosa PAGE 228

One species. Northern California to Washington.

These two genera are extremely similar. Each, however, is regionally represented by a single species separated by both appearance and distribution. *Trochosa terricola* has a distinctive pair of dark comma-shaped dashes in the broad, pale band that runs down

(continued)

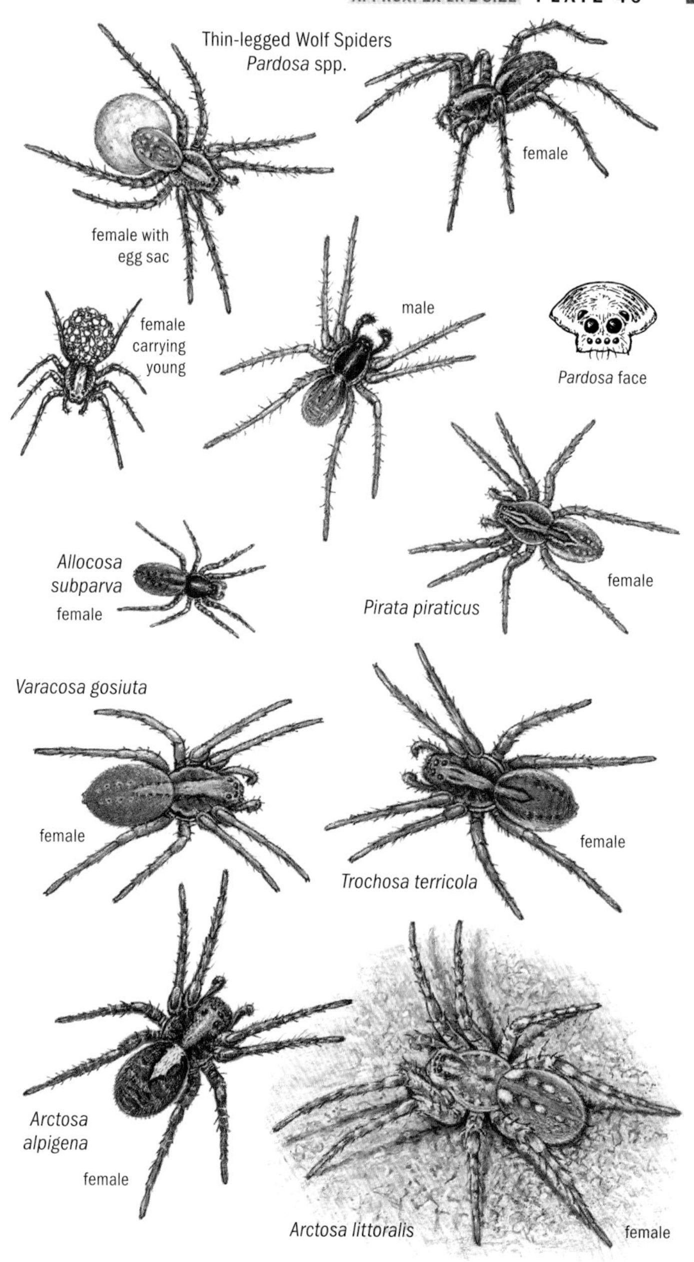
Thin-legged Wolf Spiders
Pardosa spp.
female
female with egg sac
female carrying young
male
Pardosa face
Allocosa subparva
female
Pirata piraticus
female
Varacosa gosiuta
female
Trochosa terricola
female
Arctosa alpigena
female
Arctosa littoralis
female

(continued)

the midline of its carapace. It is found in deep grass and leaf litter from Northern California through Washington. *Varacosa gosiuta* is very similar to *T. terricola* but lacks the paired, dark dashes and is found in arid portions of Southern California.

Arctosa

PAGE 227

Four species. Southern California to northern Washington.
Arctosa are medium-size, thick-legged wolf spiders with broad, low carapaces that are often either bare or only lightly dusted with setae in the middle. Of the four species that have been collected along the Pacific coast, only two can reasonably be considered common. *Arctosa littoralis* lives along beaches, streams, and lakeshores throughout the region, while *A. alpigena* is a forest and mountain field species found widely across Washington and northern Oregon. The other two species are rarely seen members of northern Washington's spider fauna.

Plates continue on the following page.

Family Lycosidae (cont.) PAGE 223

Alopecosa PAGE 227

Two species. Southern California to northern Washington.
Alopecosa are fairly large, thick-legged wolf spiders. While there is some degree of variability in their appearance, *Alopecosa* always have conspicuous broad, pale stripes down the midline of the carapace but never show lightly colored, sharply defined carapace borders.

Schizocosa PAGE 225

Four species. Southern California to northern Washington.

Hogna PAGE 225

Five species. Southern California to northern Washington.
These two genera house some of the largest wolf spiders in North America, and they can be difficult to differentiate. Also known as lanceolate wolf spiders, *Schizocosa* generally have large, sharply defined black heart marks on their abdomens. Although their ventral sides are usually reddish brown, they can also show extensive dark smudging. The most widespread regional species, *Schizocosa mccooki,* is found across most of the Pacific coast. Only one species, *H. antelucana,* can be considered regionally common. It lives in the grasslands of Southern and Central California and has a weakly defined, brownish heart mark and an entirely black venter.

Melocosa PAGE 230

One species. Washington.
Melocosa fumosa lives under high-elevation rocks and scree in Washington's Cascade Range. Unmistakable within its range, it is a large, dark spider with a hairy abdomen and eyes that are set in a particularly broad quadrangle.

Sosippus PAGE 230

One species. Southern California.
Unlike other wolf spiders, *Sosippus* weave sheet webs with a retreat extending from one side. *Sosippus californicus* is a fairly large, distinctively patterned spider most common in desert riparian strips.

Geolycosa PAGE 229

One species. Southern California.
Also known as burrowing wolf spiders, *Geolycosa* dig vertical, turret-edged burrows and hunt from their open entrances as some trapdoor spiders do. The carapaces on burrowing wolf spiders are unusually steep, sloping sharply upward from the rear edge toward the eyes. Found in the Mojave Desert, *Geolycosa gosoga* is a silvery-gray spider with black markings under its fore legs.

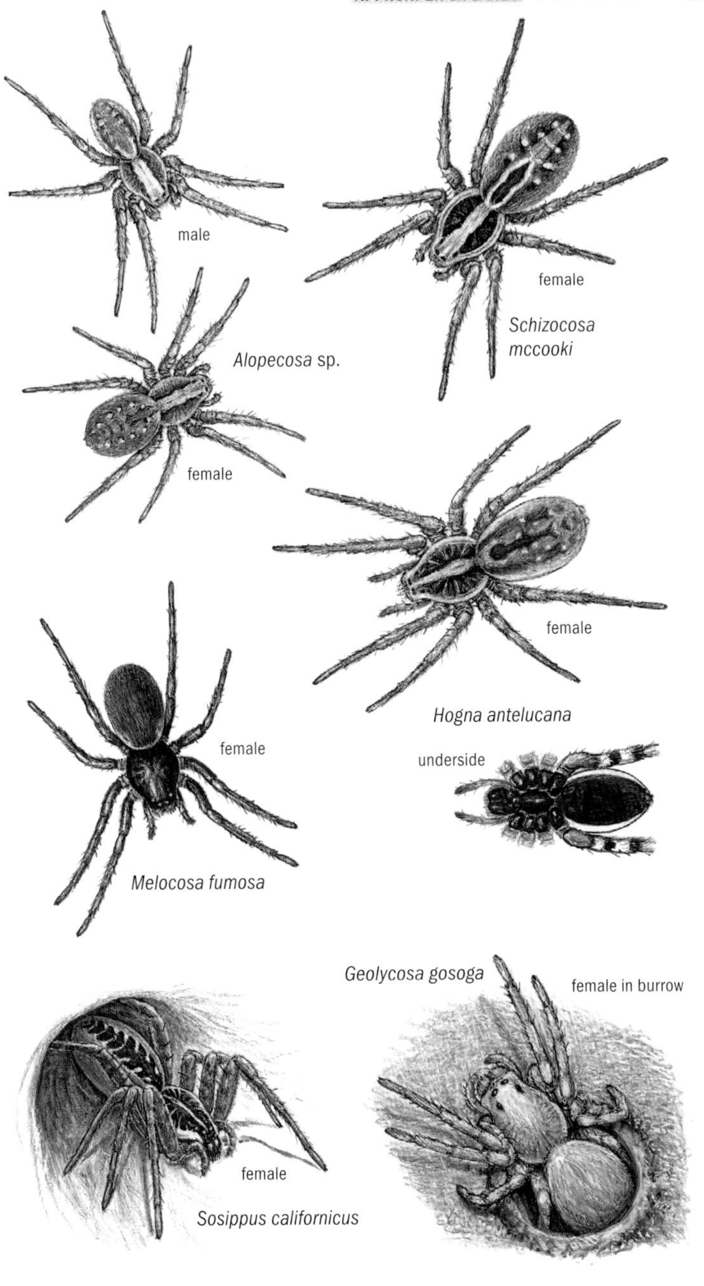
male
female
Schizocosa mccooki
Alopecosa sp.
female
female
Hogna antelucana
female
underside
Melocosa fumosa
Geolycosa gosoga
female in burrow
female
Sosippus californicus

Family Agelenidae (in part)

Funnel Web Weavers

PAGE 232

Six genera, 57 regional species. California, Oregon, Washington.

IDENTIFICATION: Spinners of webs made of a flat sheet with a tubular retreat off of one corner The webs are more frequently seen than the spiders that make them and are especially common in old buildings, grassy fields, and dense shrubbery. Funnel web weavers have long legs, feather-like setae on the body, and with the exception of *Tegenaria,* procurved eye rows.

Tegenaria

PAGE 232

Four species. Southern California to northern Washington.

Tegenaria are synanthropic spiders, especially common around barns and other buildings. Unlike other funnel web weavers, their eye rows are straight or only slightly procurved, the lateral bands on their carapaces are often indistinct, and their sterna have sharply defined patterns (fig. 16). Introduced to the Pacific coast, the region's *Tegenaria* include the Barn Funnel Weaver *(T. domestica),* the Giant House Spider *(T. duellica),* and the Hobo Spider *(T. agrestis).* The Hobo Spider has been blamed for causing necrotic wounds throughout the Pacific Northwest despite a lack of evidence linking it to the reported injuries. In Europe, it is common in rural areas and is not considered a species of medical concern.

Novalena

PAGE 235

Four species. Northern California to northern Washington.

Hololena

PAGE 235

22 species. Southern California to Washington.

Rualena

PAGE 235

Nine species. Southern and Central California.

These three genera are all extremely similar. However, they differ in the structure of the reproductive organs, as well as in the arrangement of their eyes and spinnerets, as discussed in the text. Each genus has a pair of well-defined dark carapace bands, a brighter abdominal midline stripe, procurved eye rows, and short posterior lateral spinnerets. *Novalena* is a montane genus and is generally darker than *Hololena,* the most species-rich of the region's Agelenidae genera. Despite this diversity, only two of its species, *Hololena nedra* and *H. rabana,* are found north of California. Several of Southern California's *Rualena* are island endemics.

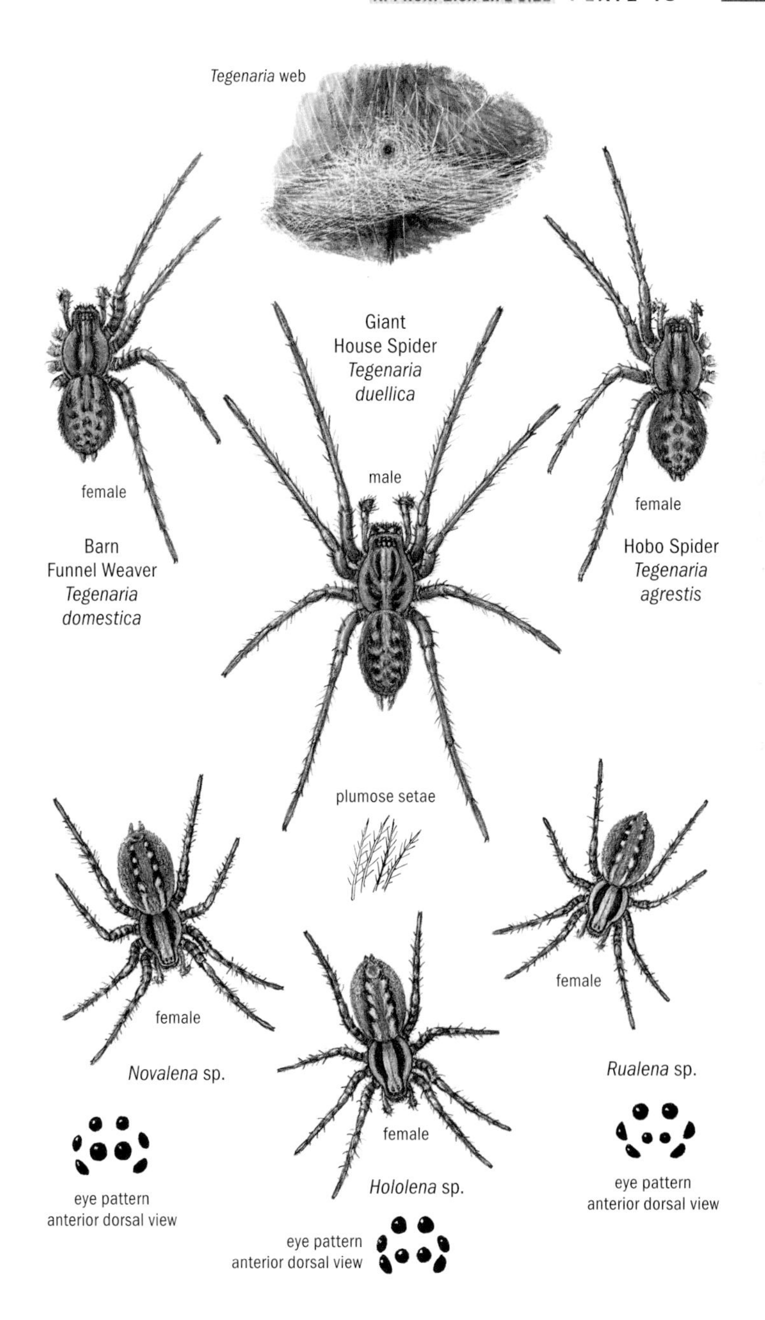
Tegenaria web
Giant
House Spider
Tegenaria
duellica
male
female
Barn
Funnel Weaver
Tegenaria
domestica
female
Hobo Spider
Tegenaria
agrestis
plumose setae
female
Novalena sp.
eye pattern
anterior dorsal view
female
Hololena sp.
eye pattern
anterior dorsal view
female
Rualena sp.
eye pattern
anterior dorsal view

Family Agelenidae (cont.) PAGE 232

Agelenopsis PAGE 234

Six species. Southern California to northern Washington.

Calilena PAGE 234

12 species. Southern California to central Washington.

These two genera of medium-size to large funnel web weavers have exceptionally long posterior lateral spinnerets, dark paramedian bands on their carapaces, and strongly procurved eye rows. Also known as grass spiders, *Agelenopsis* tend to be more yellowish brown than the grayer *Calilena,* and the extensive webs of *Agelenopsis* are often built in fields of tall grass. *Calilena* generally have more of a grayish cast to their bodies and tend to build their webs close to the ground or among piles of debris.

Family Amaurobiidae

Hacklemesh Weavers PAGE 237

Seven genera, 68 regional species. California, Oregon, Washington.

IDENTIFICATION: Secretive, minute to medium-size spiders with divided cribella. They are especially common in moist, shaded woodlands. Amaurobiidae is divided into two groups: the large-bodied genera with leg spines shown here, and tiny spiders without leg spines (fig. 17). The webs of the large-bodied genera consist of messy sheets of cribellate silk radiating out from their retreats. Although the larger hacklemesh weavers are generally similar in appearance and can demonstrate considerable variation within the genera, their pattern intensity, size, and distribution can provide useful clues for identifying an unknown individual.

Callobius PAGE 238

22 species. Southern California to northern Washington.

Amaurobius PAGE 238

23 species. California, Washington (highly local).

These genera contain the largest and most strongly patterned of the hacklemesh weavers. *Callobius* is the most widespread of the region's genera. They are commonly found under tree bark, standing and fallen, and are often boldly patterned with pale-yellow spots and chevrons on their dark gray abdomens. Although there is overlap, *Amaurobius* tend to be smaller and less contrastingly

(continued)

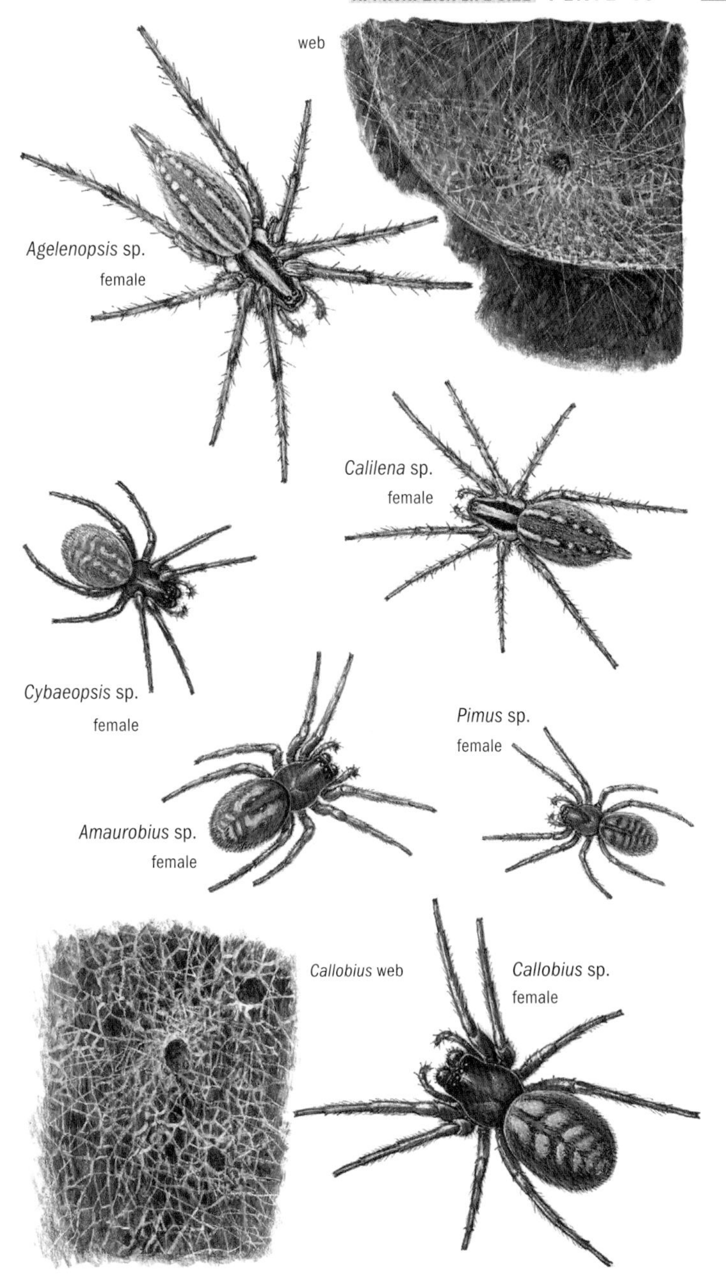
web
Agelenopsis sp.
female
Calilena sp.
female
Cybaeopsis sp.
female
Pimus sp.
female
Amaurobius sp.
female
Callobius web
Callobius sp.
female

(continued)

decorated than *Callobius*. *Amaurobius* are also found under bark, but they are more likely to be found under rocks and debris. With the exception of *Amaurobius ferox*, a European species introduced to west-central Washington, all of the Pacific coast's *Amaurobius* are found solely in California.

Cybaeopsis

PAGE 239

Three species. Oregon to Washington.

Pimus

PAGE 239

10 species. Central California to northern Oregon.

Cybaeopsis and *Pimus* tend to be smaller and less boldly patterned than either *Amaurobius* or *Callobius*. *Cybaeopsis* are small, dark hacklemesh weavers found only along a narrow strip running from southwestern Oregon to northwestern Washington. *Pimus* tend to be paler bodied and less intensely patterned than the region's other large-bodied hacklemesh weavers. Both genera are most common under debris in damp woodlands.

Plates continue on the following page.

Family Desidae

PAGE 243

One genus, one regional species. California, Oregon.

IDENTIFICATION: Builders of unique lattice-like cribellate webs, these are the Pacific coast's only Desidae. They have a divided cribellum, and the largest of their eyes are the anterior medians.

Badumna

PAGE 243

Introduced from Australia, the Gray House Spider, *Badumna longinqua*, can be exceptionally common around homes along the immediate coast from Southern California to northern Oregon. It is purplish brown with a grayish abdomen and striped legs.

Family Titanoecidae

PAGE 241

One genus, two regional species. California, Oregon, Washington.

IDENTIFICATION: Extremely similar to the cribellate amaurobiids (pl. 49), with parallel endites and rows of trichobothria that increase in length on their metatarsi and tarsi. On the females, the calamistra are nearly as long as their fourth metatarsi (but the calamistra are greatly reduced or lacking on the males). Members of the family Titanoecidae are also much more prevalent in arid areas than are the amaurobiids.

Titanoeca

PAGE 242

Two species. Southern California to northern Washington.
The Pacific coast's *Titanoeca* are essentially identical, differing mainly in their spination and reproductive structures. They have orange to brownish-red carapaces and unmarked, dark-gray to black abdomens. *Titanoeca nigrella* is found from Southern California to northern Washington. The only regional record of *T. silvicola* is of an individual collected in Northern California.

Family Amphinectidae

PAGE 244

One genus, one regional species. California.

IDENTIFICATION: Cribellate, cursorial spiders very similar to Amaurobiidae (pl. 49). Unlike any similar spider, they have five or more teeth on each cheliceral margin.

Metaltella

PAGE 245

The introduced *Metaltella simoni* is native to South America but is now extremely abundant in suburban Southern California. It has a dark-orange carapace that gets noticeably darker around the

(continued)

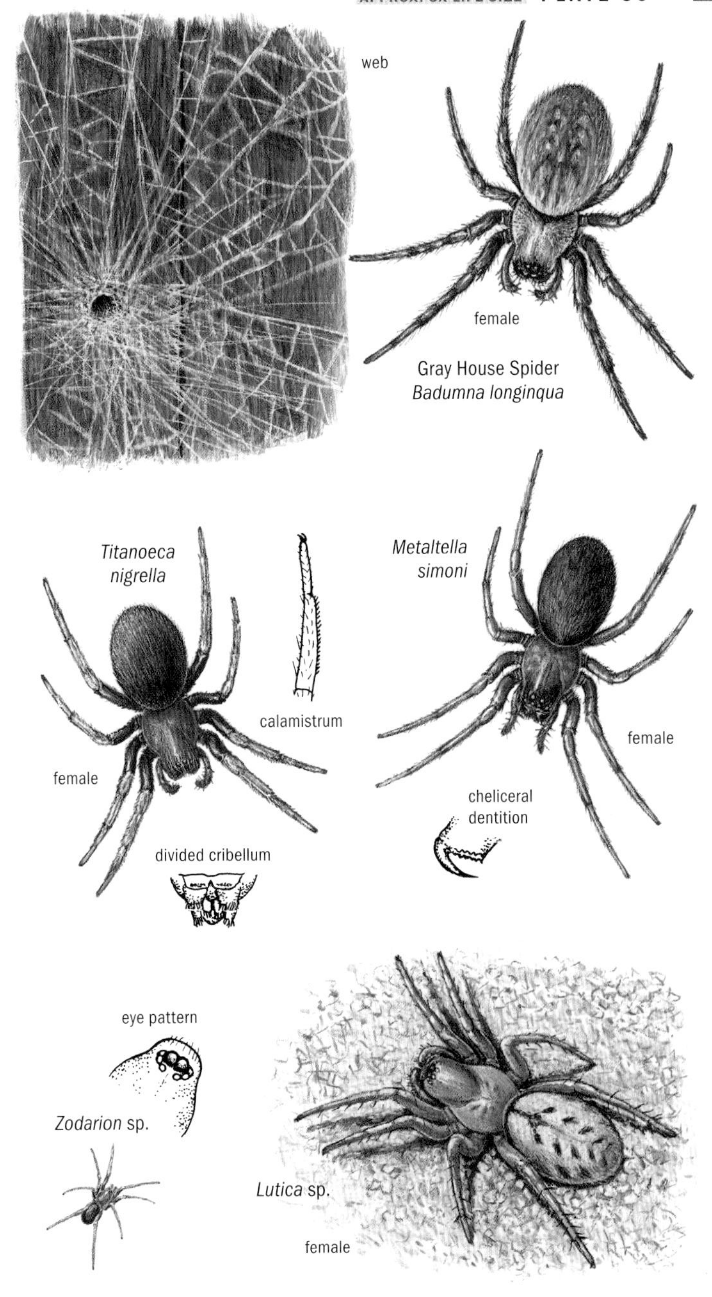
web
female
Gray House Spider
Badumna longinqua
Titanoeca
nigrella
calamistrum
female
divided cribellum
Metaltella
simoni
female
cheliceral
dentition
eye pattern
Zodarion sp.
Lutica sp.
female

(continued)

eyes and chelicerae, and its abdomen can vary from dusky yellow with chevrons to nearly black. It is especially common around well-watered gardens and lawns that have objects it can hide under.

Family Zodariidae

PAGE 214

Two genera, four described and one unidentified regional species. California.

IDENTIFICATION: Eight-eyed, ecribellate spiders with exceptionally large anterior lateral spinnerets and highly reduced posterior spinnerets.

Lutica

PAGE 214

Four species. Southern California.

Lutica are endemic to the coastal dunes and islands off of Southern California. They are fairly large, burrowing spiders with thick, heavily spined legs, dusky-yellow carapaces, and purplish markings on grayish abdomens.

Zodarion

PAGE 215

One unidentified species. Central California.

An unidentified *Zodarion* is known from the rocky, serpentine grasslands of Central California. It is a small, diurnal, fast-moving spider with comparatively large anterior median eyes and a procurved posterior eye row. No other regional spider matches this description.

The genus *Drassinella* (not illustrated) represents another group of small corinnids. They range from 2.5 to 4 mm (.1 to .16 in.) in length and are closely related to the phrurolithine genera. *Drassinella,* however, have two dorsal spines on each femur and additional ventral spines on legs II and IV. They were initially placed in the family Gnaphosidae (p. 156) because their spinnerets are conical and often separated, their endites are obliquely depressed, and they have slightly flattened posterior median eyes. Only later, when other features were looked at in detail, was it discovered that they were more closely related to members of the subfamily Phrurolithinae. Six species live along the Pacific coast, all of which are relatively rare. They have been found in scattered locations along the Coast Ranges from Southern California to the San Francisco Bay Area and in the southern Cascade Range. Only one species, *Drassinella unicolor,* has been found outside of California within the Pacific coast region. It is known from a few scattered spots across the Great Basin, and a specimen was collected in Harney County, Oregon. *Drassinella* are fairly uniform in appearance, with a brownish-orange carapace and dark-gray abdomen, which on males is almost entirely covered by a shiny scute. The genus was revised by Platnick and Ubick (1989).

NATURAL HISTORY: Corinnids are mainly terrestrial, wandering hunters, many of which imitate ants or wingless wasps. This is especially evident on the shiny, sclerotized members of the genus *Castianeira.* Many are diurnal and move in a rapid, halting fashion similar to carpenter ants, and some even walk with their fore legs held over their heads, like antennae. A few species have gone so far as to integrate themselves into ant colonies, living either around the periphery of the anthill or within the colony itself. In the deserts of southeastern California, *S. steckleri* lives among colonies of harvester ants in the genus *Pogonomyrmex.* To do this, it has evolved cuticular hydrocarbons that mimic those used by the ants to differentiate between members of the colony and intruders. Additionally, numerous corinnids in the subfamily Phrurolithinae live almost exclusively in proximity to the ants they mimic. While the ant-mimicking corinnids are frequently diurnal, genera that are not mimetic, such as *Trachelas* and *Meriola,* are primarily nocturnal. Unlike other ant-mimic sac spiders, these two genera regularly forage in vegetation and occasionally enter homes. This is particularly important as there have been reports of Trachelas bites causing mild to moderate pain and lesions in some people (Platnick and Shadab, 1974b).

Ant-mimic sac spiders don't build webs for hunting but will spin saclike cocoons for retreats. Females either hide their egg sacs within their retreats or attach them to protected spots, such as beneath rocks or under loose bark. Their egg sacs look like fried eggs, with a central mound of yellow to pinkish-orange eggs surrounded by a disk of papery white silk. In some genera, including *Castianeira,* the silk covering the egg sac has an unusual metallic sheen.

LIOCRANIDAE

Pl. 29

IDENTIFICATION: Liocranids are eight-eyed entelegynes with two tarsal claws but no claw tufts. Historically, Liocranidae's diverse genera were included in a more broadly defined version of the family Clubionidae (p. 145). Like other former (and current) clubionids, liocranids have conical anterior lateral spinnerets that are either contiguous or only moderately separated at their bases, nearly parallel endites, and a generally cursorial lifestyle. Excepting the genus *Hesperocranum,* members of the family Liocranidae lack precoxal triangles and have pseudosegmented tarsi, at least on their posterior legs. As a family, Liocranidae's diversity makes it difficult to define, although the genera within it are unusual enough that their identification can be relatively straightforward. This level of variation, coupled with a lack of strongly unifying features, predicts that future revisions will move some genera out of the family (Ubick and Richman, 2005b).

SIMILAR FAMILIES: Liocranidae is part of a complex of closely related Clubionidae-like families. The most likely candidates for misidentification are the ghost spiders (Anyphaenidae, p. 140) and members of the subfamily Phrurolithinae (Corinnidae, p. 146). While superficially similar in appearance, ghost spiders are generally more arboreal and have broad, lamelliform setae on their claw tufts. Separating liocranids from similar corinnids may require looking at their leg spination and examining their sterna for precoxal triangles. All of the genera in the subfamily Phrurolithinae have precoxal triangles, as does the liocranid genus *Hesperocranum*. However, this spider is easily identified by the brushes of bristles on its fore legs. Members of the subfamily Phrurolithinae also have four or more pairs of spines on the ventral surfaces of their anterior tibiae, and except for *Apostenus*, all of the genera in Liocranidae have fewer than four pairs of ventral tibial spines. Some liocranids might be mistaken for members of the genus *Syspira*, ground-dwelling members of the family Miturgidae (p. 143). *Syspira* are more common in deserts than liocranids and have claw tufts, a recurved eye row (rather than straight to slightly procurved as on liocranids), and elongate tips to their posterior lateral spinnerets.

PACIFIC COAST FAUNA: Four genera representing seven Pacific coast species. One of the most unusual of North America's liocranids is *Hesperocranum rothi* (pl. 29), the only known species within its genus. It is a small spider, ranging from 2.3 to 4.1 mm (.09 to .16 in.) in length, with a brownish-orange carapace and dark gray abdomen. It is distinguished from all other North American spiders by the two rows of thick bristles that line the ventral faces of the tibiae on legs I through III. This species has only been found in a narrow band of coniferous forest in the Sierra Nevada of Tulare County, California, through the Cascade Range of eastern Lane County, Oregon. At the northern end of its range, it has been found at altitudes of between 2,500 and 3,500 feet, while in the southern Sierra, it is much more common from 4,000 to 8,000 feet. *H. rothi* was described and its diagnostic structures were illustrated in Ubick and Platnick (1991).

Another exceptional liocranid is *Apostenus californicus* (pl. 29), the only representative of its genus in the Americas. It is a small spider measuring between 2.2 and 3.6 mm (.09 and .14 in.) in length. Unlike other liocranids, it has five pairs of spines on the underside of each of its anterior tibiae (the other genera have between one and three) and three pairs of spines on its metatarsi. Instead of claw tufts, *A. californicus* has a pair of long, broad, spatulate hairs that arise from below its tarsal claws. It has a brown carapace, a dark abdomen with several pale transverse bands, and light tan legs with numerous dark rings. Living predominately in oak duff between 5,000 and 7,000 feet in elevation, it has been found in the Coast Ranges and interior Transverse Ranges of Southern California from

San Diego County to San Bernardino County. An additional population is known from southwestern Kern County in the Los Padres National Forest. *A. californicus* was described and illustrated by Ubick and Vetter (2005).

Two species of *Neoanagraphis* (pl. 29) have been described, both of which live in arid portions of Southern California. They range from 3.2 to 9.1 mm (.13 to .36 in.) in length, and on their posterior legs they have highly unusual elongate, smooth-edged tarsal claws, armed with only a few teeth near their bases. In both species, the cephalothorax and legs are pale orange tan, and the abdomen is cream to tan. The anterior median eyes are dark, but the others are light. Despite the superficial similarity of the two species, Vetter (2001) found that *Neoanagraphis chamberlini* and *N. pearcei* can be differentiated by a number of easily seen features. *N. chamberlini* has only two pairs of spines on the ventral surfaces of its anterior tibiae, is relatively large, and prefers shad scale scrub flatlands. *N. pearcei* has three pairs of spines on the ventral faces of its anterior tibiae, is distinctly smaller, and is found in more montane regions with an abundance of creosote bush scrub. Additionally, *N. chamberlini* males are most active from mid-August through late September, while male *N. pearcei* look for mates from mid-September until late October. Their ranges overlap across a large swath of Southern California, from the deserts and lower Central Valley to Mono County.

Three species of *Agroeca* (pl. 29) live along the Pacific coast, two of which, *Agroeca ornata* and *A. trivittata,* are widespread but rare, with records from California and Washington. The third species, *A. pratensis,* has only been recorded regionally from central Washington. They are between 4.8 and 6.3 mm (.19 and .25 in.) in length and generally live in thick, moist leaf litter. They can have either two (as in *A. trivittata*) or three (as in *A. ornata* and *A. pratensis*) pairs of spines on the ventral surfaces of their anterior tibiae. *Agroeca* carapaces are orangish brown with dusky smudges and black marginal bands, while their abdomens have numerous black, gray, dull-yellow, and red splotches and stripes. Kaston (1938) revised the genus and illustrated the reproductive structures.

NATURAL HISTORY: Very little is known regarding the natural history of these spiders beyond their general habitat preferences. They are all terrestrial hunters living under rocks, in leaf litter, and occasionally in animal burrows. At least some European Agroeca suspend their dirt-covered egg sacs from silk stalks. Similar egg sacs have been found in North America, but with no attendant adult or juveniles reared to adulthood in captivity, their identification remains tentative (Dondale and Redner, 1982).

PRODIDOMIDAE

Pl. 29

IDENTIFICATION: Members of the family Prodidomidae are regionally very rare, small, two-clawed, entelegyne spiders with enlarged, flattened anterior lateral spinnerets. Their posterior eye rows are strongly procurved, giving them a horseshoe or circular appearance. Additionally, their posterior median eyes are triangular or oval rather than circular, and while their anterior median eyes are dark, the others are silvery gray.

SIMILAR FAMILIES: Several common spider families on the Pacific coast are similar to the prodidomids, although only Gnaphosidae (p. 156) shares their enlarged anterior lateral spinnerets. On gnaphosids, however, the spinnerets are cylindrical with short spigots, while on the prodidomids they're dorsoventrally compressed with long, fingerlike spigots. With the exception of the genus *Scopoides*, gnaphosid eyes are arranged in two well-defined rows, clearly different from the distinctly procurved eye rows of the prodidomids.

PACIFIC COAST FAUNA: Two genera, each represented by a single species. *Prodidomus rufus* (pl. 29) was first discovered in the southeastern United States (Hentz, 1847) but has since been found at numerous locations around the world and is apparently synanthropic in nature. Regionally, specimens have been collected from homes and under rocks in arid areas across Southern California (Vetter, 1996; Platnick and Baehr, 2006). *P. rufus* is easily recognized by its large, diverging chelicerae (less divergent on the male) and strongly procurved posterior eye row. Its carapace is pale yellow to orange, and the abdomen varies from grayish pink to cinnamon red with numerous short, white setae. The adult ranges from 3 to 5.5 mm (.19 to .22 in.) in length.

Neozimiris pubescens (pl. 29) is quite rare in the United States, having only been found in the desert regions of Riverside County and San Diego County, California. *N. pubescens* is generally smaller than *P. rufus,* with the adult measuring between 1.7 and 4 mm (.07 to .16 in.) in length. Its chelicerae are traditionally aligned, its carapace is light orange to yellow with dark lateral margins, and its abdomen is pale gray. The posterior eye row of *N. pubescens* is extremely procurved, creating a nearly circular pattern. Prodidomids were formerly placed in the family Gnaphosidae, and during this time, Platnick and Shadab (1976b) revised the world's *Neozimiris* species.

NATURAL HISTORY: Almost nothing is known regarding the natural history of either of these species. *P. rufus* has been collected in North and South America, Asia, and the Caribbean, although its native range remains to be identified. According to the very limited collection records available in Platnick and Baehr (2006), mature *N. pubescens* are present in Southern California's deserts in winter and early spring.

GNAPHOSIDAE — Stealthy Ground Spiders

Pls. 30–32, Figs. 12, 13

IDENTIFICATION: Stealthy ground spiders are predominately nocturnal, cursorial spiders. They are generally recognizable by the shape and size of their anterior lateral spinnerets, which are widely separated, enlarged, and cylindrical and are often clearly visible when the spider is viewed from above. Additional unifying features include their concave endites with distinct depressions across their ventral surfaces and their posterior median eyes that are often oval, triangular, or elliptical in shape. Like most running spiders, stealthy ground spiders have two claws and claw tufts. They have eight eyes, are entelegyne, and are common across the Pacific coast region. While most genera are somberly colored, a few are strikingly patterned with white stripes, spots, and iridescent scales. This is especially true in *Micaria,* a genus whose anterior lateral spinnerets are fairly small and close together. However, the placement of these ant-mimicking spiders within Gnaphosidae is supported by their eye arrangement and endite structure.

SIMILAR FAMILIES: The family Prodidomidae (p. 155) also has elongated, separated, anterior lateral spinnerets; however, they are flattened and have a multitude of long spigots. Prodidomids also have strongly procurved posterior eye rows, a feature that among the gnaphosids is only found on the genus *Scopoides*. The gnaphosid genus *Micaria* can be confused with ant-mimicking members of the family Corinnidae (p. 146), but their anterior lateral spinnerets show some basal separation, while on the corinnids they are contiguous across their bases. *Micaria* also have iridescent scales, modified posterior median eyes, and concave endites, features consistent with Gnaphosidae.

PACIFIC COAST FAUNA: Twenty-one genera containing 97 regional species. While stealthy ground spiders are relatively easy to recognize at the familial level, their identification to genus and species is often much more difficult. Many genera are somberly colored and lack conspicuous markings. Their identification normally requires a close examination of the spider's leg spination, eye arrangement, dentition, and reproductive structures as described in Ubick (2005b). Murphy (2007) made a major contribution toward our understanding of Gnaphosidae diversity by breaking the world's genera into groups based on a few fairly consistent features, an outline of which is generally followed in this text.

Before delving into the more cryptic genera, it is worth examining the more distinctive members of the family. *Micaria* (pl. 30) is superficially the least like the other stealthy ground spider genera. *Micaria* are small (1.3 to 6.5 mm, or .05 to .26 in., in length), generalized ant mimics with thin, tapering legs and a dusting of iridescent scales on their abdomens. Their ant-like appearance is further enhanced by their morphology and coloration. They vary from brick red to black, often with contrasting abdominal bars or spots, and in some species the abdomen is slightly constricted. Unlike other gnaphosids, *Micaria* are active during the day in dry, rocky, or sandy areas, frequently near the ants they resemble. They can also hold their front legs over their heads like antennae, adding another level to their mimicry. Twenty-one species have been reported from the Pacific coast, and they are common throughout the region. A key to the American *Micaria* can be found in Platnick and Shadab (1988).

According to Murphy (2007), the *Herpyllus* group contains three regional genera: *Cesonia, Sergiolus,* and *Herpyllus.* Many of the species in the *Herpyllus* group possess black and white patterned abdomens, and the males are adorned with abdominal scuta. While an examination of the spider's reproductive structures and eyes is sometimes needed to make a positive identification, quite a few species are boldly patterned and easily recognized. Spiders in the genus *Cesonia* (pl. 30, fig. 12) are especially distinctive, bearing between two and four black and white longitudinal bands across the abdomen and carapace. In the other genera, markings are limited to the abdomen, are less ornate, or are lacking altogether. *Cesonia* also differ from other members of the *Herpyllus* group in that their posterior median eyes are closer to their lateral eyes than they are to each other. All five of the Pacific coast's *Cesonia* are regionally limited to small portions of Southern and Central California. Each species also has a diagnostic color pattern, making its identification relatively straightforward. One of the more common species in California, *Cesonia josephus* (pl. 30), has a design very similar to the larger and much more widespread Western Parson Spider (*Herpyllus propinquus,* pl. 30). However, unlike the Western Parson Spider, *C. josephus* has white stripes along the sides of its abdomen. *Cesonia* range from 3.2 to 7.5 mm (.13 to .3 in.) in length and are normally found under rocks and leaf litter on sandy soils. They are fast-moving hunters that apparently prefer other spiders as prey. A summary of the genus and a key to its species can be found in Platnick and Shadab (1980b).

While the stripes on *Cesonia* are generally longitudinal, *Sergiolus* (pl. 30, fig. 13) are usually decorated with a number of pale, transverse abdominal bars. They can be either broken or complete and often stand in striking contrast to the abdomen's otherwise dark coloration. Five species of *Sergiolus* are known from the Pacific coast states. Three of these, *Sergiolus montanus* (pl. 30), *S. columbianus* (pl. 30), and *S. angustus* (see fig. 13), show a surprising degree of intraspecific variation in their abdominal patterns. In each species, there is a general trend for individuals to be darker at the northern portion of the group's range and lighter toward the south, but atypical individuals can show up anywhere. *S. columbianus* and *S. montanus* are both widespread and fairly common across the contiguous Pacific coast states, with records from a wide variety of habitats, including coastal and montane forests, chaparral, and grasslands. *S. angustus* is a California species whose abdominal bars are broken in the middle but conjoined laterally such that they

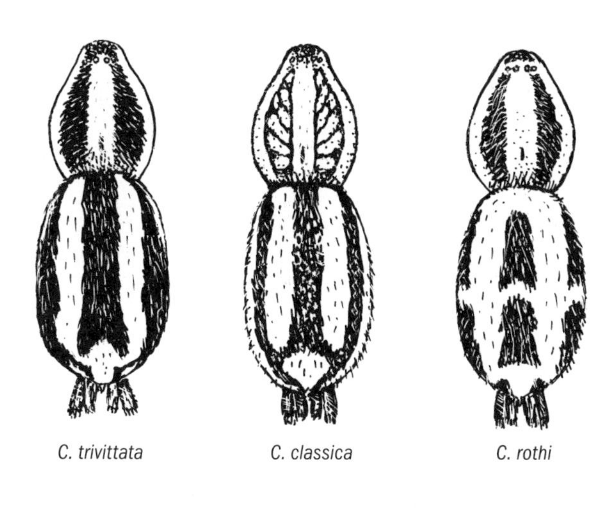

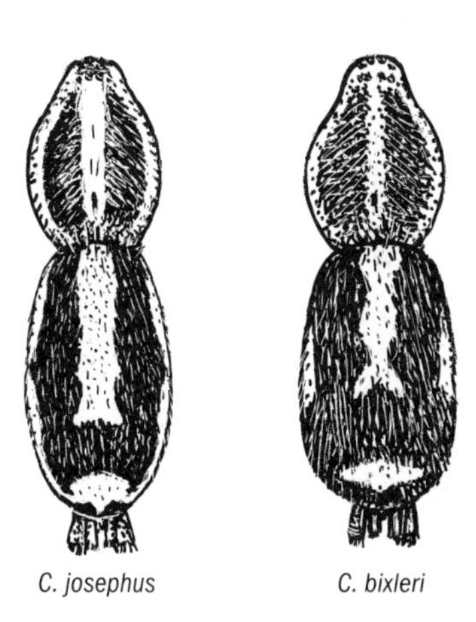

Figure 12. Carapaces and abdomens of five Pacific coast species of *Cesonia* (after Platnick and Shadab, 1980b, in part).

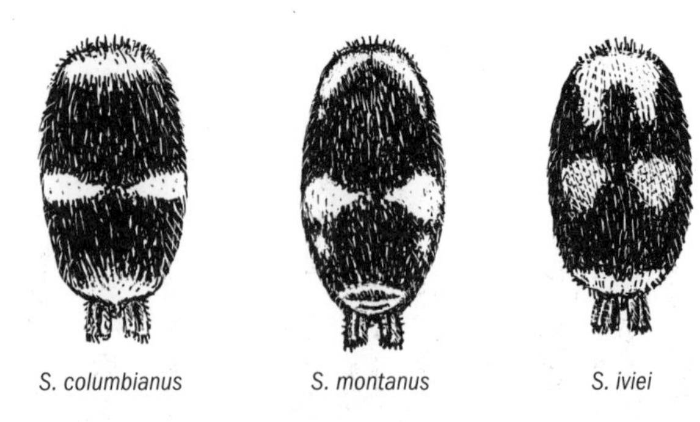

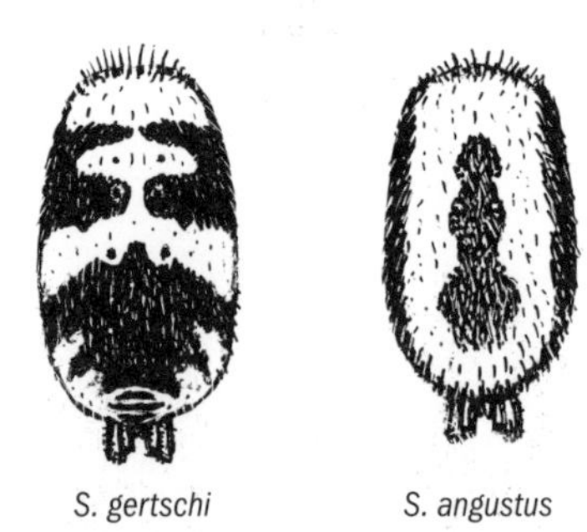

Figure 13. Representative abdominal patterns of five Pacific coast species of *Sergiolus* (after Platnick and Shadab, 1981, in part).

resemble a pair of pale longitudinal stripes divided by a dark central marking. *S. gertschi* (see fig. 13) is found in Southern California's coastal and desert lowlands, while *S. iviei* (see fig. 13) is a Great Basin species with at least one record from southeastern Oregon. An illustrated key to the genus *Sergiolus* can be found in Platnick and Shadab (1981).

Herpyllus (pl. 30) is unusual among this group of stealthy ground spider genera in that only one of the six regional species has well-defined abdominal markings. The Western Parson Spider (*Herpyllus propinquus,* pl. 30) is found throughout California and western Oregon with just a few scattered records from Washington. It is a distinctive spider with a dark body, a broad pale line through the cephalothorax, and a white, scalloped stripe along the anterior two-thirds of its abdominal midline. Often this stripe is broken, leaving an isolated white spot above the spinnerets. There are numerous records of people responding to *Herpyllus* bites with excessive tenderness and swelling, and while most records refer to the Eastern Parson Spider, *H. ecclesiasticus,* at least one is clearly attributable to the Western Parson Spider (R. Vetter, pers. comm., 2010). The five remaining regional *Herpyllus* species lack diagnostic markings and vary in color from tan to dark grayish brown. *Herpyllus* males have scuta on the dorsoanterior portion of their abdomens and range from 5 to 12 mm (.2 to .47 in.) in length. Among the unpatterned species, *H. hesperolus* (pl. 30) lives widely across the Pacific coast, but the remaining species are regionally limited to Southern and Central California. Their identification is very difficult and nearly always requires a detailed examination as described in Platnick and Shadab (1977).

Extremely similar to the unmarked *Herpyllus* are the spiders in the *Echemus* group. They lack conspicuous patterns and are generally colored in shades of gray, reddish, or tawny brown. Males have dorsal abdominal scuta, and for this reason, differentiating unmarked *Herpyllus* from genera in the *Echemus* group typically requires looking at the spider's reproductive structures. Five genera are found along the Pacific coast, one of which, *Scotophaeus,* is represented by a single, introduced European species, the Mouse Spider (*Scotophaeus blackwalli,* pl. 31). Exceptionally common in coastal communities, it is frequently found in and around homes from Southern California to northern Washington. Its common name comes from the short, gray pubescence that covers its abdomen, giving the spider a soft "mousy" look. Its cephalothorax and legs are reddish brown, and it ranges from 6.3 to 10 mm (.25 to .39 in.) in length. This species is discussed in Platnick and Shadab (1977).

Also in the *Echemus* group is the genus *Scopoides* (pl. 31). With a strongly procurved posterior eye row and greatly enlarged anterior lateral spinnerets, it is among the most readily identifiable of North America's stealthy ground spider genera. Four species live along the Pacific coast, all of which are regionally limited to Southern California's coastal scrub and desert regions. They range from 4 to 8.4 mm (.16 to .33 in.) in length and have orangish-brown cephalothoraxes and legs and brownish-gray abdomens with reduced abdominal scuta on the males. A key to this genus can be found in Platnick and Shadab (1976c). The three remaining *Echemus* genera, *Sosticus* (not illustrated), *Nodocion* (pl. 31), and *Synaphosus* (pl. 31), lack conspicuous distinguishing features, and their identification requires a microscopic examination of their legs, reproductive structures, and dentition. Each species typically has a pale to dark gray abdomen and a brownish-orange cephalothorax and legs. *Sosticus californicus* is the sole West Coast representative of its genus. The few known specimens are between 5 and 5.4 mm (.2 and .21 in.) in length and are only known from a small area of Madera County, California. It is described and its reproductive structures are illustrated in Platnick and Shadab (1976c). Five species of *Nodocion* (pl. 31) live along the West Coast, most of which are regionally limited to Southern California. Only two species, *Nodocion voluntarius* and *N. mateonus,* have been found as far north as central Washington. They range from 3.7 to 9.8 mm (.15 to .39 in.) in length and have shallow indentations on the ventrodistal margins of their trochanters. *Synaphosus syntheticus* (pl. 31) is a small, 1.8 to 2.9 mm (.07 to .11 in.) in length, synanthropic species that was introduced to Southern California from the Middle East and is the only representative of this Old World genus in western North America. *Synaphosus* have smooth edges to the ventrodistal margins of their trochanters and an unusual cluster of setae at the distal ends of their third metatarsi. Known as a preening brush, this cluster of setae is missing from *Nodocion*. Both genera are revised in Platnick and Shadab (1980a).

Genera in the *Haplodrassus* group are very similar to those in the *Echemus* group, but the males lack abdominal scuta. Their cephalothoraxes and legs are orangish to dark brown, and their abdomens are generally gray, occasionally with darker chevrons along the midline. The features that distinguish the different genera in the *Haplodrassus* group are subtle, and a detailed examination is often required to make an accurate identification. Spiders in the genus *Haplodrassus* (pl. 31) have one easily seen diagnostic feature: their oval-shaped posterior median eyes are enlarged and very close together. This feature is shared with members of the widespread genus *Drassyllus* (pl. 31), part of the *Zelotes* group; however, *Drassyllus* have preening combs on the ventrodistal edges of metatarsi III and IV.

Six species of *Haplodrassus* live on the Pacific coast in a wide variety of habitats, from chaparral and high deserts to montane forests. They range from 3.6 to 8.6 mm (.14 to .34 in.) in length, and while all are known from California, only one, *Haplodrassus signifer,* can truly be considered widespread in Oregon and Washington. Very similar to *Haplodrassus* are the genera *Orodrassus* (pl. 31) and *Parasyrisca* (not illustrated). In these genera, the posterior median eyes are smaller and separated by at least the eye's diameter. Distinguishing *Orodrassus* from *Parasyrisca* requires looking at the spider's leg spination and reproductive structures. All three of North America's *Orodrassus* species live along the Pacific coast. *Orodrassus assimilis* is widespread from eastern California north through central and eastern Oregon, *O. coloradensis* is thinly distributed across the contiguous Pacific coast states, and *O. canadensis* is regionally found almost exclusively in northwestern Washington. They range from 6 to 12 mm (.24 to .47 in.) in length. Very similar in size and color is *Parasyrisca orites,* the only New World member of its genus. Along the Pacific coast it is known only from a small area of northwestern Washington, and it ranges from 6 to 7.8 mm (.24 to .31 in.) in length. *Haplodrassus, Orodrassus,* and *Parasyrisca* are revised in Platnick and Shadab (1975b), although *P. orites* is listed under its earlier placement within *Orodrassus.*

With a powerful hand lens, spiders in the *Zelotes* group can be recognized by the presence of preening combs on the distoventral edges of metatarsi III and IV. The preening comb should not be confused with the more clustered preening brush found on members of the genus *Synaphosus.* Four genera from this group are found along the Pacific coast: *Zelotes, Drassyllus, Trachyzelotes,* and *Urozelotes.* The most common of these, *Zelotes* (pl. 31), has 35 described regional species, the vast majority of which are endemic to California. They are generally small (2.3 to 10.9 mm, or .91 to .43 in., in length), blackish spiders that can usually be separated from similar genera by the structure and position of their posterior median eyes. In most gnaphosids, the posterior median eyes are roughly oval or triangular in shape and differ in size from the posterior laterals. In *Zelotes,* the posterior median eyes are almost always round and nearly the same size as the posterior laterals.

Additionally, the four posterior eyes are equidistant from one another. *Zelotes* are common in a wide variety of habitats, from coastal beaches to montane forests. Descriptions and keys to the region's *Zelotes* can be found in Platnick and Shadab (1983), Platnick and Prentice (1999), and Platnick and Dondale (1992).

Similar to *Zelotes* is the genus *Drassyllus* (pl. 31) with 14 regional species. *Drassyllus* vary from dark greenish brown with a gray abdomen to nearly black, and they range from 1.9 to 6.9 mm (.07 to .27 in.) in length. Unlike in *Zelotes,* the posterior median eyes in *Drassyllus* are normally oval in shape, larger than the posterior laterals, and quite close together. *Drassyllus* are found in a wide variety of habitats from Southern California to northern Washington. One species in particular, *Drassyllus insularis,* can be especially common around homes throughout the region. Descriptions of the Pacific coast's *Drassyllus* can be found in Platnick and Shadab (1982) and Platnick (1984).

Urozelotes rusticus (pl. 31) is the only representative of its genus in North America. It is a highly synanthropic species that has been transported around the world. It has an eye arrangement much like that of *Drassyllus,* but its body is paler with an orangish-brown carapace and a tan abdomen. It ranges from 5.7 to 7.4 mm (.22 to .29 in.) in length and has been found near buildings throughout Southern and Central California, with at least one record from southern Oregon. Its complicated taxonomic history is reviewed and its diagnostic reproductive structures illustrated in Platnick and Murphy (1984).

The most easily recognized members of the *Zelotes* group are members of the genus *Trachyzelotes* (pl. 31). They have a thick, easily seen coating of stiff setae on the front of their chelicerae. Native to the Mediterranean area, *Trachyzelotes* have been introduced widely around the world, with three species known from the Pacific coast region, all of which have been found in California and at least two recorded from Oregon. However, these spiders are synanthropic and easily transported and could show up nearly anywhere. Established species range from 4.3 to 9.4 mm (.17 to .37 in.) in length and are generally grayish to black with a reddish-brown carapace and legs. The genus was revised and its species illustrated in Platnick and Murphy (1984).

Drassodes (pl. 32) is the only genus in its group in North America and can be easily recognized by the presence of a distinctive notch on the distoventral edge of each trochanter. The three regional species range from 5.4 to 10.2 mm (.21 to .4 in.) in length, are fairly elongate, and tend to be buffy gray in color, often with small bars or dots along the abdominal midline. *Drassodes* are found in a wide variety of habitats from Southern California to northern Washington. An illustrated key to the species can be found in Platnick and Shadab (1976a).

Talanites (not illustrated) are members of the *Anagraphis* group and are an exceptionally rare part of California's spider fauna. Two species are known from the state: *Talanites ubicki* from Marin County's serpentine grasslands, and *T. moodyae* from mountainous areas of Fresno County and Tulare County. Both are unusual in that their eyes are greatly reduced in size. The overall morphology and coloration of these species have never been described, but other *Talanites* have pinkish-brown carapaces and legs and light-gray abdomens with dark chevrons along the midline. The few collected specimens range from 4.6 to 10.4 mm (.18 to .41 in.) in length. The region's species are described in Platnick and Ovtsharenko (1991) and Ubick and Moody (1995).

The remaining three genera are part of two different Gnaphosidae groups and are defined by features of their cheliceral retromargins, structures that are nearly impossible to see except when examined under a microscope. *Gnaphosa* (pl. 32) has 11 regional species spread throughout the contiguous Pacific coast states, from intertidal marshes and desert scrub to mountain meadows. They range from 3 to 11.5 mm (.12 to .45 in.) in length and vary considerably in their overall appearance. Males have small, shiny scuta on their abdomens' dorsoanterior margins. Despite their dissimilarity, *Gnaphosa* are united by unusual serrated keels on their cheliceral retromargins. A key to the American *Gnaphosa* can be found in Platnick and Shadab (1975a).

The two remaining genera, *Callilepis* and *Eilica,* are part of the *Laronius* group. They possess one or more rounded, translucent lobes on their cheliceral retromargins. Three species of *Callilepis* (pl. 32) live along the Pacific coast, from Southern California to northern Washington. This is an uncommon genus whose members are most prevalent in dry rocky or sandy areas. The distinguishing features of this genus include a single lobe on the cheliceral retromargin, a broad sternum behind short endites, and slightly flattened posterior median eyes. *Callilepis* range in color from orange to dark brown, although some species have conspicuous silvery, scalelike setae, a feature that is especially conspicuous on the California species *Callilepis gosoga* (pl. 32). Other regional species include *C. eremella,* which has been found from Southern California through central Washington, and *C. pluto* (pl. 32), which is known from widely scattered locations throughout Oregon and Washington. The Pacific coast's *Callilepis* range from 2.7 to 6.2 mm (.11 to .24 in.) in size and are reviewed in Platnick (1975a).

Eilica bicolor (pl. 32) is the only representative of this unusual genus in North America. Regionally known only from Riverside County, California, *Eilica* are closely associated with ants, both preying on them and, on occasion, living within their colonies. *Eilica* have two to three translucent lobes on their cheliceral retromargins, narrow sterna, elongate endites, and rectangular posterior median eyes. *Eilica bicolor* ranges from 2.3 to 6.3 mm (.09 to .25 in.) in length and has an unpatterned brown abdomen and an orangish-brown carapace and legs. A revision of the genus can be found in Platnick (1975a).

NATURAL HISTORY: Most stealthy ground spiders are nocturnal wandering hunters resting by day beneath rocks, loose tree bark, and other debris. The exceptions are the ant-mimicking *Micaria* and members of the genera *Callilepis* and *Eilica* that feed on ants. Gnaphosids are especially common in rocky, open areas, generally becoming less numerous in closed-canopy, humid forests. Several species, including *Scotophaeus blackwalli* and *Urozelotes rusticus,* are highly synanthropic and are regularly found in homes. Because of the generally secretive nature of most stealthy ground spiders, many aspects of their natural history remain to be discovered. Males are known to follow pheromone-laced draglines to find females and, in some instances, will sequester themselves next to a nearly mature female's retreat in order to be the first to mate with her. Gnaphosid egg sacs are either deposited within the females' retreats or attached to the undersides of rocks or other objects, where the females may guard them or abandon them, depending on the species. In some genera the sacs look like fried eggs, while in others they are camouflaged with bits of dirt and vegetation.

SALTICIDAE — Jumping Spiders

Pls. 33–37

IDENTIFICATION: With their enlarged anterior median eyes, powerful fore legs, and stocky, often ornately decorated bodies, salticids are among the most easily recognized spiders in North America. Jumping spiders are wandering entelegynes and are common in nearly every habitat, from coastal dunes and deserts to montane forests and suburban gardens.

SIMILAR FAMILIES: With their unusual morphology and characteristic hunting behaviors, jumping spiders are unlikely to be confused with any other spider.

PACIFIC COAST FAUNA: Thirty-three genera represented by 119 described and numerous undescribed species. Identifying jumping spiders can be one of the most difficult and most gratifying of arachnological challenges. While the defining features of each genus often involve subtle differences in the spiders' leg lengths, spination, dentition, and reproductive features, their diurnal habits and bright colors make the field identification of many species relatively straightforward. In some ways, the observation of jumping spiders is more akin to bird watching than traditional collection-based arachnology. As with birds, the males are often gaudily colored while the females generally exhibit more cryptic coloration. Unfortunately, the living colors of many species have never been described because even just a few hours in preservative fluid permanently changes their appearance. For this reason, in much of the scientific literature only the reproductive structures are illustrated. With advances in digital photography, people are increasingly able to photograph the living spiders, collect them, and then match them to the descriptions of preserved individuals. This has dramatically improved our ability to field-identify many of North America's jumping spiders, although it is still often necessary to use keys such as Richman et al. (2005) to identify unfamiliar individuals.

Among the most commonly encountered salticids are those in the genus *Phidippus* (pl. 33). It contains the largest and some of the most colorful jumping spiders in North America. Nineteen species live along the Pacific coast, ranging in length from 3.3 to 18.9 mm (.13 to .74 in.), although most are between 8 and 14 mm (.31 and .55 in.) long. Many are boldly marked with red or orange abdomens while others have more-muted patterns. They are stout bodied with carapaces that are curved along their sides. *Phidippus* females have tufts of setae behind their posterior median eyes, and nearly all species have bright, iridescent chelicerae. *Phidippus* reaches its greatest regional diversity in California, although at least half a dozen species occur in Washington, including the ubiquitous Johnson Jumper (*Phidippus johnsoni,* pl. 33). This is one of the most common garden-dwelling spiders along the Pacific coast and is found throughout the region. Like many spiders, *Phidippus* often have fairly specific habitat requirements, with some species found almost solely on shrubs while others live in grasslands. One of the region's most specialized species is *P. aureus.* Endemic to Southern California, this yellow spider is found almost entirely on creosote bushes. An illustrated guide and key to the North American species can be found in Edwards (2004).

There are numerous morphologically similar genera that are closely related to *Phidippus,* including *Eris, Dendryphantes, Terralonus, Phanias, Pelegrina,* and *Metaphidippus.* Differentiating individuals of these genera often requires closely examining their eye arrangements, relative leg lengths, and reproductive structures. Unfortunately, several of these genera are poorly defined and are in need of comprehensive revisions.

Among the most common spiders in North America is the Bronze Jumper (*Eris militaris,* pl. 34), the only member of its genus whose presence has been confirmed along the Pacific coast. Its common name comes from the coating of brassy setae across the male's body that gives it a bronzy hue. Like many spiders in this collection of genera, the male *E. militaris* is brown with broad creamy stripes along the front and sides of his carapace and around the anterior and lateral margins of his abdomen. The female is gray with a series of black and white dashes on either side of her abdomen. The Bronze Jumper is between 6 and 8 mm (.24 and .31 in.) long and is frequently found in bushes and other foliage from Central California through Washington. *Eris* and *Phidippus* can be differentiated from similar genera by the position of their posterior median eyes. In both genera they are closer to the anterior lateral eyes than they are to the posterior lateral eyes. On related genera, the posterior median eyes are either directly between or closer to the posterior laterals. *E. militaris* is described in detail by Maddison (1986).

Terralonus (pl. 36) are unusual in that both the males and females are grayish brown and mottled. Three species are found along the Pacific coast, with scattered records from Southern California through northern Washington. They are ground-dwelling jumping spiders found in barren areas under rocks and debris, and *Terralonus californicus* (pl. 36) is one of the few spiders that lives mainly on beaches. *Terralonus* are generally 5 to 6.5 mm (.2 to .26 in.) in length, and while the genus was defined by Maddison (1996), it has not received an in-depth revision.

Phanias (pl. 34) is another poorly known genus whose members are generally more elongate and yellower than related salticids. There is also a great deal of variation, and identifying a suspected *Phanias* regularly requires looking at the fine details of its reproductive structures. *Phanias* are foliage-dwelling jumping spiders whose center of diversity is western North America and northern Mexico. Six species have been recorded along the Pacific coast, all of which are found in California. Only one species, *Phanias albeolus,* can be considered regionally widespread, with records extending into Washington. *Phanias* are normally between 4.5 and 6.5 mm (.18 and .26 in.) in length and are discussed in greater detail in Maddison (1996).

Fairly common along the Pacific coast are the foliage-dwelling genera *Metaphidippus* (pl. 34) and *Pelegrina* (pl. 34). Until Maddison (1996), the spiders housed in *Pelegrina* were included in *Metaphidippus,* and separating these two genera can be difficult. Most *Metaphidippus* males have distinct patches of white or pale orange setae on their chelicerae, and the forehead band does not connect the anterior median eyes. *Pelegrina* males on the Pacific coast lack distinct cheliceral patches and, with the exception of *Pelegrina aeneola* (pl. 34), have a white stripe connecting the anterior median eyes. Females are especially difficult to separate. They are clad in various shades of gray, brown, and yellow mottling, and you often need to closely examine the epigyna to distinguish between these two genera. *Metaphidippus* range from 1.6 to 6.2 mm (.06 to .24 in.) in length. Six species are found along the West Coast. *Metaphidippus mannii* (pl. 34) is common from Southern California through northern Washington, while the rest are regionally limited to California. *M. diplacis* is unusual in that it appears limited to Southern California beachfront vegetation. *Pelegrina* are similar in size to *Metaphidippus* but are much more prevalent north of California. Six species are found along the Pacific coast, one of which, *P. flavipedes,* is regionally limited to Washington. *Pelegrina* are found on a wide range of plants, from conifers and oaks to weedy fields. An illustrated key to North America's *Pelegrina* and *Metaphidippus* can be found in Maddison (1996).

With only a few scattered records from Washington, *Dendryphantes nigromaculatus* (not illustrated) is the only American representative of this otherwise Old World genus. It is a dully marked species with a brownish carapace, and on the male, a pale stripe wraps around the front and sides of his reddish-yellow abdomen. Its taxonomic situation is discussed and its reproductive structures are illustrated in Maddison (1996).

The most species-rich genus of jumping spiders in North America is *Habronattus* (pl. 35), with 30 described and several undescribed species from the Pacific coast region. *Habronattus* are predominately ground-dwelling spiders that live in nearly every environment, from coastal dunes and desert washes to suburban lawns and mountain meadows. A great majority of the region's *Habronattus* are found in California, although around a dozen species have been noted north into Washington. They are generally between 5 and 6 mm (.2 and .24 in.) long, and unlike most jumping spiders their third leg is longer than their fourth, a condition shared with only a few other genera, including the closely related genus *Pellenes.* Exhibiting an amazing diversity in decoration and color, *Habronattus* males often have brightly ornamented faces and fore legs, as well as elaborate sculptural ornamentation on their carapaces, legs, and chelicerae. These features are used during their complex courtship displays. Females are more subdued in color, and their identification often requires a close examination of their epigyna as described in Richman et al. (2005). Griswold (1987) revised *Habronattus* and provides a key to its many species.

Very similar to *Habronattus* are members of the genus *Pellenes* (not illustrated). The third leg is longer than the fourth in this genus also, but *Pellenes* tend to be duller in color and lack the ornate structures that characterize male *Habronattus.* The most noticeable feature on many *Pellenes* individuals is a pale chevron-studded stripe up the dorsal abdominal midline. *Pellenes* are ground-dwelling spiders around 5 mm (.2 in.) long. They reach their greatest diversity in the more northerly climes, and several species live among the region's high-elevation mountain chains. Five species are found along the Pacific coast, ranging from Southern California to northern Washington. Lowrie and Gertsch (1955) provide a brief overview of the North American *Pellenes*; however, many species are poorly described and a comprehensive revision of the genus is needed.

The genera *Salticus* (pls. 34, 36) and *Sassacus* (pl. 34) include several small spiders, 3.5 to 5.5 mm (.14 to .22 in.) in length, that are decorated with bold patterns or iridescent scales. Three described and several undescribed species of *Salticus* live along the Pacific coast, the most common of which is the Zebra Spider (*Salticus scenicus*, pl. 36). This presumably introduced species is found from Southern California to northern Washington and is easily recognized by the row of angled bars on either side of its otherwise dark abdomen and by the limited iridescence on its carapace. California's *S. palpalis* (pl. 36) is similar, but it has broad bands across its abdomen, and its carapace is covered in shiny green scales. *S. peckhamae* (pl. 34) is unusual in that it lacks abdominal bars and is entirely covered in green, purple, and pink iridescent scales. It has been found from Northern California through central Washington. *Salticus* has not been the subject of a modern revision, and for most species, the descriptions in Peckham and Peckham (1909) remain the most comprehensive.

Similar to *Salticus* is the genus *Sassacus* (pl. 34), which is made up of small, beetle-like jumping spiders heavily coated in iridescent scales. They are thought to be mimics of the abundant and foul-tasting leaf beetles (Chrysomelidae). The region's three *Sassacus* species are found predominately on foliage, from shrubs to agricultural fields. *Sassacus papenhoei* (pl. 34) is common and widespread across the contiguous Pacific coast states, while *S. vitis* (pl. 34) is found in Southern and Central California and in the mountains of eastern Oregon and Washington. *S. paiutus* (not illustrated) is regionally limited to southeastern California and is entirely covered with silvery-gold scales, giving it a polished, jewelry-like appearance. *Sassacus* was revised and its species illustrated in Richman (2008).

Another genus of mostly iridescent salticids, *Tutelina* (pl. 37), has been found widely, but sporadically, along the Pacific coast. *Tutelina* are foliage-loving spiders, ranging from 4.3 to 7 mm (.17 to .28 in.) in length, with longitudinal stripes on their legs. The most widespread of the three regional species, *Tutelina similis,* has been found in small numbers across both California and Washington. The male is normally pale gray-green and has a distinctive tuft of setae above his eyes. The female lacks setal tufts, is darker, and varies from metallic green to deep reddish purple. The two remaining species, *T. elegans* and *T. hartii,* have historically been reported from Washington but lack modern records. With their long, thin legs, *Tutelina* closely resemble ants, which are believed to be their main prey item. The region's *Tutelina* are discussed (under the name *Icius*) in Peckham and Peckham (1909).

Several genera have noticeably elongate bodies, including *Marpissa, Paramarpissa,* and *Hentzia,* each of which is regionally represented by a single California species. On males, the first pair of legs are exceptionally long while the remaining legs are unremarkable. *Hentzia palmarum* (not illustrated) has been reported sporadically across California. It is widespread in the eastern United States and may have been introduced to the Pacific coast on imported plants, but it is unknown whether the scattered records represent lone individuals or small but established populations. The male is brownish bronze with a broad white band that circles around the front and sides of his carapace and abdomen. He also has long, dark reddish-brown front legs and conspicuous, forward-pointing chelicerae. The female lacks the especially long fore legs and jaws and is not as dramatically patterned. *H. palmarum* ranges from 4 to 6.1 mm (.16 to .24 in.) in length. The genus was revised (although the spider's presence in California wasn't recorded at the time) by Richman (1989).

Marpissa robusta (pl. 36) lives on foliage from Southern California through the San Francisco Bay Area. It is a fairly large jumping spider around 8 mm (.31 in.) in length. The male is dark brown dorsally with scattered orange or whitish patches on his carapace. While his front legs are long and dark, his other legs are short and pale, and the sides of his abdomen are yellowish white. The female is pale gray with a pair of orange stripes running down the sides of her abdomen. The genus *Marpissa* was revised in Barnes (1958).

Paramarpissa griswoldi (pl. 36) is easily recognized by its massive front legs and, in males, by its striking color patterns.

Its enormous fore legs may be related to the fact that it lacks spines on the ventral side of tibia I. The male has a broad, dark-bordered, whitish band running down the midline of his body. The female is similar in size and shape but is generally mottled gray and lacks the pale longitudinal stripe. *P. griswoldi* is around 6 mm (.24 in.) in length and has been found in Central and Northern California. A key to *Paramarpissa* can be found in Logunov and Cutler (1999).

Two closely related regional jumping spider genera, *Platycryptus* and *Metacyrba,* are normally found in crevices under tree bark and stones. *Metacyrba taeniola* (pl. 36), the region's only *Metacyrba* species, is a dark brown to black spider with enlarged fore legs and two thin white lines (occasionally broken) on either side of its abdominal midline. It ranges from 3.7 to 7.3 mm (.15 to .29 in.) in length and has been found widely throughout Southern and Central California, with at least one record from central Washington. *Metacyrba* was revised in Edwards (2006).

Platycryptus (pl. 36) are fairly elongate, grayish-brown spiders whose foliate and chevron-decorated abdominal patterns camouflage them against their preferred habitat, tree bark and old wooden fences. Two species live along the Pacific coast: the widespread *Platycryptus californicus* (pl. 36), which is found from Southern California through northern Washington, and *P. arizonensis,* which is regionally limited to eastern California's deserts, with records from both Mono County and San Bernardino County. *Platycryptus* range from 6.2 to 9.2 mm (.24 to .36 in.) in length, and unlike *Metacyrba,* they are heavily coated in setae. *Platycryptus* was reviewed in Barnes (1958) under its old placement in *Metacyrba,* and additional informative notes can be found in Edwards (2006).

Marchena minuta (pl. 34) is a small (4.1 to 5.1 mm, or .16 to .2 in., in length) jumping spider that lives in woodlands, especially conifer forests, from Central California to northern Washington. It is found on tree bark, and while the female is grayish brown with white and black spots on her abdomen, the male is strikingly patterned with three longitudinal white lines running across his otherwise dark-brown body, the middle stripe ending near the anterior edge of his abdomen. *Marchena* also possess distinctive rows of setae-tipped tubercles on their first femora and on their carapaces. Rubbed across one another, the tubercles create a rasping sound that is almost certainly used for intraspecific communication. A detailed description and illustrations of *M. minuta* can be found in Maddison (1987).

Abundant across the eastern United States, *Phlegra hentzi* (pl. 37) has been introduced to Southern California, with records from both Ventura and Los Angeles Counties, and may be more widespread than is currently appreciated. The sexes are nearly identical, with two thick white longitudinal bands across the top of the carapace and three on the abdomen's predominately dark background. The male's clypeus is turquoise blue, while the female's is rusty brown. *P. hentzi* is between 5 and 7 mm (.2 and .28 in.) long and is described in detail in Logunov and Koponen (2002).

Introduced to North America, two species of *Menemerus* live along the Pacific coast. The Gray Wall Jumper, *Menemerus bivittatus* (pl. 36), is an Old World species that is widespread throughout Southern and Central California. The male is gray with a wide black stripe along the top of his abdomen that becomes a broad Y-shaped mark on his carapace. The female is also gray but with a distinctive thin black border around her carapace and abdomen. The female also has another white band along the edge of her carapace. The second species, *M. semilimbatus* (pl. 36), is native to the Mediterranean region and was recently discovered living in California's northern Central Valley, and the surrounding foothills. The male's carapace is dark brown with a light spot and wide, white bands along its lateral margins. He has black palps with contrasting white femora and an abdomen that is mottled brown. The female is browner than *M. bivittatus* and lacks a black abdominal border but does also have a white band around the margin of her carapace. In both species, the females' palps are covered with long, white hairs. When hunting, the female *M. semilimbatus* often fans her palps in front of her face, hiding her dark clypeus and chelicerae (T. Manolis, pers. obs., 2009). Both *Menemerus* species are common around homes and other buildings. They are generally between 5.5 and 10 mm (.22 and .39 in.) in length. Their reproductive structures are illustrated in Barnes (1958).

Another recently introduced species is the Pantropical Jumper, *Plexippus paykulli* (pl. 36). Native to the Old World tropics, its full distribution along the Pacific coast is unclear.

There are several recent records from Orange County, California, but its synanthropic habits almost ensure that it is more widespread than currently recognized. The male Pantropical Jumper has a dark body with a thick white stripe running longitudinally up the midline and between the anterior median eyes. He also has white stripes bordering his carapace that curve up and around his anterior lateral eyes, giving his "face" three white stripes on a dark background. The female is less conspicuous and varies from sandy yellow to brown. Her pale dorsal stripe is less contrasting and extends just past her posterior lateral eyes. Both sexes normally have two pale spots on either side of the posterior abdominal midline. This is a fairly large jumping spider, measuring between 9 and 12 mm (.35 and .47 in.) in length. A more detailed description of the Pantropical Jumper can be found in Edwards (1979).

Related to the Pantropical Jumper is *Evarcha proszynskii* (pl. 34). It is found from northern Asia and Alaska to the mountains of Southern California. The male *E. proszynskii* has a brown abdomen, often with four pale spots on its posterior half, and a central area is bordered by diffuse darker bands edged in white. The posterior two-thirds of the carapace is generally very dark, becoming paler near the posterior lateral eyes. There is also a thick white band that nearly encircles the spider's carapace. The female tends to be uniformly dark brown with a nearly black carapace and to lack any distinctive markings. *E. proszynskii* is between 6 and 7 mm (.24 and .28 in.) in length and is fairly common on foliage in damp areas. The North American *Evarcha* are reviewed in Marusik and Logunov (1998).

Sitticus (pl. 37) is an uncommon genus greatly in need of a comprehensive revision. At least four species live along the Pacific coast from Central California through northern Washington, and they can be exceptionally challenging to identify, as the literature on these spiders is spread across numerous publications. The one clear exception is *Sitticus dorsatus* (pl. 37). The male of this species is rusty orange and is thought to mimic velvet ants (Mutillidae). The reproductive structures used to differentiate *Sitticus* females are internal and require microdissection to see. *Sitticus* range from 4 to 7 mm (.16 to .28 in.) in length, and most species are dark to brownish gray, often with a thin white stripe along the midline of the carapace. *Sitticus* also commonly have a "spread wing" pattern on the abdomen and a diffuse pale band running up the abdominal midline. Literature on the Pacific coast species can be found in Peckham and Peckham (1909) and in Prószyn'ski (1971, 1973, and 1980).

Thiodina hespera (pl. 33) is the only member of its genus along the Pacific coast. Its defining feature are two pairs of bulbous setae on the underside of each tibia I. However, both sexes are distinctive enough that you can normally recognize them without resorting to this level of examination. The male *T. hespera* has a brown carapace with a large white eye patch that extends backward between his posterior lateral eyes. Additional white swatches wrap below and behind his posterior lateral eyes. His abdomen is pointed and pale yellow, occasionally with darker yellow lateral stripes. The female has conspicuous dark areas around her eyes that give her a look that has been described as "clown-faced." While most jumping spiders build silken cocoons as nightly retreats, *T. hespera* spends the night hanging upside down, suspended from a silk chord. *T. hespera* is a common arboreal species in suburban neighborhoods throughout most of California. The adult is between 5 and 7 mm (.2 and .28 in.) in length, and a detailed description of *T. hespera* can be found in Richman and Vetter (2004).

Among the most unusual of the jumping spiders are those that mimic ants, in some cases even holding their legs over their carapaces like antennae, further hiding their true identity. Three genera housing six species of ant-mimicking salticids live along the Pacific coast. The most distinctive of these, *Sarinda cutleri* (pl. 37), is the only regional representative of this primarily tropical genus. Both sexes have enlarged palps and are similar in their overall appearance, with a reddish-brown cephalothorax and two rows of pale spots running down the otherwise dark abdomen. *S. cutleri* has been found in Imperial County, California, and was described by Richman (1965) under its previous placement in the genus *Myrmarachne*. The other two genera, *Synageles* (pl. 37) and *Peckhamia* (pl. 37), are very similar. They range from 2.5 to 5 mm (.1 to .2 in.) in length and vary from dusky red to nearly black with small white abdominal patches. Separating these two genera generally requires looking at very subtle morphological features; however, when the known Pacific coast species are looked at in isolation, a consistent difference in the spination of tibiae

I becomes apparent. Both of the region's *Peckhamia* species have three pairs of tibial spines, while only two pairs of are found on the region's three *Synageles* species. All of the Pacific coast's ant-mimicking jumping spiders are found in California, although only *Synageles occidentalis* makes it north into Washington, and *S. idahoanus* has been found in south-central Oregon. The genus *Peckhamia* is reviewed in Peckham and Peckham (1909), and *Synageles* is revised in Cutler (1988).

There are numerous genera of small ground-dwelling jumping spiders that, despite their inconspicuous nature, can be common in appropriate habitats. Four species of *Neon* (pl. 37) are found along the Pacific coast from Southern California to northern Washington. They live in thick beds of leaf litter and are all similar in appearance, ranging from 2.2 to 3 mm (.09 to .12 in.) in length and having a yellowish-brown carapace that is darker around the eyes and a brownish abdomen decorated with grayish-yellow bands and chevrons. A key to the North American species can be found in Gertsch and Ivie (1955).

Four genera of the Salticidae subfamily Euophryinae live along the Pacific coast. United by the anatomy of their reproductive structures, they are all small, ground-dwelling jumping spiders. *Talavera minuta* (pl. 37) is tiny (2 to 2.5 mm, or .08 to .1 in., in length) and grayish brown with a coating of short white setae that gives it a "frosted" look. Its rear legs are always banded, and its front legs are darkened depending on the spider's age and sex. Its palps are white and stand in sharp contrast to its otherwise dark coloration. *T. minuta* is an uncommon leaf litter denizen of the coast and mountains from Southern California to eastern Washington. There is also a single report from the 1800s from the Olympia region of northwestern Washington (Crawford, 1988). *Chalcoscirtus diminutus* (pl. 37) is also the only regional representative of its genus. Very little is known about this small (around 2.5 mm, or .1 in., in length) ground-dwelling spider. Both sexes are shiny black with white fringes across the base of the abdomen, and the male has white on his clypei. The female has variable amounts of white on her abdomen. *C. diminutus* has been collected widely, if uncommonly, across the United States, with several records from Southern California. The genus is described in greater detail in Cutler (1990). *Euophrys monadnock* (pl. 37) is also a small (4 to 5 mm, or .16 to .2 in., in length) rarely seen member of the spider fauna found in coniferous leaf litter, and it has been collected sporadically from Canada through Central California. The male is glossy black with pale orange femora on his third and fourth legs. The female is less ornate, with a dark-brown to black cephalothorax and gray abdomen marked with several pale chevrons.

While the previous species are leaf litter specialists, the Pacific coast's two species of *Mexigonus* (pl. 37) prefer drier, rocky areas. The mottled *Mexigonus morosus* (pl. 37) has been found throughout Southern and Central California, while the more distinctly striped *M. minutus* has been collected in Los Angeles County, California. *Mexigonus* range from 4 to 8 mm (.16 to .31 in.) in length, and making a confident identification may require reviewing the fine details of their anatomy as described in Peckham and Peckham (1909). Additional information on these euophryine genera can be found in Edwards (2003).

California also houses one of the least-known jumping spiders in America. Only a few specimens of "*Pseudicius*" *siticulosus* (not illustrated) have been collected, including a female from Owens Lake in Inyo County. Its name is in quotes because it was incorrectly placed in this otherwise Old World genus by Peckham and Peckham (1909) and has yet to be reassigned. The male has a silvery-gray cephalothorax with a broad, ornate black border and a fairly elongate abdomen with a central row of pale, thick chevrons. The female is also grayish with irregular black bands on either side of her cephalothorax and large, dark lateral spots connected by chevrons along her abdomen. The one recorded female was 7 mm (.28 in.) long. An illustration of the male's palp can be found in Maddison (1996).

NATURAL HISTORY: With their diurnal habits, bright colors, and fascinating array of behaviors, jumping spiders have been the subject of many natural history studies. Unlike most spiders, which depend on vibrations to alert them to potential prey, salticids use their acute eyesight to find and capture insects and other small arthropods. Their large anterior median eyes provide exceptional acuity while the others are used for motion detection. Both stalking and ambush hunting methods are commonly employed by jumping spiders. If potential prey is found but is out of range, salticids will move into an effective attack position. This behavior, known as detouring, might take the spider out of visual contact with its prey, even leading it away before bringing it to a better location, demonstrating a surprising level of memory and problem-solving abilities for a spider. Jumping spiders anchor a silk dragline to the substrate before pouncing, and they eat mainly insects and other small invertebrates, including other spiders. In the case of aggressive insects, the jumping spider may roll on its back and clutch the prey in its legs, holding it upside down to prevent it from escaping. In the case of larger, slow-moving insects, such as caterpillars, salticids have been observed biting their prey, sometimes repeatedly, and then moving away until it is still. This may be a protective measure, hiding the spider in case the convulsions of its prey attract the attention of wasps, birds, or other potential predators.

Courtship in jumping spiders can be a very dramatic affair, often employing a combination of movement, sound production, pheromones, and touch. When a male Johnson Jumper, *Phidippus johnsoni,* encounters a female outside of her nest, his courtship involves a series of abdominal twitches, body elevations, and leg waves in hopes of gaining her favor. If he finds a female inside her retreat where visual displays would be ineffective, he pulls, probes, and vibrates against the outside of her silken chamber. If he encounters a subadult female in her retreat, the male will spin an adjacent chamber and wait in it without attempting to mate, until she has completed her final molt. Males in some other genera, such as *Habronattus,* perform elaborate dances that show off their iridescent patterns and ornamentation, while more-cryptic genera, such as *Platycryptus,* have less complicated visual displays. Sound production, by rubbing tubercles against stiff setae, is used by several genera of jumping spiders, including *Marchena* and many species of *Habronattus.* Genera such as *Phidippus, Menemerus,* and *Hentzia* indulge in male-male aggressive interactions, some of which are superficially similar courtship behaviors. These include elevating their bodies and fore legs, spreading their chelicerae, rocking and circling, and in some cases actually fighting. The bouts end when one spider is either killed or driven away.

After mating, female jumping spiders deposit their eggs in dense, silken egg sacs surrounded by cocoon-like retreats. Inside the retreats, the females guard their eggs from predators and parasites. The retreats are often hidden under rocks or behind loose bark, and some small foliage-dwelling species will even use the insides of dried dragonfly exuviae to shelter their eggs. Two species of jumping spider, *Habronattus tranquillus* and *Metaphidippus mannii,* engage in brood parasitism, an extremely rare occurrence in spiders. *Diguetia mojavea* (Diguetidae, p. 68) places its stacked egg sacs in a suspended, cone-shaped retreat. Upon the death of the protective *Diguetia* female, *H. tranquillus* and *M. mannii* invade her retreat and lay their own egg sacs among hers. When the salticid spiderlings emerge, they feed on the *Diguetia* eggs and young. Most jumping spiders appear to have an annual life cycle, although individuals of the same species but from different habitats may demonstrate different life history patterns. Those from the mountains emerge and mate during the brief spring and summer period, while those from milder, coastal climes, especially in Southern and Central California, can have an extended breeding period, with adults present for a much longer portion of each year.

With their diurnal habits and often vivid colors, some jumping spiders effectively mimic other, more aggressive or toxic animals. Several species of *Phidippus* are similar in color and movement to wingless velvet ants (Mutillidae), a group of wasps well known for their painful stings. Other species of jumping spider closely resemble colonial ants. In some cases, the mimicry is so convincing that even trained entomologists can be easily fooled. By living in close proximity to the ants, the spiders receive protection from predators; however, this also leaves them susceptible to attack from the very insects they impersonate. To compensate, some spider species use chemical mimicry, producing cuticular hydrocarbons similar to those used by the ants to recognize one another. Other genera of jumping spider are exceptionally similar to small beetles with hard exoskeletons and noxious chemical defenses.

THOMISIDAE — Crab Spiders

Pls. 38, 39, Fig. 14

IDENTIFICATION: Thomisidae is among the most distinctive spider families in North America. The spiders' fore legs are stocky and laterigrade, an alignment that angles them so that their spined ventral surfaces face out from the body. Legs I and II are also distinctly longer and stouter than legs III and IV, giving these spiders (especially the females) a rather boxy, crab-like appearance. Thomisids can be found across a wide variety of habitats, including leaf litter, tree trunks, and flower heads, and can range in color from white with vivid splashes of pink and green to mottled brown, red, and gray. They are eight-eyed entelegynes, and unlike most other two-clawed spiders, they lack claw tufts.

SIMILAR FAMILIES: The laterigrade leg condition is found in several families, including Philodromidae (p. 187) and Sparassidae (p. 192), although none show this feature as intensely as the thomisids. Spiders in the family Philodromidae are unique in that leg II is longer than leg I, and sparassids have long, thin fore legs. Additionally, both Philodromidae and Sparassidae have claw tufts.

PACIFIC COAST FAUNA: Nine genera represented by 55 regional species. Among the most unusual of the region's crab spider genera is *Tmarus* (pl. 39). They are relatively small spiders, ranging between 4.5 and 6 mm (.18 and .24 in.) in length, and are easily recognized by the tubercle on the dorsoposterior margin of the abdomen, their forward-projecting clypei, and their exceptionally long fore legs. Unlike most crab spiders, *Tmarus* cling to small branches where their unusual shape and mottled coloration disguise them as leaf buds or the nubs of broken twigs. Placing their fore legs along either side of a stem, they wait for a small insect to wander between them. Two species are known from the Pacific coast. *Tmarus angulatus* lives in forested regions throughout most of California but becomes progressively rarer in Oregon and Washington. The second regional species, *T. salai,* is only known from a small region of coastal and near-coastal Southern California. A key to the western *Tmarus* can be found in Schick (1965).

Diaea livens (pl. 38) is the only representative of its genus along the Pacific coast. It is a distinctive spider whose cephalothorax and legs are bright, translucent green and whose abdomen has a pale-yellow base color. The female abdominal pattern varies from a large, clearly defined red patch to scattered dark spots and red anterior bars. The male also has a pale yellow abdomen, often with dark reddish dots at the base of each seta. *D. livens* ranges from 4 to 6.5 mm (.16 to .26 in.) in length and is almost entirely endemic to California's coast live oak

(Quercus agrifolia) woodlands from San Diego County to Del Norte County. It is described in more detail in Schick (1965).

The remaining seven genera are divided into two groups. The first contains four genera: *Misumenoides, Misumena, Misumessus,* and *Mecaphesa*. These crab spiders are normally found on flowers and at the tips of budding branches. They are often brightly colored, and superficially the different genera can be quite similar. In addition, there is often a great deal of individual variation within any given species. Definitively separating these genera requires closely examining their eye arrangement, facial structure, and setal pattern. On *Mecaphesa* and *Misumessus,* the anterior lateral eyes are distinctly larger than the anterior medians, while on *Misumenoides* and *Misumena,* the eyes in the anterior eye row are approximately the same size (fig. 14). *Mecaphesa* and *Misumessus* also have numerous pointed setae on the carapace and abdomen, giving them a particularly bristly appearance. *Misumenoides* and *Misumena* lack conspicuous body setae except around their eyes.

The Goldenrod Crab Spider (*Misumena vatia,* pl. 38) and the White-banded Crab Spider (*Misumenoides formosipes,* pl. 38) are among the best-known spiders in North America and are the only regional representatives of their genera. They are generally either yellow or white, and adult females are capable of slowly changing their base colors to better match their environment. Both species also exhibit a great deal of variation in their coloration and pattern. The most reliable feature for distinguishing them is the presence of a clypeal ridge running along the edge of the White-banded Crab Spider's carapace, just above its chelicerae and below its eyes. The Goldenrod Crab Spider may have white in this area as well, but never on a raised ridge. Females of both species may be solidly colored or may be decorated with pink, red, or green stripes. On the Goldenrod Crab Spider, the stripes are normally limited to the sides and, occasionally, the leading edge of the abdomen. On the White-banded Crab Spider, there are regularly markings on the sides of the abdomen, in addition to a V-shaped set of stripes on its dorsal surface. Female White-banded Crab Spiders range from 5.4 to 11.3 mm (.21 to .44 in.) in length, and female Goldenrod Crab Spiders are between 6.2 and 8.4 mm (.24 and .33 in.) long.

Sexual dimorphism in crab spiders can be extreme, a feature well illustrated by the two aforementioned species. Male

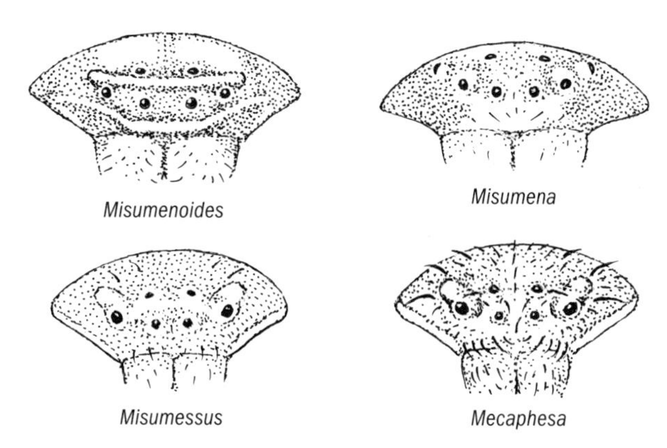

Figure 14. Faces of four crab spider genera.

Goldenrod and White-banded Crab Spiders are only 2.6 to 3.3 mm (.1 to .13 in.) in length. They have extremely long, dark fore legs and short, tan rear legs. The male White-banded Crab Spider also has the distinctive white ridge below its eyes. Its carapace is normally olive green and its abdomen is amber. The male Goldenrod Crab Spider has a much darker carapace and often shows several pink to chocolate-brown bands on its white abdomen. The Goldenrod Crab Spider is extremely common throughout Washington, in large parts of Oregon, and into Central California along the Coast Ranges and the Sierra Nevada. The White-banded Crab Spider is found from coastal Southern California through the Central Valley. It is exceptionally rare north of California, with less than half a dozen very old, questionable records from Washington (R. Crawford, pers. comm., 2010). Additional information on both species can be found in Schick (1965) and in Dondale and Redner (1978a).

Until recently, *Misumenops* was one of the most diverse Thomisidae genera in North America. Based on details of the spiders' reproductive structures, Lehtinen and Marusik (2008) revised the genus and reassigned all of the West Coast's species into one of two different genera, *Misumessus* or *Mecaphesa* (see fig. 14). *Misumessus oblongus* (pl. 38) is the only described species of its genus. It has only a few short, stiff setae on its carapace, while the *Mecaphesa* carapace is decorated with numerous long, conspicuous setae. Additionally, both sexes of *M. oblongus* have unmarked, pale-yellow to silvery-green abdomens, unlike the patterned design normally found on *Mecaphesa*. The male *M. oblongus* is between 2.5 and 3 mm (.1 and .12 in.) in length, and his carapace varies from pale green to reddish brown. The female ranges from 5.5 to 6.2 mm (.22 to .24 in.) in length, and her carapace is usually yellowish green. Regionally, *M. oblongus* is known from Southern and Central California, where it has been collected off flowers and foliage. It is further reviewed in Schick (1965) and in Dondale and Redner (1978a) under its old name *Misumenops oblongus*.

Sixteen species of *Mecaphesa* (pl. 38, see fig. 14) live along the Pacific coast, often on the surfaces of flowers and occasionally at the tips of budding branches. As in other crab spider genera, *Mecaphesa* exhibit a great deal of variation in their color pattern and intensity, even within a single species. For this reason, a close examination of the spider's reproductive structures and setal pattern is often needed to confirm a species-level identification. *Mecaphesa* are most common in Southern and Central California, where different species live across a wide variety of plant communities, from chaparral and desert scrub to riparian woodlands. They become increasingly uncommon farther north, and of the roughly half dozen species known from Washington, only one, *Mecaphesa lepidus,* can be considered widespread. Male *Mecaphesa* range from 2.5 to 4 mm (.1 to .16 in.) in length, while the larger females are between 3.5 and 7 mm (.14 and .28 in.) long.

The second major group of Thomisidae consists of three genera of cryptically patterned crab spiders: *Coriarachne, Xysticus,* and *Ozyptila*. Two of these genera, *Xysticus* and *Ozyptila,* are common in leaf litter and grassy fields, but *Coriarachne* live under loose tree bark and in tight crevices on wooden buildings. *Coriarachne* (pl. 39) carapaces are exceptionally flat, especially when compared with those of the other genera. Two species are known from the Pacific coast states. *Coriarachne brunneipes* (pl. 39) is common from Central California north through western Oregon and Washington and has an extremely flat carapace decorated with thin, sharply pointed setae. *C. utahensis* (pl. 39) is a widespread species with a slightly more rounded carapace and thicker, blunt-tipped setae. Male *Coriarachne* range from 3.8 to 6.3 mm (.15 to .25 in.) in length, and females are between 4.5 and 11.2 mm (.18 to .44 in.) long. Both species are reviewed in Bowling and Sauer (1975). Occasionally *C. utahensis* is placed in the genus *Bassaniana*; however, numerous reviewers have found that *C. utahensis* and other North American *Coriarachne* should be considered members of the same genus (Dondale, 2005b).

With 26 species, *Xysticus* (pl. 39) is the most diverse of the area's crab spider genera. These spiders range from 3.8 to 8.7 mm (.15 to .34 in.) in length and are generally larger and more abundant than members of the very similar genus *Ozyptila* (pl. 39). There are six species of *Ozyptila* along the Pacific coast, and they measure between 2.8 and 4 mm (.11 and .16 in.) in length. Both genera vary in color from dull yellow and pink to black with patches and chevrons of beige, gray, and orange. Most also have a broad, pale dorsal stripe down the midline of the carapace. Except for *Xysticus punctatus*, a rare resident of northwestern Washington that lives on tree trunks and branches, *Xysticus* and *Ozyptila* spend the majority of their lives in thick leaf litter and grassy fields and under rocks and fallen logs. The two genera are best separated by the structures of their first legs. *Xysticus* have three or more pairs of spines on the ventral surfaces of tibiae I, and their femora are long and slender (about four times longer than they are wide). *Ozyptila* have thicker first femora (about three times longer than wide) and only two pairs of spines on the ventral sides of tibiae I. *Ozyptila* are also generally smaller than *Xysticus*. *Xysticus* are found throughout the Pacific coast states in nearly every habitat. *Ozyptila* are much less common and are generally limited to forest leaf litter in Central California's Sierra Nevada and coastally from Mendocino County, California, through western Oregon and Washington. There is also at least one record of *Ozyptila conspurcata* from near Spokane, Washington. Gertsch (1953) revised several of the North American thomisid genera, including *Xysticus*, *Ozyptila*, and *Coriarachne*. Each genus has also received regional coverage in Schick (1965) and Dondale and Redner (1978a). *Ozyptila* was the subject of a detailed revision by Dondale and Redner (1975b).

NATURAL HISTORY: Crab spiders are sit-and-wait predators, relying on their cryptic coloration to hide them from predators and conceal them from prey. Adult female Goldenrod Crab Spiders and White-banded Crab Spiders are even able to change their color by either secreting or reabsorbing yellow epidermal pigments, a process that can take a few days to several weeks.

Surprisingly, experimental research by Brechbühl et al. (2010) has shown that in the Goldenrod Crab Spider, color had no significant effect regarding prey capture. When spiders were placed on flowers of different colors than themselves, they did not catch any fewer prey than spiders hunting in matching flowers did. Bees also did not discriminate between the empty and spider-inhabited flowers, helping explain why bees make up such a large portion of the crab spiders' diet (in addition to butterflies and moths). There was also no evidence that matching coloration provided any greater protection from predators. These counterintuitive results may be explained by the phenomenon that what appears to be cryptic coloration to humans may be highly contrasting to an insect that sees in the ultraviolet range, thus leaving the adaptive benefits of color changing unresolved.

The crab spider's potent bite quickly immobilizes its prey. Rather than wrapping the prey in silk, the spider feeds directly from the insect's body, leaving an empty, but otherwise undamaged, husk. *Xysticus* eat a wide range of soil-dwelling invertebrates, including other spiders, harvestmen, insects, and on occasion even earthworms. Infrequently, *Xysticus* will climb into foliage and attempt to capture flying insects.

There is some evidence that chemical cues are used by adult males to track down females. Based on observations of *Xysticus ferox*, when a male encounters a female, they engage in several bouts of sparring with their fore legs. The male then attempts to climb on the female's back, and after a few minutes of trying to dislodge him, she relaxes. He then spins a bridal veil over her body. This thin silk net appears to have a pheromonal effect, keeping her still while the much smaller male crawls beneath her to mate. Afterward, the female is easily able to tear free. Because so few species have been studied in detail, it's unknown how widespread the use of bridal veils is across the crab spider genera.

The female Goldenrod Crab Spider hides her lens-shaped egg sac behind the tip of a folded leaf. She then stitches the nest closed with a thick coating of silk and spends the next four to six weeks guarding it until all of the young have hatched. Some *Xysticus* females seal themselves within their nests rather than guard them from the outside. Once the young hatch, *Misumena* spiderlings climb to the tips of their host plants and balloon away. In contrast, *Xysticus* young rarely balloon. Instead, they drop into the leaf litter and crawl away from their natal sites (Morse, 1992).

PHILODROMIDAE — Running Crab Spiders

Pl. 40

IDENTIFICATION: The running crab spiders are two-clawed entelegynes and make up an exceptionally abundant family within North America's spider fauna. The different genera are relatively flat bodied, and their outlines vary from short and round *(Titanebo)* to long and thin *(Tibellus)*. The running crab spiders' most distinguishing feature is that leg II is longer, sometimes significantly, than leg I. In most genera, the fore legs are laterigrade. In this alignment, the leg's spined ventral surface faces outward from the front of the body and allows the spider to more effectively grab struggling prey. Nearly all philodromids also have both claw tufts and scopulae.

SIMILAR FAMILIES: The giant crab spiders (Sparassidae, p. 192) and the crab spiders (Thomisidae, p. 180) are anatomically and behaviorally similar to the philodromids. However, in neither of these families is leg II longer than leg I. Thomisids also lack claw tufts. Running crab spiders in the genus *Tibellus* may be mistaken for long-jawed orb weavers (*Tetragnatha,* Tetragnathidae, p. 127), but philodromids never build orb webs, and *Tetragnatha* do not have clearly defined dark bands running the length of the body.

PACIFIC COAST FAUNA: Sixty regional species representing six genera. While the generic placement of some philodromids is fairly straightforward, a careful examination of the eyes, legs, and carapace dimensions of many individuals is required for an accurate identification. *Ebo* (pl. 40) and *Titanebo* (pl. 40) are among the most distinctive of the region's running crab spider genera, as their second legs are nearly twice as long as their first. Separating them from each other, however, is more challenging. Both have fairly broad, rounded abdomens, vary in color from yellowish gray to dark brown, and often have dark bands or mottling across their dorsal surfaces. *Ebo* are small, ranging from 1.9 to 4.9 mm (.07 to .19 in.) in length. They lack lateral spines on their tibiae and metatarsi, and their clypei are shorter than the space between their anterior median eyes. *Ebo* are most commonly found in leaf litter and among the roots in weedy fields but have also been found in caves, ant nests, and wood rat middens and are thinly scattered across the Pacific coast, with four regional species, all of which are described in Sauer and Platnick (1972).

The area's ten recorded species of *Titanebo* range from 2.8 to 6.4 mm (.11 to .25 in.) in length. In each, there are spines on the lateral faces of the tibiae and metatarsi, and the clypeus is distinctly higher than the width between the anterior median eyes. Although *Titanebo* have been collected in all three of the contiguous Pacific coast states, they are most diverse in Southern and Central California, becoming progressively rarer farther north. Only one species, *Titanebo parabolis,* is known from Washington. *Titanebo* are also much more likely to be found in bushes, creosote scrub, and other standing foliage than *Ebo.* The different *Titanebo* species are described in greater detail in Saurer and Platnick (1972) as a subgenus of *Ebo.*

Also known as slender crab spiders, *Tibellus* (pl. 40) are elongate and yellow or tan in color, with a dark stripe running the length of the body. Their fore legs are long and often held pressed together as the spider clings to a plant stem. *Tibellus* range from 6 to 10 mm (.24 to .39 in.) in length and are especially common in grassy fields and brushy meadows, where their coloration and habits afford them excellent camouflage. There are six regional species, and representatives are common in all three of the contiguous Pacific coast states. Keys to their identification can be found in Gertsch (1933), Schick (1965), and Dondale and Redner (1978a).

The three remaining genera are similar in appearance, and their identification often requires a detailed examination of their eyes, legs, and overall body structures. When identifying an unfamiliar running crab spider, you should first look at its eyes. On *Philodromus* (pl. 40) the posterior median eyes are significantly closer to the posterior laterals than they are to each other. In both *Apollophanes* and *Thanatus,* the posterior median eyes are nearly equidistant from the posterior laterals.

With 33 described regional species, *Philodromus* is the most species-rich of the Pacific coast's running crab spider genera. *Philodromus* are common across the region, from Southern California's chaparral to northern Washington's montane meadows. They live in leaf litter, under tree bark, and on foliage. *Philodromus* are between 2.5 and 8 mm (.1 and .31 in.) in length and vary in color from pale gray to dark reddish brown, often with a great deal of mottling, striping, or other patterns on their abdomens and carapaces. In some species there is a black diamond-shaped mark on the dorsal abdominal midline, similar to that found on the region's *Thanatus*. Based on the fine details of the spiders' internal and external reproductive structures, numerous "species groups" of *Philodromus* have been defined. To identify an adult *Philodromus,* you should begin by discovering which species group it belongs to, using Dondale and Redner (1976b). You can then look to regional guides such as Schick (1965) or Dondale and Redner (1978a) or the major taxonomic works of Dondale and Redner (1968, 1969, and 1975a) to make a species-level identification.

The differences between *Apollophanes* (pl. 40) and *Thanatus* (pl. 40) are subtle, and there is some overlap in the features normally used to distinguish these two genera. *Apollophanes* are fairly flat, long-legged spiders with speckling on the abdomen and legs. While this can be difficult to judge in the field, their carapaces are also noticeably longer than they are wide. *Thanatus* are comparatively shorter legged and less speckled, and their carapaces are scarcely longer than wide. Both genera often show a dark diamond-shaped heart mark, but in *Apollophanes* its outline tends to be diffuse, while in *Thanatus* it is more sharply demarcated. This feature is also found on several species of *Philodromus*. Two species of *Apollophanes* live along the Pacific coast. *Apollophanes margareta* is an uncommon species associated with conifer forests across Washington, Oregon, and Northern California, while *A. texanus* is regionally limited to Southern California, where it has been found in juniper scrub and chaparral. *Apollophanes* are between 5 and 9 mm (.2 and .35 in.) in length and are revised in Dondale and Redner (1975c).

There are five regional species of *Thanatus* (pl. 40), and while not particularly common, representatives are found throughout the Pacific coast states. They are grayish yellow to reddish brown, often with a crisply defined brown or black diamond-shaped heart mark. *Thanatus* range from 4 to 8 mm (.16 to .31 in.) in length and have the least laterigrade leg arrangement of any of the Pacific coast's running crab spider genera. A key to their identification can be found in Dondale et al. (1964).

NATURAL HISTORY: There has only been limited research on the Philodromidae of the western United States, although several studies have looked at the life histories of eastern species, and many of the facts they have revealed are broadly relevant across the genera. *Philodromus* and *Titanebo* are most common in herbaceous foliage and woody scrub but can also be found on the ground. *Tibellus* prefer grassy fields and marshes with sedges and reeds, while *Ebo* are most commonly found amid plants' roots. *Apollophanes* and *Thanatus* are most common in leaf litter, mountain meadows, and desert scrub.

Like other running crab spiders, *Philodromus* are able to move rapidly when hunting, but they also often take a sit-and-wait ambush approach. They are mainly tactile and visual hunters, triggered by the movement of their prey. Resting on a twig with its fore legs spread open, a *Philodromus* will quickly grab any small insect that touches it or walks nearby. Once a prey item is captured, the spider rapidly scoops it in with its fore legs and bites it, delivering a mix of venom and digestive fluids. *Philodromus* neither wrap their prey in silk nor crush it, but feed on it whole, sucking the prey's liquefied remains through the bite wound. Similar hunting techniques have been observed in *Tibellus*.

As the most species-rich genus, *Philodromus* has been the subject of most philodromid natural history studies. When a male finds a female, he normally begins courtship by steadily tapping on the substrate, a signal that tells the female that he is a courting male and not necessarily prey. If she allows him to approach, he taps on her legs and abdomen until he is in a position where mating can occur. A *Philodromus* egg sac consists of a single silk sheet below the eggs and two slightly separated sheets above. At least one species, *Philodromus alascensis*, places its egg sac in a small, rocky recess, where it is covered with grains of sand, wood flakes, and other debris. The female then hides the egg sac beneath her body. In a similar manner, a *Tibellus* female builds her egg sac either on blades of grass or in the folds of a leaf. She then covers it with a thick coating of silk and guards it until she dies.

Once *Philodromus* spiderlings emerge from their egg sac, they climb to the tips of nearby twigs and balloon away. Some *Philodromus* have an annual life cycle, in which they overwinter in their penultimate stage. After spending the cold months hidden in protected spots, they emerge in late spring or early summer and molt into adulthood. Other *Philodromus*, especially those from more-northern climes with shorter growing seasons, have a biennial life cycle, taking two years to fully develop (Dondale 1961).

SELENOPIDAE **Flatties**

Pl. 41

IDENTIFICATION: Selenopids are dorsoventrally compressed entelegynes with only two eyes in the posterior eye row and six in the anterior eye row, although the most lateral of the anterior eyes are exceptionally small and difficult to see. Selenopid bodies are mottled grayish yellow or brown with dark spotting on the abdomen and carapace. Their legs are heavily banded, armed with numerous large spines, and in a laterigrade arrangement, a position that further magnifies their already flattened appearance.

SIMILAR FAMILIES: No other regional spider family shares Selenopidae's unique eye arrangement, although several families are superficially similar in form and habit. The giant crab spiders (Sparassidae, p. 192) have two very distinct, relatively straight eye rows with four eyes in each, and their legs are rarely as boldly patterned as those of our regional selenopids. Some running crab spiders (Philodromidae, p. 187) also resemble flatties, but philodromids have two clearly visible eye rows (recurved rather than straight) with four eyes in each, and leg II is longer than the other legs, a situation unmatched in the selenopids.

PACIFIC COAST FAUNA: One genus, *Selenops* (pl. 41), with a single described Pacific coast species. *Selenops actophilus* was described by Chamberlin (1924) based on a male collected in Sonora, Mexico, and a female from Los Angeles County, California. Muma (1953) provided additional records from Santa Barbara County, California, to western Texas. He also described several other *Selenops* species from the southwestern United States and northern Mexico. Recent work involving both the physical examination and genetic analysis of large numbers of specimens has found that there is a great deal of variation in *Selenops* across the southwestern United States (Crews, 2011). The spider currently known as *S. actophilus* is part of a complex that Muma (1953) referred to as the *debilis* group, but the characteristics currently used to differentiate the *debilis* species are variable even within populations. A revision that combines both genetics and morphology will be necessary before the true identity of the Pacific coast's *Selenops* can be confirmed. Regionally, *Selenops* are found in deserts and rocky chaparral from Santa Barbara County through southeastern California. According to Muma (1953), the male *S. actophilus* is 8 to 8.3 mm (.31 to .33 in.) in length while the larger female spans 9.2 to 13 mm (.36 to .51 in.) in length.

NATURAL HISTORY: *Selenops* are nocturnal spiders that live in the cracks of boulders and other narrow fissures. They don't build webs, relying instead on their stealthy habits and exceptional speed to capture prey. Selenopids are nocturnal ambush predators. Resting just outside the entrances of their retreats, they can snatch any small arthropods wandering past. This also allows them to rapidly return to their retreats when threatened, where they are protected by the retreats' narrow apertures. Very little has been written regarding the natural history of the region's *Selenops*, although studies from other populations (Crews et al., 2008; S. Crews, pers. comm., 2009) have revealed aspects of their biology that seem relatively consistent throughout the genus. Acts of cannibalism have been observed, and in the Caribbean, *Anolis* lizards feed on *Selenops* disturbed from their lairs. Along the Pacific coast, native geckos (Gekkonidae) and other nocturnal lizards may play a similar role, but this has yet to be confirmed.

Selenops are fairly long-lived, taking two or more years to reach maturity and possibly surviving well beyond that. For this reason, both adult and immature spiders may be found year-round. When an adult male encounters a mature female, he initiates courtship by moving his body in a shuddering manner, and if she accepts his advances, she lets him crawl on top of her into a mating position. The female then builds a flat, disk-shaped egg sac containing between 40 and 50 eggs, which she guards until the spiderlings hatch and disperse.

SPARASSIDAE — Giant Crab Spiders

Pl. 41

IDENTIFICATION: Giant crab spiders are dorsoventrally flattened spiders whose laterigrade leg arrangement allows them to hide in narrow crevices. They are eight-eyed, two-clawed entelegynes with conspicuous claw tufts and scopulae. Their most distinctive feature is a trilobed extension on the dorsal surface of each metatarsal-tarsal joint that allows the spider to lift the tarsus above the axis of the metatarsus, much like a person resting a forearm on a tabletop can raise the hand at the wrist. Regionally, giant crab spiders range from 6.2 to 48 mm (.24 to 1.89 in.) in length.

SIMILAR FAMILIES: Spiders in the family Selenopidae (p. 191) and running crab spiders (Philodromidae, p. 187) are the most likely candidates for misidentification. Selenopids are extremely flat in profile, and the selenopid's six-eyed anterior eye row is unique. The giant crab spiders have carapaces that are more robust in profile and four eyes in each of their two eye rows. Philodromids are generally smaller than sparassids (under 10 mm, or .39 in., in length) and have second legs that are clearly longer than their first, a feature exclusive to running crab spiders. Additionally, giant crab spiders have numerous teeth along the cheliceral retromargin, while running crab spiders lack retromarginal teeth.

PACIFIC COAST FAUNA: Three genera represented by five Pacific coast species, all of which are regionally limited to California. Two of the genera, *Olios* and *Macrinus,* are native, while the third, *Heteropoda,* is introduced but not established. With three species, *Olios* (pl. 41) is the most diverse of the region's giant crab spider genera. Two species, *Olios peninsulanus* and *O. naturalisticus,* range from 6.2 to 11.3 mm (.24 to .44 in.) in length and are found in Southern California's deserts and chaparral. The third, *O. giganteus,* is an exceptionally large species, with the adult ranging from 11.3 to 48 mm (.44 to 1.89 in.) in length. In California, it has been reported as far north as Yolo County in the Coast Ranges and along the Sierra foothills to Tulare County. Additional *Olios* records, presumably representing this species, have been reported from as far north as Shasta County, California. *Olios* vary in color from yellowish gray to chestnut brown, often with a thin Y-shaped stripe along the abdomen's dorsal midline. They regularly have spotting on their legs, and their chelicerae are normally dark brown to black (occasionally pale orange). North America's *Olios* were revised in Rheims (2010b). Formerly included in *Olios* was the extremely similar genus *Macrinus* (not illustrated), which is represented by a single species, *Macrinus mohavensis*. Rheims (2010a) split *Macrinus* from *Olios* based on the fine details of their reproductive organs. Only the female *M. mohavensis* has been described and was collected from the Mohave National Preserve in San Bernardino County, California.

The Huntsman Spider (*Heteropoda venatoria*, pl. 41) is found in tropical areas worldwide. In the United States it is established only in Florida, but it has been found numerous times around California's commercial ports, presumably shipped in with loads of imported fruit. It is an exceptionally long-legged spider with records of females up to 28 mm (1.1 in.) long and leg spans of 12 cm. Both sexes are beige with a multitude of large, black spines on the legs. The female has a tan carapace with creamy-yellow bands across its posterior margin and, at times, around its sides. The male is smaller with proportionately longer legs, a large dark, divided patch in the center of his carapace, and a pale triangular area behind his eyes. Due to the Huntsman Spider's size, it often causes a great deal of unnecessary alarm. Its venom is harmless to people, but if mishandled, it can give a painful bite.

NATURAL HISTORY: Little has been published regarding the natural history of our native giant crab spiders. They use their incredible speed to capture prey and often hunt across vertical surfaces. They are regularly found crawling up the sides of walls, on the bark of trees, and even across ceilings. They live in a wide variety of habitats, from riparian forests to deserts and suburban parklands. While not entirely nocturnal, they often spend the day hidden in narrow crevices, such as under loose bark, or even on walls behind hanging artwork. Because they readily enter buildings, giant crab spiders can inspire panic, but they are harmless insect eaters and pose no threat to people.

Olios and *Heteropoda* differ dramatically in how they care for their eggs and young. In protected locations, *Olios* females build silken retreats, completely enclosing themselves in alongside their egg sacs. They then open the retreats only once the spiderlings have hatched and are ready to disperse. The female *H. venatoria* builds a large, disk-shaped egg sac that she carries beneath her with her palps and chelicerae. In Florida, all stages of development are present throughout the year. This life cycle pattern is almost certainly applicable across the tropical distribution of *H. venatoria*, so it is possible for individuals of any age to be transported at any time. In the wild, *H. venatoria* is a voracious predator of cockroaches and other large insects.

HOMALONYCHIDAE

Pl. 41

IDENTIFICATION: Juveniles and adult female homalonychids are among the most distinctive and least conspicuous spiders in North America. Small grains of sand adhere to stiff setae on their abdomens, carapaces, and legs, camouflaging them against the background soil. Although adult males lack this applied camouflage, they can be recognized by the combination of a strongly recurved posterior eye row, a broadly rounded carapace with a W-shaped band along the posterior margin, and a roughly pentagonal abdomen. Homalonychids are eight-eyed entelegynes endemic to the deserts of the southwestern United States and northwestern Mexico.

SIMILAR FAMILIES: With their unusual habits, distinctive body form, and recurved posterior eye row, even adult males, which aren't covered in sand, are unlikely to be confused with any other regional spider family.

PACIFIC COAST FAUNA: One genus, *Homalonychus* (pl. 41), with two regional species. *Homalonychus theologus* is found west of the Colorado River in desert regions and chaparral from southern San Diego County northeast through central Inyo County. The second species, *H. selenopoides,* predominantly lives east of the Colorado River but has also been collected along the border of southern Nevada, where it extends west into the Grapevine Mountains in Death Valley National Park, Inyo County, California. The two species are extremely similar, differing mainly in their geographic distribution and in the fine details of their reproductive organs. An illustrated key to *Homalonychus* can be found in Roth (1984) where he discusses the exceptionally variable epigynal structure of *H. selenopoides.* Adult *Homalonychus* range from 6.8 to 12.5 mm (.27 to .49 in.) in length, and as in other spiders, the females are generally larger than the males.

NATURAL HISTORY: Homalonychids are nocturnal, cursorial spiders adapted to some of the hottest, driest areas in North America. They feed mainly on insects but will also prey upon other spiders. They don't build capture webs and are often found under rocks, beneath dead vegetation, and in rodent burrows. One of the most remarkable features of these spiders is the veneer of fine-grained soil covering the bodies of juveniles and adult females that renders them nearly invisible against the ground. After molting, the immature or adult female *H. theologus* digs a shallow pit and, starting with its abdomen, rolls backward until it's lying upside down. From this position, it slowly rock its body back and forth, allowing the fine-grained sand to stick to its setae. Then it flexes its legs, dorsally, then ventrally, across the sand, eventually covering its entire body, a process that may be done multiple times to ensure a complete coating.

While they are often found under rocks and other debris, *Homalonychus* can also conceal themselves by rapidly digging small holes with their front legs and flipping dirt over their bodies. When relaxed, *Homalonychus* rest with their legs splayed out, but when distressed they bring their front and rear legs together into an X shape or "paired-leg formation" (Vetter and Cokendolpher, 2000), a rigid stance they will maintain even when picked up. By holding this position, they may be mimicking a broken twig or dropped cactus spine, a feature that is certainly much less appetizing than a living spider to predators.

Adult males and juveniles wander at night, but adult females are nearly always found near their retreats. It is not uncommon to find *Homalonychus* under a rock along with a series of progressively larger exuviae, implying that the spiders spend a considerable amount of time in the same location. They are long-lived for araneomorphs, with lab-raised individuals taking three years to reach maturity and adult females living at least two years beyond that. Courtship in Homalonychidae involves the male alternately drumming his palps on the ground in front of the female and attempting to climb on top of her. If receptive, she allows him to crawl over her. She then tightly pulls her legs up alongside her body so they nearly touch above her carapace. The male then ties them together with a silk thread wrapped around her patellae and tibiae. In this position, the spiders mate. Throughout the mating process, the male rapidly vibrates his legs against the female's body, possibly providing a sensory stimulus that helps maintain the female's immobility. Afterward, the male rapidly departs, and within a few seconds the female breaks the noose around her legs and cleans off the remnant scraps of silk.

Homalonychus egg sacs are extremely well camouflaged and nearly indistinguishable from the surrounding substrate. They are roughly spherical and attached to the female's retreat by a mesh of silk threads, the whole of which is covered with fine sand. It is also surrounded by a sand-covered, multilayered silk collar. In addition to protecting the egg sac from predators and parasites, the sandy covering may help in temperature and humidity control, a function of special importance in the arid regions where these spiders live.

ZORIDAE

Pl. 29

IDENTIFICATION: Zorids are small, entelegyne spiders with a straight anterior eye row and a strongly recurved posterior eye row. They are two-clawed, and tibiae I and II are armed with six to eight pairs of long, stout spines on their ventral surfaces.
SIMILAR FAMILIES: The Zoridae eye arrangement is highly unusual. A similar pattern is shared by the wolf spiders (Lycosidae, p. 223), but lycosids lack the heavily armed fore legs of the Zoridae.
PACIFIC COAST FAUNA: One genus, *Zora,* with a single regional species, *Zora hespera* (pl. 29). The male *Z. hespera* averages around 3 mm (.12 in.) in length, and the female is around 5 mm (.2 in.) long. Along the Pacific coast, *Z. hespera* can be identified by its overall coloration and eye arrangement and by the heavy spination of its fore legs. Its carapace is pale yellowish brown with a pair of wide, dark bands running from its eyes to its posterior margin along with thin, dark lines on its lateral edges. Its abdomen has grayish-brown mottling on a pale background, and its sternum has a dark patch at the base of each coxa. *Z. hespera* has been collected from San Diego County, California, to Vancouver Island, British Columbia (Corey and Mott, 1991; Bennett and Brumwell, 1996). It lives in a wide range of habitats, from chaparral and mountain forests to coastal fields and oak woodlands, within which it prefers open, sunny areas.
NATURAL HISTORY: *Zora* are diurnal, actively pursuing their prey both along the ground and in foliage. In the whole *Zora* range, at least some adults can be found year-round, although

courtship and breeding are concentrated in late spring and summer. The female hides several egg sacs behind a flat sheet of silk, which in turn she places under a rock or other object. She then stays with them for an extended period, guarding them from predators and parasites.

DICTYNIDAE — Mesh Web Weavers

Pl. 42

IDENTIFICATION: Dictynidae is a poorly defined, morphologically diverse family of three-clawed spiders whose members include arboreal, cribellate web weavers as well as ecribellate, cursorial hunters. Additionally, many genera are poorly defined and in desperate need of revision. Based on current knowledge, the specific and sometimes even generic placement of many individuals, especially adult females, is extremely difficult, and in some cases impossible.

Cribellate mesh web weavers are generally less than 4 mm (.16 in.) in length, have eight eyes (with one very rare exception), and build netlike webs around the tips of dried weeds or in the niches of stone walls and wood fences. Their calamistra extend across more than half the length of metatarsi IV, and their cribella are usually undivided. They normally lack spines on their legs and have at most only a few tarsal trichobothria. Adding to the difficulty of identification is that on the males of some species, the calamistrum and cribellum may be significantly reduced.

While there is overlap, the ecribellate mesh web weavers are often larger, ranging up to 8 mm (.31 in.) in length. The ecribellate species live close to the ground, including, in some cases, deep inside caves. Unlike the cribellate dictynids, their legs are conspicuously spined, and they have rows of trichobothria on both their tarsi and metatarsi that get progressively longer toward the distal end of each leg segment. In both cribellate and ecribellate groups, the anterior lateral spinnerets are clearly separated and are either shorter than or equal in length to the posterior lateral spinnerets. While many ecribellate dictynids have eight eyes, a significant number in several genera have only six.

SIMILAR FAMILIES: The diversity of Dictynidae in form and ecology make it inevitable that many of its genera are challenging to identify. Ecribellate species are most similar to members of the family Cybaeidae (p. 205) and a subset of the family Hahniidae (p. 209). However, on cybaeids the anterior lateral spinnerets are either close together or contiguous at their bases and are longer than the posterior lateral spinnerets. Distinguishing ecribellate dictynids from the cryphoecine hahniids is more difficult, often requiring a close examination of the spider's dentition and leg structure. Two genera in Hahniidae, *Ethobuella* and *Dirksia*, have four or more pairs of spines along the undersides of their anterior tibiae, while the ecribellate dictynids never have more than three pairs of spines there. Of the Hahniidae genera with fewer ventral tibial spines, *Calymmaria* have proportionately longer, slender legs and build very distinctive cone-shaped webs. Members of the Hahniidae genus *Cryphoeca* are very similar to some ecribellate dictynids, differing mainly in the pattern of their cheliceral dentition. The Hahniidae have between two and four teeth of similar size on their cheliceral retromargins, while the ecribellate Dictynidae have either a mix of teeth and denticles or more than four retromarginal teeth.

The cribellate species are most likely to be mistaken for a member of the family Amaurobiidae (p. 237) or for the introduced Gray House Spider, *Badumna longinqua* (Desidae, p. 243). However, in each of these families the cribellum is divided, while in Dictynidae it is almost always entire.

PACIFIC COAST FAUNA: Fifteen genera with 107 described and several undescribed regional species. A key to the genera can be found in Bennett (2005b). Regionally, there are four ecribellate and 11 cribellate genera. One of the region's most unusual spiders is the ecribellate *Saltonia incerta* (pl. 42). It lives almost entirely under salt crusts along the shorelines of lakes, lake beds, and streams in the Mojave Desert and Imperial Valley regions of Southern California and northern Mexico. Its legs and cephalothorax are pinkish red and its abdomen is gray. The adult *S. incerta* is generally between 3 and 8 mm (.12 and .31 in.) in length and has a large, broad colulus and fore legs armed with numerous long, conspicuous spines. For many years after its initial description, very few additional specimens of *S. incerta* were found, but more recent searches have revealed that in the right microhabitats it can be fairly common. It differs from other ecribellate mesh web weavers genetically, in its spinneret morphology, and in its cheliceral dentition, differences that have led to serious questions regarding the appropriateness of its inclusion within Dictynidae (Spagna et al., 2010).

The remaining ecribellate mesh web weaver genera, *Yorima, Blabomma,* and *Cicurina,* are exceptionally similar and can be difficult to distinguish. They overlap widely in size, ranging from 2.1 to 7.7 mm (.08 to .3 in.) in length, live in leaf litter, and are found in wooded areas throughout the Pacific coast region. *Yorima* (pl. 42) contains five species of six-eyed spiders that are endemic to the California Coast Ranges from San Diego County to Sonoma County. Adding to the complexity, at least three species of six-eyed *Blabomma* live within the distribution of *Yorima,* and the anterior median eyes of the remaining *Blabomma* species may be minute and difficult to see. The structure of the spinnerets can be helpful in distinguishing *Yorima* from other mesh web weavers. In *Yorima,* the distal segment of a posterior lateral spinneret is half as long to nearly as long as the basal portion. In the other two genera, the distal segment is only about a third as long as the basal. *Blabomma* and *Cicurina* also tend to be larger than *Yorima*. Overall, *Yorima* are fairly consistent in their appearance, with a yellowish cephalothorax and legs and a solid gray or lightly marked abdomen. A key to the *Yorima* can be found in Roth (1956).

Ten species of *Blabomma* (pl. 42) have been described from the Pacific coast states, with records extending from San Diego County, California, through western Washington to southern British Columbia. Additionally, several museums house numerous undescribed species in their collections. *Blabomma* normally have eight eyes, although the anterior medians are often very small and difficult to discern. In a few species, they have disappeared entirely, making those species look very similar to *Yorima*. *Blabomma* legs and cephalothoraxes range from yellow to brownish orange, and their abdomens are normally dark gray with a smattering of light gray spots, streaks, and smudges. The recognized species are described in Chamberlin and Ivie (1937) under the generic names *Blabomma* and *Chorizommoides*.

Cicurina (pl. 42) is the most diverse ecribellate dictynid genus in North America, with over 100 described species, only 14 of which are known from the Pacific coast states. All of the region's species have eight eyes, with their anterior medians normally between one-half and two-thirds the size of their anterior laterals, distinctly larger than those found on *Blabomma*. Additional differences between the two genera can be found in the structure of their reproductive organs, as outlined in Bennett (2005b). *Cicurina* are found widely throughout the Pacific coast states, and a key to their identification can be found in Chamberlin and Ivie (1940).

Eleven genera of cribellate Dictynidae live along the Pacific coast and can be divided into two groups. Four genera have at least one tarsal trichobothrium and have clypei that are no higher than the width of the spiders' anterior lateral eyes. These four are occasionally referred to in the literature as the "low-clypeus" genera. The remaining seven "high-clypeus" genera lack tarsal trichobothria and have clypei that are at least one and a half times the height of the spiders' anterior lateral eyes. Among the high-clypeus Dictynidae are some of the family's most common or readily identified genera, including the primarily Neotropical genus *Mallos* (pl. 42). There are four regional species, each of which has a chocolate-brown carapace with diagnostic wide, white bands along the margins of the thoracic portion. Their abdomens are grayish white and decorated with species-specific chevrons, stripes, and other markings. They range from 1.3 to 4.3 mm (.05 to .17 in.) in length. *Mallos* are found widely throughout California, with two species, *Mallos pallidus* and *M. niveus,* occurring as far north as northern Washington. An illustrated key to the genus can be found in Bond and Opell (1997).

Similar to *Mallos* but larger and missing the white thoracic borders is *Mexitlia trivittata* (pl. 42), the only regional representative of its genus. It is one of the largest mesh web weavers in North America, ranging from 4 to 8 mm (.16 to .31 in.) in length. Its carapace and legs are dark reddish brown, and its abdomen is chalky gray with large, dark anterior and posterior spots and two or three pairs of smaller lateral spots. This unusual species is communal, and colonies several hundred strong have been recorded. Unlike truly social spiders, *M. trivittata* individuals are territorial within their own webs, although their support lines and web edges may be interconnected. *M. trivittata* is found in mountains throughout the southwestern United States, with regional records limited to Inyo County, California. It is reviewed in detail in Bond and Opell (1997).

A unique member of the region's mesh web weaver fauna is *Nigma linsdalei* (pl. 42). It has a pale, translucent green abdomen with several rows of irregular whitish spots and often a prominent black or red central spot. *N. linsdalei* is around 3 mm (.12 in.) long and is found in California's Coast Ranges from Los Angeles County to Marin County, where it has been repeatedly found on the budding tips of oak branches (D. Ubick, pers. comm., 2010). It is described and illustrated in Chamberlin and Gertsch (1958) under its previous name, *Heterodictyna linsdalei*.

The four remaining high-clypeus genera, *Dictyna, Emblyna, Phantyna,* and *Tivyna,* contain the majority of the region's Dictynidae fauna. Unfortunately these genera are poorly defined and difficult to separate. In nearly all cases, a detailed examination of the males' reproductive structures is needed to differentiate them. While female *Tivyna* have an unusually large, spatula-shaped scape on the epigynum, the remaining females are essentially indistinguishable. Initially described by Chamberlin (1948), *Tivyna, Phantyna,* and *Emblyna* were reassigned into *Dictyna* by Chamberlin and Gertsch (1958). The genera were then resurrected by Lehtinen (1967) in a move that is still considered controversial (Crawford, 1988; Bennett, 2005b). As a group, these cribellate mesh web weavers range from 1.2 to 4 mm (.05 to .16 in.) in length. Their posterior median eyes are fairly large and close together, and in males the chelicerae are laterally bowed. They can be concolorous or patterned, pale or dark, and they can be found on leafy foliage and tree bark, under rock piles, and in leaf litter. Fifty-eight species representing the four "*Dictyna*-type" genera have been reported from within the contiguous Pacific coast states. *Tivyna moaba* (not illustrated) is the only regional representative of its genus. It is a pale, unpatterned spider known along the Pacific coast only from the desert areas of Imperial County, California. Members of each of the remaining genera can be found in all three of the Pacific coast states. *Emblyna* (not illustrated) is the most diverse with 33 reported regional species, followed by *Dictyna* (pl. 42) with 21 species, and *Phantyna* (not illustrated) with only three species. A key to these genera can be found in Bennett (2005b), while each species is more fully described under its previous placement within *Dictyna* in Chamberlin and Gertsch (1958).

The region's four genera of cribellate low-clypeus dictynids all have eight eyes, with the possible exception of *Lathys delicatula* (not illustrated). It is the only member of its genus recorded from the Pacific coast, and according to Chamberlin and Gertsch (1958), its anterior median eyes are either minute or missing entirely. The remaining eyes form two compact triads that nearly touch at the carapace's midline. Chamberlin and Gertsch (1958) mention an unconfirmed 1928 report from Friday Harbor, San Juan County, Washington, that they choose to disregard because of its distance from other records from the southwestern United States. Crawford (1988) accepted this record because several other species have disjunct populations within the Olympic Mountains rain shadow. Since this initial record, no additional specimens have been found (R. Crawford, pers. comm., 2010). The female *L. delicatula* runs between 1.4 and 2.1 mm (.05 and .08 in.) in length and has a yellowish cephalothorax and a gray abdomen that can be either unmarked or decorated with dark chevrons. *L. delicatula* is described and illustrated in Chamberlin and Gertsch (1958).

Eight described species of *Tricholathys* (pl. 42) live along the Pacific coast, and while widespread in California, they become scarce north into Washington. They are small spiders, ranging from 2 to 5.5 mm (.08 to .22 in.) in length. Their carapaces are pale yellow to reddish brown and are often decorated with hazy patches and dusky lines. Abdomens on *Tricholathys* vary from pale and unmarked to dark brown with numerous light and dark chevrons, although there is often variation even within the same species. Definitively separating *Tricholathys* from the other low-clypeus cribellate mesh web weavers requires looking at their cheliceral fang furrows. In *Tricholathys* the promargin is armed with four or five teeth while the retromargin has three or four. The other low-clypeus genera have fewer teeth along their cheliceral margins. The region's *Tricholathys* are described and their diagnostic reproductive structures are illustrated in Chamberlin and Gertsch (1958).

The region's two remaining low-clypeus genera are represented by tiny, rarely collected, montane species. *Brommella* (not illustrated) are usually less than 2 mm (.08 in.) in length, with short calamistra limited to the basal half of metatarsi IV. Two species have been found on the Pacific coast: *Brommella bishopi* from California's Yosemite National Park, and *B. monticola* from Washington's Mount St. Helens National

Volcanic Monument. Each has a yellowish-brown carapace and an unmarked, white to gray abdomen. Members of the genus *Argenna* (not illustrated) are larger than *Brommella,* ranging from 2 to 3.3 mm (.08 to .13 in.) in length. They also have longer calamistra. Two species live along the Pacific coast: *Argenna yakima* in Washington's Mt. Rainier National Park, and *A. obesa,* which while widely distributed across North America has only been regionally reported from Umatilla County, Oregon. Descriptions of each of these four species, supported by illustrations of their diagnostic features, can be found in Chamberlin and Gertsch (1958).

NATURAL HISTORY: Dictynidae can be broadly divided into either ecribellate cursorial hunters or cribellate web-building predators, although some cribellate species are found in thick beds of leaf litter. In a situation common to many spider families, most Dictynidae genera have received little attention from ecologists or arachnologists, and many aspects of their natural history remain unknown. This is especially true of the ecribellate genera, most of which have received essentially no attention beyond their initial descriptions and a few taxonomic revisions. The one exception is *S. incerta.* Several studies have focused on this enigmatic species, particularly on its unusual morphology, genetics, and distribution (Roth and Brown, 1975; S. Crews, pers. comm., 2010).

Cribellate species often live in vegetation, their tentlike webs draped over the tips of dry grasses, flower heads, and tree branches. The mesh wraps around the foliage with a tubular retreat near its center. The genera that have received the most attention have been *Mallos, Mexitlia,* and *Dictyna* because of the various degrees of sociality, from solitary to communal, that these spiders express. In each of these genera, flies make up the majority of the diet (Jackson, 1977). When a fly or other insect is captured and killed, it is not wrapped up. Instead, after feeding, the spiders leave the empty husks of their prey littering the outsides of their webs, possibly helping camouflage the spiders and acting as lures for additional prey.

Of the cribellate species that have been examined, there is a fair degree of diversity in courtship behaviors, although most species generally follow a common outline (Jackson, 1979). On encountering a female's web, a *Dictyna* male explores its outer surface, laying down strands of silk and slowly working his way toward her retreat. The female initially responds aggressively, but within a few moments, she generally becomes more sedate. Both the male's initial behaviors and the female's actions strongly imply a pheromonal component to the silk used in this courtship routine. In *Dictyna volucripes,* a species common across the eastern United States, the male is thought to lightly bind the female within her retreat (Starr, 1988). Facing her, he then lightly strokes her with his palps and fore legs. If receptive, she raises her body so mating can occur. After mating, a *Dictyna* male may stay with the female for an extended period, and during the breeding season, it's not uncommon to find both sexes together along with several egg cases in the same web.

CYBAEIDAE

Pl. 43

IDENTIFICATION: Cybaeidae is a family of three-clawed, entelegyne spiders that is difficult to define. As with the other two members of the Dictynoidea superfamily, Dictynidae (p. 198) and Hahniidae (p. 209), Cybaeidae's genera have historically been shuffled into and out of several families. While most cybaeids have eight eyes, some have only six and a few are completely eyeless. There are, however, several features that, when looked at as a group, unite Cybaeidae's dissimilar genera. They have rows of trichobothria on their tarsi and metatarsi that become increasingly longer as they progress toward the distal end of each leg segment. Their anterior lateral spinnerets are very close together or touching at their bases and are also thicker and often longer than the posterior lateral spinnerets. Additionally, their coluli have been reduced to between one and 10 pairs of setae. Cybaeids are common in wooded areas throughout the western United States. Many live under leaf litter on the forest floor, while other populations have adapted to life in caves, and in some cases they are fully troglobitic.

SIMILAR FAMILIES: Members of the family Cybaeidae are most similar to the ecribellate Dictynidae and to several genera in the family Hahniidae. All three families are generally yellowish gray to dark olive brown with either unmarked or patterned abdomens and live in cool, moist areas near the forest floor. Among those groups most likely to be confused with cybaeids, the anterior lateral spinnerets are clearly divided, thinner, and shorter than the posterior lateral spinnerets. With their heavily spined legs and small size, two Cybaeidae genera, *Cybaeota* and *Cybaeina*, might be mistaken for members of an ant-mimic sac spider family (Corinnidae, p. 146), but ant-mimic sac spiders are two-clawed and lack the Cybaeidae tarsal trichobothria pattern.

PACIFIC COAST FAUNA: Four genera represented by 42 regional species are currently recognized. In an unpublished thesis, Bennet (1991) described four additional genera and over 30 new species. However, because this work was never published, these changes are not officially recognized under the rules governing international zoological nomenclature. Bennet (2005a) includes these undescribed genera and species, and while this may lead to some confusion, it is the most accessible key available to North America's Cybaeidae fauna. The four currently recognized genera can be divided by the spination pattern on the underside of tibia I. *Cybaeota* and *Cybaeina* have four or five pairs of spines under this leg segment, while *Cybaeozyga* and *Cybaeus* have only two or three.

Three species of *Cybaeota* (not illustrated) live along the Pacific coast. They are all less than 2.5 mm (.1 in.) in length, with two to five small teeth on their cheliceral retromargins. *Cybaeota* have pale yellowish legs, grayish-yellow carapaces with dark pigmentation around the eyes, and abdomens that vary from concolorous to heavily patterned. They are forest spiders, and specimens have been found in leaf litter, in wood rat middens, and under thick moss growing on tree trunks from Southern California through northern Washington. Rare in collections, they can be fairly common in the right microhabitats. Bennet (1988) revised the genus and provides a key to its component species.

Very similar to *Cybaeota* are the four described species of *Cybaeina* (pl. 43). Like *Cybaeota, Cybaeina* have yellowish gray carapaces and legs but their abdomens are often paler. They are larger than *Cybaeota*, ranging from 3.5 to 4.7 mm (.14 to .18 in.) in length, and have between five and 15 teeth of various sizes along their cheliceral retromargins. *Cybaeina* are leaf litter specialists endemic to the Pacific Northwest's coastal forests. Their greatest diversity is in western Oregon, where three of the four described species have been recorded. Only one species, *Cybaeina sequoia*, has been found in California. Roth

(1952) collected multiple individuals from redwood duff in Humboldt County. *Cybaeina minuta* is the most widespread of the described species, with coastal records from both northern Oregon and central Washington. No comprehensive revision of *Cybaeina* has been published, and any attempt to identify a specimen will require comparing it with the original descriptions in Chamberlin and Ivie (1932, 1937, and 1942b) and Roth (1952). Additionally, numerous undescribed species exist in museum collections that match the definition of *Cybaeina* as currently understood (Bennet, 2005a).

The two remaining Cybaeidae genera, *Cybaeozyga* and *Cybaeus,* are extremely similar in appearance, although *Cybaeus* are generally larger than *Cybaeozyga.* Confidently differentiating them requires examining their cheliceral dentition. *Cybaeozyga* have between five and seven small teeth of similar size along their cheliceral retromargins and three large, widely spaced teeth on their promargins. *Cybaeus* have a mix of teeth and denticles along their cheliceral retromargins and three large, closely grouped teeth along their promargins. Adult male *Cybaeus* also have an unusual collection of peg setae on the apophysis of each palpal patella, a feature missing on male *Cybaeozyga.*

Cybaeus (pl. 43) is by far the most species-rich of the Cybaeidae genera, with 34 described and nearly as many undescribed Pacific coast species (Bennett, 2005a). Even within the same species the spiders are extremely variable in size, with regional records ranging from 3.9 to 10 mm (.15 to .39 in.) in length. *Cybaeus* carapaces and legs vary from pale yellow to dark reddish brown with duskier pigmentation around the eyes. Their abdomens are gray and can be either unmarked or decorated with pale chevrons. With the exception of *C. signifer,* the anterior median eyes on *Cybaeus* are distinctly smaller than the others. While most *Cybaeus* build loose, unassuming webs in leaf litter and debris, *C. signifer* is a nocturnal hunter that prowls the mossy trunks of trees in search of prey. Its anterior median eyes are quite large. It is found from Monterey County, California, north to the Queen Charlotte Islands, British Columbia. There are no published, comprehensive keys to the region's *Cybaeus* species, although Copley et al. (2009) breaks the genus into eight groups, all of which occur along the Pacific coast, based on fine details of their reproductive structures. Studies on the genetic and morphological diversity within *Cybaeus* have revealed that the genus likely evolved from two separate stocks. One is a Holarctic clade with representatives in both Eurasia and North America, and the other clade is found strictly in western North America (Copley et al., 2009). Known as the Californian clade, its members reach their greatest diversity in California and southern Oregon. If future research confirms these findings, then only those species that are part of the Holarctic clade will retain the generic name *Cybaeus* while those that are part of the Californian clade will be reassigned to another genus.

Cybaeozyga heterops (not illustrated) is the only described species in its genus. It is 2.7 to 2.9 mm (.11 in.) in length, with yellowish-brown legs, a matching carapace, and a dark gray abdomen with indistinct light markings. *C. heterops* is a rare spider that has been collected from Douglas Fir leaf mold in Klamath County, Oregon. Since it was described by Chamberlin and Ivie (1937), three additional undescribed species of eyeless *Cybaeozyga* have been discovered living in caves in Northern California (Roth and Brame, 1972).

NATURAL HISTORY: Cybaeidae are most common under rocks and rotting logs and in moist leaf litter, although Bennett (2005a) reported finding specimens in habitats as diverse as beach debris and alpine meadows. Cybaeid webs are inconspicuous and are used predominately as retreats, not hunting snares. They are squarish, loosely woven baskets with signal lines extending from openings at all the corners and are built in damp, protected nooks. Depending on the species and location, mature adults are found between late spring and early fall. While most die soon after maturing, some female *Cybaeus* appear to live for a second year, as evidenced by an increased thickening and sclerotization of their epigynal walls (Bennett, 2006). Few egg sacs have been described, but those of *Cybaeus reticulatus* are lens shaped, covered by a thin sheet of silk and soil, and attached to the underside of a sheltered recess (R. Bennett, pers. comm., 2010).

Most cybaeids have highly restricted distributions, and numerous species with nonoverlapping ranges can often be found across a relatively small area. Unlike many juvenile spiders, juvenile cybaeids don't appear to balloon but disperse by wandering. In this situation, rivers, arid valleys, and mountain ranges can act as nearly impenetrable barriers to gene flow, thus isolating populations and greatly increasing the likelihood of speciation. This same highly localized population structure also makes some members of the family Cybaeidae especially susceptible to disturbance or even extinction.

HAHNIIDAE

Pl. 43, Fig. 15

IDENTIFICATION: The Hahniidae are three-clawed, eight-eyed entelegynes. Like several other spider families, they have rows of trichobothria on their metatarsi and tarsi that increase in length as they progress down each leg segment. As members of the superfamily Dictynoidea (along with Dictynidae, p. 198, and Cybaeidae, p. 205), Hahniidae's genera have been split apart and regrouped numerous times. As currently recognized, Hahniidae contains two subfamilies, Hahniinae and Cryphoecinae. These subfamilies differ conspicuously in the arrangement of their spinnerets; however, similarities in their reproductive structures led to lumping Cryphoecinae within Hahniidae by Lehtinen (1967), a move that more than 40 years later is still considered controversial.

Members of the subfamily Hahniinae are easily recognized by their spinnerets. They are uniquely aligned in a transverse row across the abdomen's posteroventral surface. Spiders in the subfamily Cryphoecinae are much more difficult to define and are readily confused with the members of several spider families. Their spinnerets are more traditionally arranged, with the anterior laterals clearly separated and shorter than the posterior laterals. The Cryphoecinae genera vary considerably in size, leg spination, and cheliceral dentition. As a group, they lack any easily defining characteristics, as illustrated in Ubick (2005c), in which the cryphoecine spiders are keyed out in three separate locations. Most Cryphoecinae are pale gray to reddish brown, nocturnal forest dwellers that live in leaf litter and under old logs. Members of the genus *Calymmaria* are the easiest to recognize, not because of their anatomy, but by the unusual cone-shaped webs they build in protected, moist alcoves. When attempting to identify a member of the subfamily Cryphoecinae, you have to look at several features in conjunction. In addition to their trichobothrial pattern and spinneret design, they lack feather-like (plumose) setae on their bodies and have narrow coluli. Any spider meeting these requirements with four or more pairs of ventral spines on its anterior tibiae is almost certainly one of the Cryphoecinae. If it has fewer ventral spines on its anterior tibiae, then a careful examination of its cheliceral dentition, leg structure, and size is generally needed to confidently distinguish it as one of the various Cryphoecinae genera rather than a morphologically similar spider.

SIMILAR FAMILIES: No other family displays the Hahniinae spinneret pattern, making Hahniinae some of the region's easiest spiders to recognize. Cryphoecinae contains numerous morphologically diverse genera, making it much more challenging to identify. Its genera are likely to be confused with either members of the family Cybaeidae (p. 205) or one of the ecribellate Dictynidae (Dictynidae, p. 198). In this situation, you should check the spider's spinnerets. On cybaeids, the anterior lateral spinnerets are thicker and longer than the posterior laterals and either are connected at their bases or rest very close together. In both Cryphoecinae and Dictynidae, the anterior lateral spinnerets are either the same size or shorter than the posterior laterals and are clearly separated at their bases.

Distinguishing between the ecribellate Dictynidae and members of the subfamily Cryphoecinae requires carefully examining the spination on their anterior tibiae, their overall sizes, and their cheliceral dentition. Two genera in Cryphoecinae, *Ethobuella* and *Dirksia,* stand out because they have either four or five pairs of spines on the ventral sides of their anterior tibiae. No other hahniid or dictynid has more than three pairs of spines at this location. Several cheliceral dentition patterns occur on the remaining genera of ecribellate Dictynidae and Cryphoecinae. If a spider has either three or four equal-size teeth and no denticles on its cheliceral retromargins, it is a member of one of the region's small Cryphoecinae genera, *Willisus* and *Cryphoeca*. However, if the spider has both a mix of teeth and denticles on its cheliceral retromargins and long, slender legs, then it is part of the species-rich Cryphoecinae genus *Calymmaria,* a genus also recognizable by its member's distinctive webs. Some ecribellate dictynids demonstrate a similar cheliceral dentition pattern, but they have distinctly stockier legs than *Calymmaria*.

PACIFIC COAST FAUNA: Seven genera containing 40 regional species. Bennett (2005c) provides an illustrated key to North America's Hahniidae genera, including both the Hahniinae and the Cryphoecinae subfamilies. Two Hahniinae genera live along the Pacific coast: *Hahnia* (pl. 43) and *Neoantistea* (pl. 43). They can be separated by the structure of their spinnerets and anterior median eyes. On *Hahnia* the anterior median eyes are distinctly smaller than the anterior laterals, while on *Neoantistea* they are nearly the same size or larger than the anterior laterals (fig. 15). Also, the distal and median segments of the lateral spinnerets are of nearly equal length on *Neoantistea*, but on *Hahnia* the distal segment is significantly shorter than the median. Adult *Hahnia* range from 1.2 to 2.4 mm (.05 to .09 in.) in length while *Neoantistea* are slightly larger, running from 2.5 to 4.6 mm (.9 to .18 in.) in length. Both genera have brown carapaces and abdomens that are decorated with several pale, anteriorly pointed chevrons. Occasionally, these chevrons are broken, forming paired bars on either side of the abdominal midline. *Hahnia* and *Neoantistea* are each represented by four regional species, and representatives of both genera are found in all three of the contiguous Pacific coast states, reaching their greatest diversity in Washington. An illustrated key to the North American Hahniinae can be found in Opell and Beatty (1976).

Cryphoecinae has never been comprehensively reviewed, and most of its members have received little attention beyond their initial descriptions, although the most species-rich of its genera, *Calymmaria*, was revised by Heiss and Draney (2004). Excepting *Calymmaria* in their webs, the simplest Cryphoecinae genera to identify are *Ethobuella* (pl. 43) and *Dirksia* (pl. 43). *Ethobuella* have four pairs of spines on the undersides of tibiae I while *Dirksia* have five pairs. All other regional Cryphoecinae genera have two or fewer pairs of spines on the ventral side of tibia I. The two known species of *Ethobuella* both have dark-rimmed, yellowish carapaces with dusky splotches, and yellowish-gray abdomens with jagged black lateral margins. Additionally, their anterior median eyes are minute in comparison to their anterior laterals. *Ethobuella* range from 2.5 to 3 mm (.1 to .12 in.) in length, and while *Ethobuella tuonops* has been collected widely from southwestern British Columbia through the mountains of western Washington (Crawford, 1988), *E. hespera* has been reported only from San Mateo

Hahnia

Neoantistea

Figure 15. Eye patterns of two Hahniidae genera, *Hahnia* and *Neoantistea*.

County, California. Illustrations and brief descriptions of both species can be found in Chamberlin and Ivie (1937), although the male *E. hespera* has yet to be described. *Ethobuella* have been collected from leaf litter, on bushes and trees, and from moss mats growing on deciduous trees in old riparian forests (Bennett, 2005c).

Dirksia cinctipes (pl. 43) is the only *Dirksia* in North America and is similar in appearance to *Ethobuella* but has five pairs of spines on the underside of tibia I and anterior median eyes that are slightly larger than its anterior laterals. The adult ranges from 2.3 to 4.4 mm (.09 to .17 in.) in length and can be found from Northern California to the mountains of coastal Washington, where it has been collected off the leaves of shrubs and trees (Bennett, 2005c; Roth and Brame, 1972). *D. cinctipes* is described and its reproductive structures are illustrated in Chamberlin and Ivie (1942b).

The most diverse Hahniidae genus in North America is *Calymmaria* (pl. 43) with 27 regional species. *Calymmaria* are among the most common spiders in the coastal and near-coastal woodlands from northwestern Baja California to southern British Columbia, with a separate population in California's Sierra Nevada. *Calymmaria* range from 2 to 9.8 mm (.08 to .38 in.) in length and nearly always require a detailed examination of their reproductive structures to be identified to the species level. They have relatively long legs and are dusky yellow or grayish brown, often with a series of dark V-shaped dashes along the thoracic region of the carapace. The abdomen normally has a dark heart mark and an irregular amalgamation of spots, smudges, and chevrons. While *Calymmaria* are fairly inconspicuous, their conical webs are easily recognized and are often found suspended beneath ledges, in caves, and in other protected locations. *Calymmaria* can be differentiated from other Cryphoecinae by having only a single pair of spines on the ventral surface of each tibia I (rarely two pairs) and a combination of both teeth and denticles on the cheliceral retromargin. An illustrated revision of *Calymmaria* can be found in Heiss and Draney (2004).

One of the least known of the region's hahniids is *Willisus gertschi* (pl. 43). Around 2.0 mm (.08 in.) in length, it lacks ventral spines on tibiae I and demonstrates an unusual form of autospasy. When its legs are entangled, they break along a cleavage plane within the patella, a feature shared with spiders in the genus *Calymmaria*. *W. gertschi* has a shiny, dark carapace, and its abdomen is pale gray with numerous dark transverse bars. It lives under large rocks in the San Gabriel Mountains of San Bernardino County, California, and was described and illustrated by Roth (1981).

Cryphoeca (pl. 43) contains several species of small (2.9 to 4.5 mm, or .11 to .18 in., in length), leaf litter–dwelling spiders with two pairs of spines on the ventral faces of their anterior tibiae and three to five equal-size teeth on their cheliceral retromargins. While North America's *Cryphoeca* have never been revised, the defining characteristics of the genus are discussed in Roth and Brame (1972). Along the Pacific coast, *Cryphoeca exlineae* is very common in Washington, and unidentified *Cryphoeca* have been found as far south as Central California.

NATURAL HISTORY: Discussing the natural history of Hahniidae is best accomplished by looking at the two subfamilies, Hahniinae and Cryphoecinae, separately. Among the region's Hahniinae, only *Neoantistea* are known to make webs. They weave small sheet webs, rarely more than a few inches across, over damp depressions in the ground or over beds of moss. No webs are known from *Hahnia*, and it is unclear whether this is because they don't build them or because they are tiny and constructed in leaf litter where they would easily be missed or destroyed during the collecting process. Hahniinae egg sacs have a mound shape, are several millimeters across, and are attached to the undersides of rocks or beneath loose bark. Nothing is known regarding Hahniinae courtship behaviors or feeding preferences.

Little is known regarding the life histories of most of the Cryphoecinae genera except that they are mainly nocturnal, forest-dwelling spiders. Only the webs of *Calymmaria* have been described, and while it is possible that the other genera don't build hunting webs, in some cases they may just be so small and flimsy that they have been overlooked. The cone-shaped basket of *Calymmaria* is made by attaching support strands to the bottom of a sheet web and pulling it down in the center. A spider will often hang from a thin platform above the web, although it will also move about on the outside of the cone. Because they prefer damp areas, *Calymmaria* can be especially common on shaded cliffs above streams. Their egg sacs are flat, irregularly shaped, coated with a tough outer layer of soil structures, and suspended around the perimeters of the females' webs.

ZODARIIDAE

Pl. 50

IDENTIFICATION: Two very dissimilar genera of Zodariidae live along the Pacific coast. Both are three-clawed, eight-eyed entelegynes with exceptionally large anterior lateral spinnerets and highly reduced posterior spinnerets. Additionally, the genus *Lutica* is distinctive in both habitat and coloration, living only along a narrow strip of Southern California's coastal dunes and on the Channel Islands.

SIMILAR FAMILIES: No other spider families share the gross morphological features and habitat preference of *Lutica*. Some Corinnidae (p. 146) might be mistaken for members of the genus *Zodarion*, but none have its greatly enlarged anterior median eyes and highly procurved posterior eye row.

PACIFIC COAST FAUNA: Two genera with four described species and one undetermined species. All four described regional Zodariidae species are in the genus *Lutica* (pl. 50). They range from 10 to 14 mm (.39 to .55 in.) in length and all have a yellowish-gray cephalothorax, a pale-gray abdomen with purplish stripes radiating along the midline and sides, and robust, heavily spined legs. *Lutica* are endemic to the coastal dunes and islands from central Baja California, Mexico, to Santa Barbara County, California, where they build burrows under the sand. Gertsch (1961) described the four extant species based on individuals collected from Southern California's Channel Islands and coastal dunes. Ramirez and Beckwitt (1995) later discovered that based on genetic data, a different set of species limits should be defined that more accurately reflect the spiders' evolution and biogeographic history. They did not, however, specifically change the species names or provide new descriptions, so the names designated by Gertsch (1961) remain in use.

Zodarion (pl. 50) are small spiders, about 2 mm (.08 in.) in length, with long, unspined legs, massive anterior median eyes, and a strongly procurved posterior eye row. A purplish-brown diurnal species has been found in a small area of rocky, serpentine grasslands in Central California (Ubick and Craig, 2005); however, the identity of this spider is yet to be determined.

NATURAL HISTORY: The two genera are so different in morphology, habitat, and behavior that their natural histories are best dealt with separately. Nearly everything known regarding the ecology of *Lutica* was published by Ramirez (1995). They are almost entirely endemic to coastal dunes, where they dig delicate, silk-lined burrows among the roots of native dune plants. The *Lutica* that live on Santa Barbara Island off Southern California make their burrows in the sandy soil beneath vegetation-covered sea cliffs instead. Most *Lutica* burrows are angled and between 2.5 and 15 cm (1 to 6 in.) in length. Occasionally they are built horizontally just beneath the surface. In this situation, *Lutica* may hunt by tracking the movements of insects walking above. When an insect is directly on top of a spider, the spider rips through the burrow's roof, grab its prey, and drags it back into the retreat. *Lutica* will also ambush prey by pouncing from behind almost perfectly camouflaged trapdoors. Breeding apparently occurs in late summer and early fall, and like most burrowing spiders, *Lutica* juveniles do not balloon. Instead, they wander through the dunes until they find a suitable patch of sand for burrowing. The only known *Lutica* predator is *Aptostichus simus* (Euctenizidae, p. 39), a trapdoor spider that shares the same coastal dune habitat.

Besides their diurnal habits, almost nothing is known regarding California's undetermined *Zodarion*. Other members of the genus in other regions specialize in hunting ants and build small dome-like shelters of sand on the undersides of rocks.

TENGELLIDAE

Pl. 44

IDENTIFICATION: Members of the family Tengellidae are ground-dwelling, nocturnal hunters with several features that in combination distinguish them from other entelegyne spiders. All of the Pacific coast genera have a prograde leg arrangement, deep notches on the underside of each trochanter, and a vestigial third claw hidden within the claw tufts. They also have at least five pairs of spines on the ventral surface of each anterior tibia. Tengellids range from 7 to 21 mm (.28 to .83 in.) in length. Their eight small eyes are round and similar in size and occur in two transverse rows. Additionally, their anterior lateral spinnerets are conical and nearly contiguous at their bases.

SIMILAR FAMILIES: No other regional spider family shares Tengellidae's collective morphological features, which include claw tufts, deeply notched trochanters, a prograde leg arrangement, conical spinnerets, and a large number of ventral spines on the anterior tibiae.

PACIFIC COAST FAUNA: Three genera represented by 34 species, all of which are regionally limited to California. With 16 species, *Titiotus* (pl. 44) includes the largest tengellids on the West Coast, ranging from 8 to 21 mm (.31 to .83 in.) in length. *Titiotus* have unpatterned tan to reddish-brown bodies and legs and reach their greatest diversity in the mountains and forests of Central and Northern California. In the Sierra Nevada they prefer rocky streamside canyons, but in the lower-elevation Coast Ranges, they are most abundant in forests and rocky grasslands. Very similar to *Titiotus* are members of the genus *Socalchemmis* (pl. 44), another species-rich genus whose center of diversity is Southern California and the northern Baja California peninsula. *Socalchemmis* range from 7.1 to 15.2 mm (.28 to .6 in.) in length, and the majority of the region's 14 species are endemic to Southern California's deserts and chaparral scrub. Only two species, *Socalchemmis monterey* and *S. arroyoseco,* live north to Monterey County. Like *Titiotus, Socalchemmis* are wandering nocturnal hunters, and separating these two genera often requires looking at their reproductive structures. Fortunately, due to the large size of many individuals, these features can often be seen with a hand lens or in a decent macrophotograph. *Titiotus* males have complex retrolateral tibial apophyses (RTAs, see fig. 2A) with three or more prongs, and females have median epigynal lobes that are nearly as broad as the epigynum. *Socalchemmis* males have only two prongs on their RTAs, and the females' median epigynal lobes are less than two-thirds as wide as the epigynum. A key to the *Titiotus* species can be found in Platnick and Ubick (2008), while information on *Socalchemmis* can be found in Platnick and Ubick (2001, 2007).

The third regional genus, *Anachemmis* (pl. 44), contains four species and ranges from 7.4 to 13.6 mm (.29 to .53 in.) in length. Two species are widespread: *Anachemmis sober* lives in oak and coniferous forests across southwestern California and in the southern Sierra Nevada, while *A. linsdalei* is found in oak, pine, and redwood forests along the California coast from San Luis Obispo County to Humboldt County. Both species are common in leaf litter and may show light patterning on their abdomens. The remaining species, *A. jungi* and *A. aalbui*, are known only from a few desert caves in Inyo and San Bernardino Counties, respectively. They are paler than the more widespread species, with unmarked grayish white abdomens. Confirming an *Anachemmis* identification often requires looking at the fine details of the reproductive features. Male *Anachemmis* usually have only single-pronged RTAs, while females have more distinctly triangular median epigynal lobes than the other tengellid genera. These features are described and illustrated in Platnick and Ubick (2005).

NATURAL HISTORY: Very little is known regarding the natural history of the region's Tengellidae fauna. They are nocturnal hunters, and while several species in each genus are found in caves, none are fully troglobitic. Their saclike retreats are woven deep in the leaf litter or under fallen debris. Among the Pacific coast genera, only *Titiotus* egg sacs have been described. They are about 1.5 cm (.6 in.) in diameter and are suspended from short strands of silk. When dissected, an egg sac of a *Titiotus flavescens* had a thick layer of shredded leaf litter between the outer, papery covering and the blanket of fluffy silk.

PISAURIDAE — Nursery Web Spiders, Fishing Spiders

Pl. 44

IDENTIFICATION: Nursery web spiders are three-clawed, eight-eyed entelegynes with deeply notched trochanters, although it is often easier to recognize them by their unusual behaviors than by their anatomy. They are closely tied to ponds and streams where they hunt aquatic invertebrates, small fish, and tadpoles either from the shoreline or directly from the water's surface. Females carry their large, spherical egg sacs under their bodies, held by their chelicerae and attached to their spinnerets. Close to the time of hatching, the females suspend their egg sacs in protective "nursery webs," normally built in some nearby vegetation. The region's Pisauridae range from 5 to 19.3 mm (.2 to .76 in.) in length but often appear larger because they regularly hold their long legs outstretched from their bodies. Their posterior eye rows are moderately recurved, and the males have conspicuous retrolateral apophyses on their palpal tibiae.

SIMILAR FAMILIES: Fishing spiders are similar to the much more common wolf spiders (Lycosidae, p. 223), whose members also occasionally move across still bodies of water. However, the posterior eyes of wolf spiders are arranged in a distinct rectangular shape, very different from the smooth recurve of the fishing spider's posterior eye row. Additionally, male wolf spiders lack retrolateral tibial apophyses on their palps, and most females carry their egg sacs behind them by their spinnerets. No wolf spiders build nursery webs, and the newly hatched young of most wolf spider genera travel by clinging to their mothers' backs.

PACIFIC COAST FAUNA: Two genera with one regional species each. The largest and most widespread of the region's fishing spiders is the Six-spotted Fishing Spider (*Dolomedes triton,* pl. 44). It ranges from 9.5 to 19.4 mm (.37 to .76 in.) in length and is generally greenish brown above with broad white stripes on both sides of its body and a series of small white spots on either side of its abdominal midline. Its common name comes from the six dark spots that decorate its sternum. In the Pacific Northwest, these spots are regularly fused into broad, comma-shaped bands. The Six-spotted Fishing Spider lives along the edges of still ponds, lakes, and slow-moving streams, and if disturbed, it can hide for up to 90 minutes amid the submerged vegetation, breathing through a bubble of air clinging to its body. It has been found sporadically throughout western Washington and south to Shasta County, California. A detailed revision of the nearctic *Dolomedes* is presented in Carico (1973).

Tinus peregrinus (pl. 44) is regionally limited to Southern California, where it lives on permanent desert ponds and in backwater pools along the banks of the Colorado River. It ranges from 4.3 to 7.8 mm (.17 to .31 in.) in length and has a brown, beige-bordered carapace. Its abdomen has a large, dark, posteriorly pointed patch that is decorated with thin, sinuous, white lines along its midline. Very little is known about the biology of *T. peregrinus,* but what is available is summarized in Carico (1976).

NATURAL HISTORY: The vast majority of the research on North America's nursery web spiders has focused on the Six-spotted Fishing Spider, *D. triton.* This species has well earned its title as a "fishing spider," readily chasing small fish and invertebrate prey underwater. It often lies in wait with its fore limbs resting lightly on the water's surface, its body and rear legs planted on a piece of floating wood or emergent vegetation. From this position, it can sense both the movement of insects on the water's surface and that of fish and aquatic insects below. When attacking a fish, it dives into the water and bites it just behind the head before carrying it back to the surface for eating. The Six-spotted Fishing Spider is able to run across the water's surface. Additionally, it employs a technique known as "sailing." By either raising its fore legs or lifting its entire body above the water's surface, it is able to catch the wind and glide with little energy expenditure across large bodies of water.

The male Six-spotted Fishing Spider is receptive to female pheromones left both on the water's surface and on strands of silk. When a male encounters a high concentration of female pheromones, he responds by waving his fore legs in the air and by rapidly flexing them on the water's surface. This produces a series of waves whose regular intervals are very different from those made by a struggling insect. If receptive, the female drums on the water with her palps and performs slow, circling motions with her elevated fore legs. The male has to be cautious when approaching the female, as cannibalism is well documented in this species. If courtship is successful, the two spiders entwine their legs and lightly tap on each other's bodies before mating, after which the male rapidly retreats. Later, the female constructs a large egg sac that she carries beneath her, held both by her chelicerae and by silk from her spinnerets. Just before hatching time, she constructs an elaborate nursery web in some waterfront vegetation (rarely among rocks) and suspends the egg sac within an enclosure of leaves and tightly woven silk. She then guards the nursery web until the spiderlings hatch and disperse.

ZOROPSIDAE — False Wolf Spiders

Pl. 44

IDENTIFICATION: Zoropsidae is represented on the Pacific coast by a single introduced species, the False Wolf Spider, *Zoropsis spinimana*. This six-eyed, cribellate spider is easily recognized by its large size (10 to 17 mm, or .39 to .67 in., in length), distinctly patterned body, strongly recurved posterior eye row, and straight anterior eye row. Its carapace is orangish brown with a thin, black V behind its eyes and dark, wavy bands along its sides, while the anterior half of its abdomen is decorated with a dark-brown heart mark. The False Wolf Spider has a divided cribellum, and its calamistrum is an oval patch on metatarsus IV. The female is slightly larger than the male and has proportionately shorter legs.

SIMILAR FAMILIES: *Z. spinimana* (pl. 44) is a distinctive species, unlikely to be confused with any other North American spider. In homes, it may be mistaken for a large wolf spider (Lycosidae, p. 223) or funnel web weaver (Agelenidae, p. 232), but its eye arrangement immediately distinguishes it from both of these families. Additionally, its large size and unique dorsal coloration separate it from all other North American cribellate spiders.

PACIFIC COAST FAUNA: Native to the Mediterranean region, *Z. spinimana* has been present in California's San Francisco Bay Area since at least 1992 (Griswold and Ubick, 2001). By the end of 2008, it had established itself in San Mateo, Santa Clara, Alameda, and San Francisco Counties (D. Ubick, pers. comm., 2008), where it has been found exclusively around homes, greenhouses, and other structures.

NATURAL HISTORY: *Z. spinimana* is a wandering hunter with adults most frequently reported from September through May. The female lays her eggs in spring, walling off the egg sac under a tent of fluffy cribellate silk. Because this spider is large and synanthropic, it can attract a fair bit of attention. It is, however, harmless to both people and pets.

OXYOPIDAE — Lynx Spiders

Pl. 45

IDENTIFICATION: Lynx spiders are easily recognized three-clawed entelegynes with a unique hexagonal eye arrangement.

They also have tapered abdomens and long, thin legs with an impressive array of prominent spines.

SIMILAR FAMILIES: No other North American spider family shares the lynx spiders' unique eye arrangement. With their darting manner and willingness to jump when disturbed, they are similar to some jumping spiders (Salticidae, p. 167) but are quickly differentiated by their eye structure and thin, spine-covered legs.

PACIFIC COAST FAUNA: Three genera with six regional species. The largest of the Oxyopidae are the green lynx spiders (*Peucetia viridans*, pl. 45) and *P. longipalpis* (pl. 45). Ranging from 8.1 to 21.6 mm (.32 to .85 in.) in length, they live in brushy fields, gardens, and grasslands across much of California. Both have a vivid-green base color with dark speckling on the legs, but despite their similarities, with experience you can recognize many individuals in the field. *P. viridans,* the more common and heavily patterned of the two, has distinctly longer legs in proportion to its body. *P. viridans* is also quite variable in color, varying from light green with a few white abdominal chevrons to conspicuously marked with white, black, and red patches extending across most of its body. *P. longipalpis* tends to be more uniformly green with more diffuse, white abdominal decorations. A female also may have a pair of dusky white bars running the length of her abdomen, but they are not as sharply defined as those on *P. viridans*. Both species are discussed in depth in Brady (1964).

Three species of *Oxyopes* live along the Pacific coast, where they range from 4.6 to 9.6 mm (.18 to .38 in.) in length. The most widespread species is the Western Lynx Spider, *Oxyopes scalaris* (pl. 45). It is common throughout the region in a wide variety of brushy habitats, from coastal chaparral to mountain meadows. While coloration is variable, the female is usually dark brown with a pair of dusky-white stripes at the front of her abdomen that meld together and continue to her spinnerets. Some individuals also have buffy midline stripes on the carapace. The smaller male Western Lynx Spider normally has a narrow-tipped, silver-blotched abdomen and brownish carapace with a wide, silvery to buff stripe down its midline. The second most widespread regional *Oxyopes* is the Striped Lynx Spider (*Oxyopes salticus,* pl. 45). In contrast to the darker Western Lynx Spider, the female Striped Lynx Spider has a cephalothorax that is generally creamy white to buff with several dark stripes behind her eyes. The female's abdomen is also pale with a dark heart mark and lateral stripes. The male's abdomen is unmarked and often has a metallic purplish or golden sheen. The Striped Lynx Spider also has distinctive thin, sharply defined black lines on the ventral surfaces of the femora on its fore legs. It lives in tall grasses and shrubs throughout California and along much of the Oregon coast. The third Pacific coast *Oxyopes* is the boldly patterned Trident Lynx Spider (*O. tridens*, pl. 45). It has a broad, buffy stripe running the length of its body that splits into a three-pronged fork near its eyes. The male and female are similarly patterned. This spider can be found in dry, rocky areas in Southern California. A revision of North America's *Oxyopes* can be found in Brady (1964).

Hamataliwa grisea (pl. 45) is a cryptically colored species normally found on trees and woody shrubs. Similar in size to the *Oxyopes* (between 5 and 7.5 mm, or .2 and .3 in., in length), it lacks their distinctive patterning. Instead, *H. grisea* is grayish brown above and pale yellow or tan below, with indistinct black and white markings on both its abdomen and cephalothorax. The female is recognizable by the long fringes of hair on her legs. *H. grisea* is an uncommon spider that has been found in coastal Southern California north through Los Angeles County. Brady (1964) discusses this species in his revision of the North American lynx spiders.

NATURAL HISTORY: Lynx spiders use both ambush and stalking behaviors when hunting. While most are associated with vegetation, the Trident Lynx Spider is a cursorial predator, most often found on walls or along the ground. The two most intensely studied species are *P. viridans* and the Striped Lynx Spider, *O. salticus*. *P. viridans* has an exceptionally broad diet that includes bees and wasps, butterflies, grasshoppers, flies, and even members of its own species. The Striped Lynx Spider has an equally diverse diet and can be quite common in agricultural fields. Lynx spiders often wait at promising locations, including the tops of flower heads, for insects to land. Alternatively, they may chase their prey through the vegetation.

Adult *Oxyopes* generally begin to appear in May, while the larger green lynx spiders take longer to mature and often do not appear until late June or early July. *P. viridans* is the only regional species of lynx spider whose courtship behaviors have been described. They include abdominal vibrations, leg lifts, palpal drumming, and light caresses. If receptive, the female will briefly touch the male's fore legs with her own, and then, attached to a thread of silk, she will drop off the edge of a leaf. While she hangs upside down, the male rotates her until her ventral side faces him. Then, hanging from his own thread, the male follows her off the leaf and gently taps on her abdomen, causing her to arch her body into a shallow U-shaped mating position. During mating, male *Peucetia* produce hard plugs that often contain the tips of their paracymbia. Lodged in a female's epigynum, a plug was previously thought to block subsequent males from mating. It is now known that it is not entirely effective, and *P. viridans* broods have been found that are the results of matings with multiple males. *Peucetia* females attach their egg sacs to the stems of large flowers or woody shrubs and protect them by resting their bodies on top of them. The female *P. viridans* also has an exceptional defensive behavior that she exhibits while guarding her eggs. When threatened, she can spray venom from her fangs in a stream up to 20 cm (7.9 in.) long. This makes *P. viridans* the only spider known to spray pure venom as a defensive response (Fink, 1984).

LYCOSIDAE — Wolf Spiders

Pls. 46, 47

IDENTIFICATION: Wolf spiders are among the most ubiquitous and easily identified spiders in North America. Except for members of the genera *Sosippus* and *Geolycosa,* wolf spiders are all cursorial hunters and are common in a wide variety of habitats, including grassy fields, suburban lawns, marshes, forests, coastal dunes, and deserts. Females are especially conspicuous when their egg sacs are attached to their spinnerets and when their bodies are covered with newly hatched young. One of their most distinguishing features is their highly unusual eye arrangement. While their anterior eyes are small and run in a straight line across the front of the carapace, their posterior median eyes are large and forward looking. Their posterior lateral eyes rest farther back, along the carapace margin. They are long-legged entelegynes and, as a family, range from 3 to 35 mm (.12 to 1.4 in.) in length. Most are patterned in shades of brown, dull yellow, gray, and black, although a few species are decorated with touches of pink or orange.

SIMILAR FAMILIES: While lycosids are readily recognized by their unusual eye arrangement, several families, superficially similar in color and behavior, may be mistaken for wolf spiders. Although funnel web weavers (Agelenidae, p. 232) rarely leave their conspicuous sheet webs, displaced individuals or wandering males may be confused for wolf spiders. However, funnel web weavers have a posterior eye row that is either straight or strongly procurved. Another spider likely to be mistaken for a wolf spider is *Zoropsis spinimana* (Zoropsidae, p. 220). This introduced species is well established around California's San Francisco Bay Area. It is an exceptionally large, six-eyed spider, and while its posterior eye row is recurved, it does not have the rectangular arrangement diagnostic of the wolf spiders. *Zora hespera,* the only Pacific coast member of the family Zoridae (p. 197), has an eye arrangement similar to that of the wolf spiders. However, *Z. hespera* is only around 3 mm (.12 in.) in length and distinctively patterned, and it has between six and eight pairs of spines on the undersides of its anterior tibiae, more than any wolf spider.

PACIFIC COAST FAUNA: Twelve genera represented by 64 regional species. Lacking bold colors or other readily distinguishing features, many genera and species regularly require a close examination of their reproductive structures for identification. However, once you become familiar with an area's resident species, each genus's particular physical and behavioral gestalt becomes increasingly obvious, helping you notice when you encounter something novel. By far, the most common of the region's lycosids are the thin-legged wolf spiders in the genus *Pardosa* (pl. 46). Forty described and numerous undescribed species have been reported from the contiguous Pacific coast states. They range from 3 to 12 mm (.12 to .47 in.) in length and vary considerably in color and pattern, from nearly black to pale sandy gray, and from solidly colored to conspicuously patterned. Their legs are exceptionally long and slender, and when running, they often hold their bodies high off the ground. While subtle, the shape of the carapace can be informative. When viewed head-on, the sides of the anterior portion of the carapace are nearly vertical, giving *Pardosa* a particularly "narrow-faced" appearance. In other genera, this area is either angled or slightly bulged out like lightly puffed cheeks. Found nearly everywhere, thin-legged wolf spiders are generally most abundant in open habitats, such as grassy fields, marshes, and river banks. Vogel (2004) provides the most comprehensive key to the region's thin-legged wolf spiders. Additional information can be found in Dondale (2007). Despite its focus on the spiders of Canada, Dondale and Redner (1990) includes a great deal of relevant information, especially regarding those species from Washington.

The pirate wolf spiders, *Pirata* (pl. 46), represent another genus of small wolf spiders. They are normally between 4 and 8 mm (.12 and .31 in.) in length and can be recognized by the presence of a V-shaped or tuning fork–shaped mark on their otherwise shiny carapaces. Only three species are known from the Pacific coast, the most common of which, *Pirata piraticus* (pl. 46), is found widely throughout the region from Southern California to northern Washington. *P. sedentarius* has been collected mainly in Southern California, with a single very old record from western Washington. The third species, *P. insularis,* has been found sporadically in southeastern Washington. *Pirata* are most common near wetlands, where they move rapidly across, and sometimes under, the water's surface. Both *P. sedentarius* and *P. piraticus* are also known from coastal brackish and saltwater marshes. A key to North America's *Pirata* can be found in Wallace and Exline's (1978) revision.

While *Pirata* and *Pardosa* house some of the region's smallest wolf spiders, *Hogna* and *Schizocosa* contain the largest. Although individual species can be fairly distinctive, as genera *Hogna* and *Schizocosa* can be challenging to differentiate. You may need to closely examine the spiders' reproductive structures as described in Dondale (2005a). *Schizocosa* (pl. 47) are common, generally nocturnal spiders that, when running, often carry their bodies fairly high off the ground. Along the West Coast, they range from 5.5 to 28.4 mm (.22 to 1.2 in.) in length and are occasionally referred to as lanceolate wolf spiders because of the sharply demarcated black or brown heart marks on the abdomens of most species. While many wolf spiders have heart marks, those on *Schizocosa* are often exceptionally well defined. Four species live along the Pacific coast, although only two, *Schizocosa mccooki* (pl. 47) and *S. maxima,* can be considered widespread. *S. mccooki* is common in grassland, chaparral, and desert habitats from Southern California to northern Washington. It is generally brownish gray with black parallel bands on either side of its carapace, which in turn are bordered by pale yellow marginal bands. It ranges from 9.1 to 22.7 mm (.36 to .89 in.) in length. Generally larger (16.1 to 28.4 mm, or .63 to 1.2 in., in length) is *S. maxima*. This impressive spider has been found along nearly the entire length of coastal and near-coastal California, with one record from Wasco County, Oregon. While the undersides of both *S. mccooki* and *S. maxima* are normally reddish brown, they can have conspicuous black markings, especially under their abdomens. These markings, however, are never as extensive as those found on the region's commonly occurring *Hogna*. The two remaining Pacific coast *Schizocosa* are predominately eastern species that have been found on very rare occasions in Washington (*S. minnesotensis* and *S. communis*) and Oregon (*S. minnesotensis*). A key to the North American *Schizocosa* can be found in Dondale and Redner (1978b).

Similar in appearance to *Schizocosa*, *Hogna* (pl. 47) has been in disarray for many years and would benefit greatly from a thorough revision. Five species of these medium to very large spiders (9.2 to 35 mm, or .36 to 1.4 in., in length) have been recorded from along the Pacific coast of the United States, although only one, *Hogna antelucana* (pl. 47), is relatively common. It has been found in grasslands throughout Southern and Central California. Similar in appearance to *Schizocosa,* it has a large heart mark that is generally browner and less defined around its edges. Its carapace is also comparatively broader across its anterior margin. What quickly differentiates this species from most of the region's other large wolf spiders is that its entire underside is black, a feature shared with the regionally very rare *H. carolinensis*, one of the largest spiders in North America (18 to 35 mm, or .71 to 1.4 in., in length). It ranges from pale gray to reddish brown and has an indistinct or partial heart mark on its abdomen. It has an entirely black venter and long, thick legs. *H. carolinensis* has been found several times in Southern California, where it excavates and builds turret-ringed burrows, similar to those of the burrowing wolf spiders (genus *Geolycosa*). While its underside is similar to that of *H. antelucana,* the two species differ considerably in the patterns of their upper parts. The remaining Pacific coast *Hogna* are regionally extremely rare. *H. labrea* is a poorly described, enigmatic species from Los Angeles County, California, that is most likely a misidentified *H. antelucana* (D. Bixler, pers. comm., 2011). *H. coloradensis* and *H. frondicola* are regionally known only from a few scattered records from Washington. Additional information on *H. antelucana* can be found in Montgomery (1904) and Gertsch and Wallace (1935); look to Dondale and Redner (1990) for information on *H. carolinensis* and *H. frondicola*; *H. coloradensis* was redescribed by Slowik and Cushing (2007); and Chamberlin and Ivie (1942b) described *H. labrea*.

Two species of *Alopecosa* (pl. 47) live within the contiguous Pacific coast states, both of which exhibit a fair degree of variability in their color and pattern and can either have or lack heart marks. *Alopecosa* range from 6.6 to 16 mm (.26 to .63 in.) in length and have stout bodies and long, robust legs. Females are generally gray to reddish brown and have broad, pale bands down the midlines of their carapaces. Adult males are often more strikingly colored with creamy white bands and black margins on their carapaces and an orangish dorsal abdominal wash. When an *Alopecosa* has a well-defined black heart mark, it might be mistaken for a *Schizocosa*. However, unlike *Alopecosa*, the region's *Schizocosa* have sharply defined pale margins to their carapaces. *Alopecosa kochi* is one of the most common wolf spiders in the western United States. It lives in forests and grassland from Southern California to northern Washington. Similar in color and pattern is *A. aculeata*. A distinctly more northern species, it has been found sporadically across Washington and in the mountains of northeastern Oregon. A key to the North American *Alopecosa* can be found in Dondale and Redner (1979).

With records extending from far above the Arctic Circle to Central America, *Arctosa* (pl. 46) is among the most widely distributed wolf spider genera in North America. Four species have been collected along the Pacific coast, and they range from 4.8 to 16 mm (.19 to .63 in.) in length. *Arctosa* are especially common in damp, sandy areas and to a lesser degree in bogs, tundra, and forests. Most appear to be nocturnal, and unlike other wolf spiders, female *Arctosa* care for their egg sacs in burrows rather than carrying them around. Like many genera, the features that define *Arctosa* involve subtleties of the spiders' reproductive structures. However, they have a fairly distinctive, broad, low carapace, the middle of which is either bare or only sparsely coated with short setae. Only two of the Pacific coast's *Arctosa* can be considered regionally common. *Arctosa littoralis* (pl. 46) is found widely throughout North and Central America, with regional records from Southern California to northern Washington. Its preferred habitat is sandy beaches, riverbanks, and desert streams. The coloration of *A. littoralis* varies dramatically to match its surroundings, from nearly pure white to heavily mottled. It also differs from all other regional *Arctosa* in that its anterior median eyes are nearly twice the size of its anterior laterals. The other fairly common Pacific coast species is *A. alpigena* (pl. 46). This dark species has an unusual pale heart mark and a creamy median band on its carapace that broadens anteriorly. Unlike *A. littoralis,* which is tied almost exclusively to sandy shores, *A. alpigena* is a spider of coniferous forests, alpine meadows, and bogs. It lives widely across Washington, and there are several records from northeastern Oregon. It is most common from early June through mid-September. The two remaining regional *Arctosa* species, *A. raptor* and the introduced *A. perita* (not illustrated), are extremely rare members of the northern Washington arachnofauna (R. Crawford, pers. comm., 2011). A revision of the North and Central American *Arctosa* can be found in Dondale and Redner (1983a).

Similar to *Arctosa* is the region's only species of *Allocosa, A. subparva* (pl. 46). It has a dark, shiny carapace, and its abdomen is mottled in brownish gray and yellow. Ranging from 4.8 to 9.3 mm (.19 to .37 in.) in length, *A. subparva* is found from Southern California to northern Oregon. It is most common along the banks of streams and beaches, where it spends the day hiding beneath driftwood and other debris. It has an indistinct pale median carapace band, and around each femur and tibia are two black rings. The most important features that differentiate *Allocosa* from *Arctosa* are the fine structures of their reproductive organs. However, because of the very small numbers of species found within the contiguous Pacific coast states and the limited areas of overlapping distribution, once you becomes familiar with one species, the differences between it and similar spiders should become increasingly apparent. A key to the North and Central American *Allocosa* can be found in Dondale and Redner's (1983b) revision.

Two extremely similar genera for many years were lumped

together are *Trochosa* and *Varacosa*. Broadly speaking, the two genera are differentiated by the fine details of their reproductive structures. However, because each genus is regionally represented by only a single species, and the two species are separated by both habitat and range, the identification of these genera is made much more straightforward. *Trochosa terricola* (pl. 46) is found across Eurasia and northern North America, with numerous records from Washington and Oregon into northernmost California (Siskiyou and Modoc Counties). It ranges from 7.2 to 11.2 mm (.28 to .44 in.) in length and has a thick body and legs. It is generally reddish brown in color with a broad, diffuse, pale band down the midline of its carapace and a pale heart mark. What immediately distinguishes *T. terricola* is the presence of two dark dashes within the median carapace band. It is most common in deep grass and leaf litter around woodland edges and is discussed in greater detail in Brady (1980). Also covered by Brady (1980) under its previous placement within *Trochosa* is the Pacific coast's only species of *Varacosa, V. gosiuta* (pl. 46). It lives in Southern California's dry foothills, in the Mojave Desert, and possibly in the Great Basin along Central California's eastern border. Similar in size and appearance to *T. terricola, V. gosiuta* is more yellowish brown in color and lacks the paired dashes on its carapace.

Among the most unusual of North America's Lycosidae are the burrowing wolf spiders, genus *Geolycosa* (pl. 47). From a side view, their carapaces rise sharply from their rear edges to their posterior lateral eyes. Burrowing wolf spiders do not readily wander but instead dig deep, silk-lined burrows around which they construct short turrets of sand, pebbles, and plant debris. Locally common throughout the Mojave Desert, especially around Joshua Tree National Park, is *Geolycosa gosoga* (pl. 47), the Pacific coast's only described species. Despite its unusual habits and large size (the adult is between 13 and 18 mm, or .51 and .71 in., in length), very little is known about this species. Its carapace is brownish gray, its abdomen is silvery, and while its underside is pale, most of the ventral portions of its fore legs are black. At night it hunkers at the edge of its burrow with its fore legs resting over the edge of its turret. When an insect wanders past, the spider lunges out and drags it back into its retreat for feeding. The burrowing wolf spider is a sedentary hunter, only leaving its burrow when forced to by external circumstances or as a dispersing juvenile or, if it is a male, when seeking out a mate. One other very rare regional Lycosidae, the Carolina Wolf Spider, *H. carolinensis,* builds a similar turret-ringed retreat, but its burrows are fairly shallow and angle off rather than descending vertically. Wallace (1942) revised *Geolycosa,* although only the female *G. gosoga* has been formally described.

Only a single species of funnel web wolf spider is found along the Pacific coast, *Sosippus californicus* (pl. 47). It builds a sheet web with a tubular retreat in one corner, much like those of funnel web weavers (Agelenidae, p. 232), although *S. californicus* is immediately identifiable by the characteristic Lycosidae eye arrangement. In addition to its unusual web, this relatively large spider (12.8 to 15 mm, or .5 to .6 in., in length) has a distinctive color pattern. Dorsally, it is dark brown with a creamy border to both the abdomen and cephalothorax. There is also a pale line down the center of its carapace and numerous small, white dashes on either side of its abdomen. In Southern California, *S. californicus* lives in riparian areas surrounded by desert scrub. Its web is built low to the ground and its retreat often extends into a pile of rock and debris. A revision of the genus can be found in Brady (2007).

One of the most specialized of the region's wolf spiders is *Melocosa fumosa* (pl. 47). It lives only under rocks and scree on mountain peaks above the tree line. It is a fairly large spider, ranging from 11 to 17 mm (.43 to .67 in.) in length, and within its habitat, it is very distinctive. It has a brown to black carapace that is shiny, low, and exceptionally broad. It also has a very hairy abdomen, and its posterior eyes are arranged in an especially wide quadrangle. All of the region's records are from Washington's Cascade Range. The most comprehensive discussion of *M. fumosa* is in Dondale and Redner (1990).

NATURAL HISTORY: Wolf spiders are among the most frequently encountered spiders in North America. Many species are diurnal, and just a few minutes in a summer field will often turn up numerous individuals moving through the grass. Others are nocturnal, and a night hike with a flashlight can reveal a great many wolf spiders, their eyes reflecting the light and sparkling like tiny stars skittering across the ground. With a few unusual exceptions (*Geolycosa* and *Sosippus*) wolf spiders are wandering hunters, using their keen eyesight and sensitivity to vibrations and chemical cues to find and capture prey. Insects are quickly dispatched with a venomous bite, often behind the head. If the prey is exceptionally large or aggressive, the spider may flip over on its back and, holding the prey against its chest, leave the insect struggling helplessly until the venom takes effect. *Geolycosa* are ambush predators that wait at the lip of the burrow for insects to wander by, while *Sosippus* feed on those that stumble onto their sheet webs.

Wolf spiders have been the focus of many behavioral studies looking at courtship and communication. As they wander, lycosids leave a dragline behind them. Chemical cues on the silk telegraph information about the spider that left it, including its species, gender, age, and how long ago the silk was left behind. When a wolf spider encounters the silk or feces of a larger, potentially predatory species, it may respond by reducing its own activity or by climbing off the ground. Its defensive response is especially vigorous if the predatory spider has been feeding on other wolf spiders. When adult males encounter draglines, they "test" them by rubbing chemosensitive palpal hairs across them. If they determine that a dragline was recently laid down by a mature female, the males may try to track her down. Courtship displays vary widely across the family and can include palpal drumming and other vibratory signals, leg waving (especially if the males have tufts or other "ornaments" on their fore legs), dance-like movements, and finally, caresses and light body tapping. If the females are receptive, they will respond with their own suite of behaviors, including reciprocal leg movements and body tapping. Mating is generally a rapid affair and the males usually die soon after.

One of the most distinctive aspects of wolf spider biology is their maternal care. Excepting members of genus *Arctosa*, female wolf spiders carry their egg sacs with them by their spinnerets. After the spiderlings hatch, a female tears the sac open and the young spiders swarm over her abdomen, clinging to the spiny, knobbed hairs that carpet her back. The spiderlings stay with their mother for up to a week without eating (possibly longer in the larger species). Over time they steadily drop off, and some further disperse by ballooning. The major exception is *Sosippus* spiderlings. They spend an extended period of time in their mother's web, feeding together on her kills and illustrating one of the rare instances of subsocial behavior in spiders.

Eventually they crawl away, creating regions with a high prevalence of related individuals. A similar situation exists for the burrowing wolf spiders, *Geolycosa*. Because the adult females are sedentary, the spiderlings disperse by crawling away from their natal burrows, although longer-distance dispersals can occur through ballooning. Most juvenile wolf spiders develop through fall, overwinter in a subadult state, and mature the following year. During winter, most hibernate or go into a temporary torpor, although a few species do remain active, surviving the coldest months under a thick blanket of snow.

AGELENIDAE — Funnel Web Weavers, Grass Spiders *(Agelenopsis)*

Pls. 48, 49, Fig. 16

IDENTIFICATION: Although funnel web weavers themselves are generally inconspicuous, their webs are often a prominent part of the summer landscape, especially in grassy fields, on shrubs, and in the corners of old buildings. Agelenids build webs that combine a flat sheet with a tubelike retreat at one corner, earning them their common name, funnel web weavers. They are eight-eyed entelegynes, and in every genus but *Tegenaria,* the eye rows are strongly procurved. Their posterior spinnerets are fairly long, often extending past the spiders' abdomens, and under magnification, a peppering of plumose hairs can be seen on their legs and bodies. While most funnel web weavers are pale yellow, brown, and gray, some have fairly bright pink or orange bands and dark chevrons on their abdomens. As a family they range from 5.3 to 18 mm (.21 to .71 in.) in length.

SIMILAR FAMILIES: Wolf spiders (Lycosidae, p. 223) are the spiders most likely to be mistaken for a funnel web weaver; however, only one lycosid genus, *Sosippus,* makes a sheet web, while the rest are wandering or burrowing hunters. The families also differ in their eye arrangements, with the wolf spiders' posterior eyes arranged in a distinctive rectangular pattern.

PACIFIC COAST FAUNA: Six genera represented by 57 regional species. The Pacific coast's funnel web weavers can be divided into several groups based on their eye arrangement, color pattern, and spinneret length. The first group, consisting of the genus *Tegenaria* (pl. 48, fig. 16), is the most distinctive of the region's Agelenidae. Unlike other genera, *Tegenaria* have eyes that are arranged in two straight or only slightly procurved

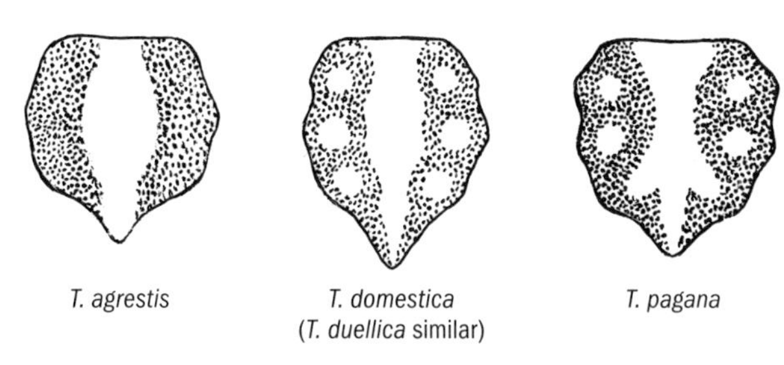

T. agrestis *T. domestica* (*T. duellica* similar) *T. pagana*

Figure 16. Sternal patterns of *Tegenaria* species.

rows. While other Agelenidae have two broad, well defined, dark bands across their carapaces, on *Tegenaria* the bands are fairly diffuse. The sterna on *Tegenaria* have sharply demarcated dark lateral borders, pale centers, and, on some species, two to four pairs of light spots (see fig. 16). Other genera of funnel web weavers can have similar sternal patterns, but they are rarely as sharply contrasting as those of *Tegenaria*. Although none of the region's four *Tegenaria* are native to North America, they are among the most common synanthropic spiders along the Pacific coast. The regional species range from 6.2 to 18 mm (.24 to .71 in.) in length.

One species that has received a great deal of media attention is the Hobo Spider (*Tegenaria agrestis*, pl. 48). Its bite is alleged to have caused necrotic lesions across the Pacific Northwest, an accusation that has certainly been greatly exaggerated. The name *agrestis* refers to the agricultural, rural nature of the spider's habitats in Europe, not to any particularly aggressive nature. Studies by Binford (2001) have also shown that there is no difference in venom composition between the *T. agrestis* spiders in Europe (where it is not a species of any medical concern) and those in the United States. Many of the same causative agents (including bacterial, viral, and fungal infections; chemical burns; assorted skin diseases; and media hype) that have led to a massive overdiagnosis of Brown Recluse bites across the United States are almost certainly falsely inculpating the Hobo Spider as well. It is found widely across Washington and Oregon and is likely present in rural Northern California. Also found across western Washington through west-central Oregon is the Giant House Spider (*T. duellica*, pl. 48) (occasionally known by its former name *T. gigantea*). In urban and suburban areas west of the Cascade Range, this exceptionally large spider is often more abundant than the generally more rural *T. agrestis*. In addition to the two species discussed above, there are the cosmopolitan Barn Funnel Weaver (*T. domestica*, pl. 48), which is exceptionally common around homes and barns throughout the Pacific coast states, and *T. pagana*, which appears to be regionally limited to coastal and near-coastal areas of California north to the San Francisco Bay Area. Although the reproductive structures are diagnostic for each species, a close look at the spider's sternal pattern can be useful when identifying a *Tegenaria* individual to the species level (see fig. 16). *T. pagana* has a pale, backward-pointing, central "trident" and two pairs of light spots in the dark lateral borders. The sterna of the Barn Funnel Weaver *(T. domestica)* and much larger Giant House Spider *(T. duellica)* have three or four pale spots in the dark lateral bands. Occasionally the spots on these two spiders are quite faint, and in this situation, either could easily be mistaken for a Hobo Spider, whose sternum has a broad, yellowish central band and solid dark borders. A key to the American *Tegenaria* can be found in Roth (1968), while more-detailed information on separating the Hobo Spider from the Giant House Spider can be found in Vetter and Antonelli (2002).

The remaining five genera of regional funnel web weavers have strongly procurved posterior eye rows and pairs of well-defined dark bands on their carapaces. One pair of genera, *Agelenopsis* and *Calilena*, share extremely long posterior lateral spinnerets in which the distal segments are about twice as long as the basal ones. *Agelenopsis* tend to have a yellowish-brown base color, and *Calilena* are generally grayer, although there is overlap and a confident identification often requires examining a spider's reproductive structures as described in Bennet and Ubick (2005). Also known as grass spiders, *Agelenopsis* (pl. 49) are found widely across North America, with six species along the Pacific coast, five of which are known from Washington, Oregon, and northwestern California. Only one species, *Agelenopsis aperta*, is found in Southern and Central California. *Agelenopsis* webs are commonly seen in the late summer and early fall blanketing grassy fields and other low vegetation and can measure up to half a meter across. *Agelenopsis* range from 6.7 to 17 mm (.26 to .67 in.) in length. The genus was revised in Chamberlin and Ivie (1941b), and Ayoub et al. (2005) used genetics to reveal the grass spiders' recent evolutionary history and mapped the distribution of North America's *Agelenopsis* species. Although the bites of very few spiders are capable of harming people, on rare occasions *A. aperta* can cause envenomation symptoms, including headaches, tenderness, swelling, and nausea (Vetter, 1998).

Very similar to *Agelenopsis* is *Calilena* (pl. 49). Ranging from 6.2 to 13 mm (.24 to .51 in.) in length, the vast majority of the region's twelve described *Calilena* species are known only from Southern California through the San Francisco Bay Area. Only two species are known from north of California: *Calilena stylophora* has been collected sporadically from Southern California to Tillamook County, Oregon, and *C. umatilla* has been found in Benton County, Washington. *Calilena* normally build their webs close to the ground with their retreats extending under debris and into cracks in the topsoil. Although several *Calilena* have been collected in chaparral and rocky grasslands in Central California, almost nothing is known about the habitat preferences of the genus as a whole. *Calilena* and its component species were described and their reproductive structures illustrated in Chamberlin and Ivie (1941a).

The final group of regional funnel weavers consists of three extremely similar genera, *Hololena, Novalena,* and *Rualena.* These spiders vary from pale yellowish to dark brown with two broad, dark bands across their carapaces, and many have a grayish-pink to yellowish-orange central stripe with dark scalloped edges on their abdomens. They build small sheet webs topped by loose tangles of threads, and at times localized populations can become very dense. While examining the structures of the different genera's reproductive organs is the most commonly cited means of separating them, other features, including eye size and spinneret arrangement, can be extremely useful. The most easily recognized of the three genera is *Rualena* (pl. 48). Nine regional *Rualena* species have been described, all of which are endemic to Southern and Central California, with several species known only from islands off the Southern California coast. *Rualena* tend to be smaller than other similar genera, ranging from 5.3 to 9.5 mm (.21 to .37 in.) in length, and their anterior median eyes are distinctly smaller than the anterior laterals. Most *Rualena* were described by Chamberlin and Ivie

(1942a), although a single species, *Rualena balboae* from San Diego County, was described by Schenkel (1950). Of the two remaining genera, *Hololena* (pl. 48) is the most diverse, with 22 regional species, only two of which have been collected north of California. *Hololena nedra* has been found from Southern California to northern Washington, and *H. rabana* is known from Jackson County, Oregon. *Hololena* range from 7 to 13.5 mm (.28 to .53 in.) in length and are grayish yellow to tan in color with a pinkish-orange to brown dorsal abdominal stripe. Their eyes are all about the same size, but the distal segments of their posterior lateral spinnerets are distinctly shorter than the basal segments. On members of the very similar genus *Novalena* (pl. 48), the posterior lateral spinnerets' distal and basal segments are about the same length, and the anterior eyes are slightly larger than the posteriors. *Novalena* also tend to be darker than *Hololena,* with a brownish carapace and grayish-brown abdomen, and they range from 9 to 11.4 mm (.35 to .45 in.) in length. Three of the region's four species are known only from Northern California's Sierra Nevada, while the fourth, *Novalena intermedia,* has been found from Northern California through northern Washington. Both *Hololena* and *Novalena* are revised in Chamberlin and Ivie (1942a).

NATURAL HISTORY: Agelenid webs are fairly uniform in appearance, each consisting of a flat sheet of nonsticky silk leading to a tubelike retreat. In many cases, there is also a tangle of dry threads above the sheet that act as a baffle to insects flying above. When an insect either is knocked down or stumbles onto the web, the spider rushes out across the web's upper surface, bites the insect, and carries it back into the retreat. Because they don't use sticky silk in their webs' construction, funnel web weavers depend on speed and the potency of their venom to capture prey. Some genera, especially *Agelenopsis,* can be extremely common in grassy fields, where their webs may blanket large swaths of the available habitat. *Tegenaria* are most prevalent in old buildings with lots of insects and crevices for their retreats. Other genera are most common in thick shrubs or low to the ground. While funnel web weavers are mainly nocturnal, they can regularly be seen during the day just inside their retreats, ready to pounce on any insects that fall onto their webs.

Most regional agelenids mature in the late summer and early fall, and during this period males often wander into homes, searching for a female's pheromone-laced web. Male funnel web weavers have a variety of courtship behaviors, including rapid leg pumping, tugging on the female's web, and abdominal rocking, all of which signal the male's presence as a potential mate and not as prey. If receptive, the female responds to his courting by folding her legs against her body and becoming effectively nonresponsive, a state normally referred to as cataleptic. The male then pulls the inert female back into her retreat, where mating occurs. The female lays egg sacs in her retreat and, in numerous species, covers them with a silk sheet, which is then decorated with bits of dirt and other debris. The Hobo Spider has an unusual egg sac that encapsulates a layer of dirt between two layers of silk. This may help explain why it is more common in rural areas than the Giant House Spider (R. Crawford, pers. comm., 2011).

AMAUROBIIDAE — Hacklemesh Weavers

Pl. 49, Fig. 17

IDENTIFICATION: The hacklemesh weavers are secretive, nocturnal spiders whose fairly distinctive webs consist of collars of tangled, bluish-gray cribellate silk radiating from protected crevices. Nearly every hacklemesh weaver has eight eyes in two transverse rows (the exception being one uncommon, minute California species) and a divided cribellum. Amaurobiidae are generally robust spiders with relatively thick legs, ranging from 1.3 to 22 mm (.05 to .87 in.) in length. Other defining features of the family include a calamistrum that is never more than half the length of metatarsus IV, rows of trichobothria on the tarsi and metatarsi that increase in length as they progress down each leg segment, and four or fewer teeth on the cheliceral retromargins.

SIMILAR FAMILIES: Several families may be mistaken for Amaurobiidae, including Amphinectidae (p. 244), which is regionally represented by a single introduced species, *Metaltella simoni*. Along the Pacific coast, this synanthropic spider is limited to Southern California, where it can be extremely common. It looks like a dark, medium-size Amaurobiidae, but hacklemesh weavers are never as abundant as *M. simoni* in Southern California's suburban neighborhoods. *M. simoni* also has three pairs of ventral spines on its anterior tibiae and metatarsi and between five and seven teeth on each cheliceral margin, more than any North American amaurobiid. Members of the family Titanoecidae (p. 241) are also similar to some amaurobiids, but on females the calamistrum is around 70 percent as long as metatarsus IV. Male titanoecids, like many male hacklemesh weavers, have significantly reduced or absent calamistra. Members of the family Titanoecidae also prefer more arid regions, in contrast to the amaurobiids' penchant for humid areas. Along the immediate coast, from Southern California to northern Oregon, the introduced Gray House Spider (*Badumna longinqua,* Desidae, p. 243) can be very common around buildings and in dense ornamental plantings. This spider makes a distinctive lattice-like cribellate web and, unlike the region's amaurobiids, it is purplish gray and has anterior median eyes that are larger than the others. On the hacklemesh weavers, the anterior median eyes are never the largest and are often significantly reduced in size.

PACIFIC COAST FAUNA: Seven genera containing 68 regional species. California is the epicenter of Amaurobiidae diversity in North America, with many species represented by small, isolated populations. The region's hacklemesh weavers can be divided into two groups. The first contains four genera: *Callobius, Amaurobius, Pimus,* and *Cybaeopsis.* These are larger spiders (greater than 3 mm, or .12 in., in length with one rare exception) with leg spines. This assemblage also includes the region's most frequently encountered hacklemesh weavers. The differentiation of these genera often requires examining their reproductive structures as outlined in Ubick (2005a). However, body size, location, and color pattern can provide important clues. The most widespread of these genera is *Callobius* (pl. 49), with 22 species extending from Southern California to northern Washington. These spiders are especially prevalent under tree bark, both standing and fallen. *Callobius* also contains the largest of the region's hacklemesh weavers, with West Coast representatives ranging from 7 to 22 mm (.28 to .87 in.) in length. They are often boldly patterned with dark reddish-orange cephalothoraxes and legs and gray abdomens decorated with pale spots and chevrons. The most comprehensive guide to their identification is Leech (1972).

Very similar to *Callobius* is *Amaurobius* (pl. 49) with 23 regional species. While the introduced species *Amaurobius ferox* is established in King County, Washington (R. Crawford, pers. comm., 2011), all of the region's native species are endemic to California. *Amaurobius* are widespread in wooded areas across the state, although individual species often have highly restricted distributions. While there is some overlap in size, *Amaurobius* tend to be smaller than *Callobius,* ranging from 2.9 to 14 mm (.11 to .55 in.) in length, although most species are less than 9 mm (.35 in.) long. They also generally have more-muted patterning on their abdomens. *Amaurobius* live under the bark of fallen trees, under rocks, and in rotting logs. A key to the region's species can be found in Leech (1972).

The two remaining large-bodied genera, *Pimus* and *Cybaeopsis,* tend to be less contrastingly colored than either *Amaurobius* or *Callobius.* Ten species of *Pimus* (pl. 49) occur along the Pacific coast from Central California through southwestern Oregon, both along the coast and in the Sierra Nevada. The one geographical exception is *Pimus salemensis,* a species known only from Marion County, Oregon. *Pimus* normally have yellowish-tan carapaces and legs and pale gray abdomens that are occasionally marked with gray chevrons. They range from 4.8 to 8 mm (.19 to .31 in.) in length and are most common beneath rocks and logs on the forest floor. The genus was revised in Leech (1972).

Regionally, *Cybaeopsis* (pl. 49) are limited to a strip running from southwestern Oregon to northwestern Washington. They are fairly dark spiders and range between 4 and 8.5 mm (.16 and .33 in.) in length. Three species are known from the contiguous Pacific coast states. Two of these, *Cybaeopsis macarius* and *C. spenceri,* are regionally limited to western Washington, while the third, *C. wabritaska,* is found from southern Oregon to northern Washington. Both *C. macarius* and *C. wabritaska* have pinkish foliate marks on their otherwise dark grayish-brown abdomens, while the abdomen of *C. spenceri* has pale, indistinct blotches on its anterior portion and chevrons on its posterior section. All three species are discussed in Leech (1972) under their old placement in the genus *Callioplus.*

The second division within the Pacific coast's Amaurobiidae fauna consists of three genera of tiny, sexually dimorphic, leaf litter–dwelling spiders without leg spines: *Zanomys, Parazanomys,* and *Cavernocymbium.* Males have round carapaces while those of females narrow anteriorly, giving them a distinctly

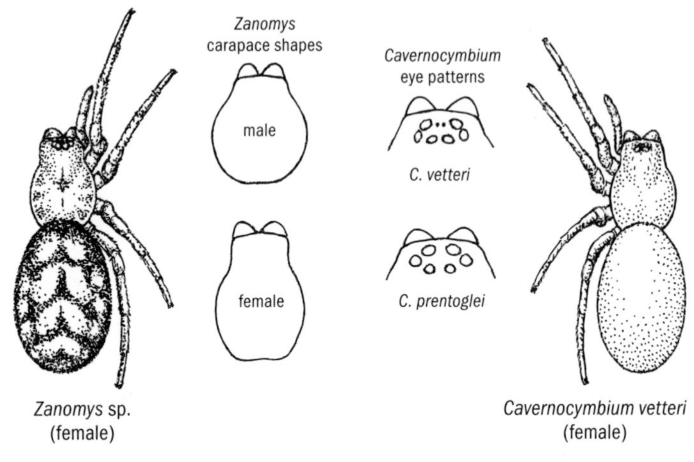

Figure 17. Small Amaurobiidae spiders in the genera *Zanomys* and *Cavernocymbium*.

pear-shaped appearance. The most widespread and species-rich of these genera is *Zanomys* (fig. 17). All seven of the world's species live along the Pacific coast, five in California and two in Oregon and Washington. They range from 1.5 to 2.5 mm (.06 to .1 in.) in length and are dusky yellow with gray abdomens that are often decorated with transverse bars, diffuse chevrons, or pale spots. They can be common in both mountain and coastal woodland leaf litter, where they build small sheet webs between fallen leaves, under rocks, and inside rotting logs. A key to the *Zanomys* can be found in Leech (1972).

Closely related to *Zanomys* are *Parazanomys* and *Cavernocymbium,* both of which are endemic to Southern California. Only a single species of *Parazanomys* (not illustrated) has been described, *Parazanomys thyasionnes*. Known from Tulare County, it ranges from 2. to 3.3 mm (.08 to .13 in.) in length and has a grayish cephalothorax and a dusky-yellow to greenish-black abdomen that is often decorated with rows of broad chevrons. *P. thyasionnes* lives in a wide variety of habitats but is most common in mixed oak and broadleaf forests. Quite similar to *Zanomys,* it differs in the arrangement of its reproductive structures. *Cavernocymbium* (see fig. 17) are diminutive spiders ranging from 1.2 to 2.6 mm (.05 to .1 in.) in length. With their unmarked whitish abdomens, they differ considerably from related, more ornately patterned genera. While the anterior median eyes in both *Parazanomys* and *Zanomys* are small, they are either especially minute or absent in *Cavernocymbium*. Two species have been described: *Cavernocymbium vetteri*, which is known from a small area of San Bernardino County, and the six-eyed *C. prentoglei*, which lives in coastal and (rarely) desert scrub leaf litter in both Riverside County and Imperial County. *Parazanomys* and *Cavernocymbium* are described in detail in Ubick (2005d).

NATURAL HISTORY: A great deal remains to be learned regarding the natural history of these often cryptic, nocturnal spiders. California is one of the world's centers of Amaurobiidae diversity. Given the spiders' limited dispersal abilities and California's complex geological history, it is not surprising that many species are known only from small, segregated populations. The courtship behaviors of many hacklemesh weavers are unknown; however, the egg sacs of several genera have been described. *Callobius* cover their eggs with a thick coating of cottony, white silk, often with bits of dirt stuck to the outside. The eggs are then hidden under loose bark and rocks. Female *P. thyasionnes* suspend their lens-shaped egg sacs in loose, silk tents sequestered beneath rocks and other protective covers.

After her young hatch, the female *A. ferox* lays a second batch of eggs as food for the newly emerged spiderlings. Several days later, the spiderlings devour their mother, making *A. ferox* one of the very few animals known to engage in obligatory matriphagy. The young spiders then stay together in their mother's web for up to four weeks, where collectively they are capable of capturing prey much larger than any one spiderling could do alone. Experimental evidence has shown that by feeding on their mother, the young spiders grow more rapidly and have a greater survival rate at the time of dispersal than individuals whose mother is removed but otherwise are well fed (Kim et al., 2000).

TITANOECIDAE

Pl. 50

IDENTIFICATION: Titanoecidae is a small family of eight-eyed, cribellate spiders. Because of their similarity to some of the region's other cribellate families, several morphological features must be used in conjunction to confirm a titanoecid's identity. Their endites are nearly parallel, and their cribella are longitudinally divided into two adjacent oval plates. While female titanoecids have calamistra that extend nearly the entire length of their fourth metatarsi, on males the calamistra are either vestigial or lacking entirely. They have numerous leg spines but lack long, conspicuous trichobothria. The region's Titanoecidae range in size from 4.5 to 8 mm (.18 to .31 in.) in length.

SIMILAR FAMILIES: Members of the family Titanoecidae have historically been housed within the family Amaurobiidae (p. 237), but unlike titanoecids, amaurobiids have rows of trichobothria that increase in length distally on both their tarsi and metatarsi. On amaurobiids, the calamistra are also only around half the length of their fourth metatarsi. Titanoecids generally have entirely dark abdomens, while many amaurobiid abdomens are decorated with light chevrons, spots, or stripes. Although there is broad overlap in the distributions of the two families, titanoecids are more common in arid regions, while amaurobiids prefer cooler, more humid areas.

In Southern California, the introduced *Metaltella simoni* (Amphinectidae, p. 244) may be confused with a titanoecid. While the *M. simoni* carapace is brownish orange with a distinct darker area around the eyes and clypeus, the carapaces on the Pacific coast's titanoecids tend to be more uniformly colored, darkening only a little near their anterior margins. Members of the family Titanoecidae also have only a few distal metatarsal spines, while *M. simoni* has three pairs on the underside of each metatarsus.

PACIFIC COAST FAUNA: One genus with two regional species. *Titanoeca nigrella* (pl. 50) is widespread along the Pacific coast from Southern California to southern British Columbia, although it is much more common in the dry interior regions than in the humid coastal forests. Its carapace is orange to brownish red, and its abdomen is oval and unpatterned and ranges from dark gray to black. The second regional species, *T. silvicola,* is essentially identical to *T. nigrella*. Found widely across western North America, the only Pacific coast record is an individual collected in Plumas County, California. To distinguish these two spiders, you must look at their legs and reproductive structures. The female *T. nigrella* has only three spines on the distal margins of the undersides of metatarsi I and II, while the female *T. silvicola* has four. Separating the males requires a detailed examination of their palpal features. A key and descriptions of the region's Titanoecidae can be found in Leech (1972) as part of his massive revision of the North American Amaurobiidae.

NATURAL HISTORY: Little is known regarding the natural history of North America's Titanoecidae. They are more common in arid regions, where they weave tangled webs of cribellate silk under rocks and fallen wood. *T. nigrella* is most common at elevations below 5,500 ft, while *T. silvicola* is generally found at elevations above that. Adults of both species are most frequently encountered in late June and July. After mating, female titanoecids suspend their spherical, cream-colored egg sacs in their webs and cover them with thin crusts of dirt.

DESIDAE

Pl. 50

IDENTIFICATION: The Pacific coast's only Desidae species is the introduced Gray House Spider, *Badumna longinqua*. It is a locally common, synanthropic spider that has a divided cribellum and anterior median eyes that are larger than the others. It is purplish brown with a gray abdomen and builds very distinctive lattice-like cribellate webs.

SIMILAR FAMILIES: The Gray House Spider was initially thought to be a hacklemesh weaver (Amaurobiidae, p. 237), but its coloration, unusual web, and large anterior median eyes all stand in contrast to those of the native Amaurobiidae.

PACIFIC COAST FAUNA: One genus containing a single introduced species. The Gray House Spider (*B. longinqua*, pl. 50) ranges from 5 to 12 mm (.2 to .47 in.) in length and has a divided cribellum and exceptionally large anterior median eyes. Its abdomen is purplish gray and its carapace is purplish brown, darker around the eyes and jaws, and is frosted with silvery-gray hairs. *B. longinqua* is an exceptionally common spider in coastal urban and suburban areas, where its lattice-like webs can be found around buildings and ornamental plantings. Based on specimens from San Francisco, it was first thought to be a native member of California's spider fauna (Gertsch,

1937). Only later was it identified as an Australian spider that had been introduced to western North America, New Zealand, and several South Pacific islands. It is found almost exclusively along the immediate coast, becoming rare even 15 miles inland, and it ranges from northern Baja California to northern Oregon (R.J. Adams, pers. obs., 2009). The morphology of the Gray House Spider is discussed in greater detail in Leech (1972) under its previous name, *Ixeuticus martius*.

NATURAL HISTORY: The Gray House Spider spins easily recognized sheets of lattice-like cribellate silk. They extend out from a tubular retreat that is hidden either in dense foliage or in a small opening, especially along windowsills and fencerows. In California, the Gray House Spider seems to have a special affinity for the small juniper bushes commonly used in coastal gardens and can become so numerous that most of the plant's surface is covered by their distinctive webs. On the coast of Central California, the Gray House Spider readily eats a wide variety of flies. It doesn't wrap its prey, but instead bites the entangled insect until it quits struggling. From summer through early fall, the adult male wanders in search of females, although the particulars of *B. longinqua* courtship have never been described. After mating, the female constructs a white, disk-shaped egg sac inside her retreat, which she guards until the spiderlings hatch and disperse. Interestingly, wandering Gray House Spiders along the Pacific coast are often preyed upon by another common introduced species, the Long-bodied Cellar Spider (*Pholcus phalangioides*, Pholcidae, p. 73).

AMPHINECTIDAE

Pl. 50

IDENTIFICATION: The region's only amphinectid, *Metaltella simoni,* is a cribellate, eight-eyed, three-clawed entelegyne whose most diagnostic feature is its dentition. It is the only cribellate spider in North America with five or more teeth on each cheliceral margin. It is also exceptionally common around homes with well-watered lawns and gardens across a large portion of Southern California.

SIMILAR FAMILIES: Amphinectidae is most similar to the families Amaurobiidae (p. 237) and Titanoecidae (p. 241). All three families share numerous features, including a divided cribellum, and Amphinectidae and Amaurobiidae share a relatively short calamistrum. However, neither Amaurobiidae nor Titanoecidae ever have more than four teeth on their cheliceral margins. Additionally, no other similarly colored amaurobiid or titanoecid is as abundant in Southern California's urban parks and suburban neighborhoods as *M. simoni*.

PACIFIC COAST FAUNA: Native to eastern South America, *M. simoni* (pl. 50) has been introduced to both the Gulf Coast of the United States and to Southern California. It ranges from 6 to 10 mm (.24 to .4 in.) in length and has a burnt-orange to mahogany-brown carapace that is noticeably darker around the eyes and chelicerae. The abdomen varies from dusky yellow with gray smudging to nearly black, and it may show several pale gray chevrons along the midline. First collected regionally in 1994 in Riverside County (Vetter and Visscher, 1994), it has since been found across suburban Southern California from San Diego County to northern Los Angeles County, with additional records from western San Bernardino County and from Palm Springs, Riverside County (Vetter et al., 2008).

NATURAL HISTORY: Native to Uruguay and Argentina, *M. simoni* quickly became an established member of Southern California's arachnofauna. In some parts of San Diego it has become so abundant that pest control measures have been initiated (Vetter et al., 2008). It is a ground-dwelling spider that builds cribellate webs under trash cans, flower pots, and other materials. While tolerant of a wide range of temperatures, it appears to require fairly humid conditions, such as those around landscaped gardens and lawns. This propensity for damp areas with ample low-lying retreats, a condition commonly found in nurseries, almost certainly contributed to its rapid spread throughout Southern California and likely played a role in its initial introduction to the state. Regionally, the female produces the majority of her globular egg sacs in midwinter, suspending them within her web.

GLOSSARY

Abdomen The posterior portion of the spider's body. The plural is *abdomens.*

Apophysis A stiff or membranous outgrowth most commonly found on the palpal segments but which can also be found on the chelicerae or legs of some adult male spiders. The plural is *apophyses.*

Araneomorph A member of the infraorder Araneomorphae, the more abundant and more evolutionarily derived of the two North American spider suborders. Araneomorphs are occasionally referred to as the "true" spiders.

Araneophagy Specializing in the eating of spiders.

Autospasy A defensive condition in which a spider is capable of abandoning a leg when it is grabbed or ensnared.

Balloon A means of dispersal used mainly by juvenile spiders in which they release strands of silk to be caught in the wind. The spiders are then carried aloft and deposited elsewhere.

Calamistrum A row of stout, curved setae on the dorsal edge of metatarsus IV on cribellate spiders that is used to card out silk from the cribellum. The plural is *calamistra.*

Carapace The dorsal plate covering the cephalothorax. The plural is *carapaces.*

Cephalothorax The anterior portion of a spider's body including the carapace, mouthparts, and legs. The plural is *cephalothoraxes.*

Chelicera The spider's "jaw." The section of a spider's mouthparts consisting of a large basal portion tipped by a hinged fang. The plural is *chelicerae.*

Clade A term used to define the entirety of a group of related organisms consisting of an ancestral species and all of its decedents.

Claw tuft A dense brush of setae between the tarsal claws.

Clypeus The area between the leading edge of the carapace (above the chelicerae) and the anterior eyes. The plural is *clypei.*

Colulus A nonfunctional cribellum that is often reduced to a fleshy lobe or collection of setae between the anterior lateral spinnerets. The plural is *coluli.*

Coxa The most basal segment of the legs and palps. The plural is *coxae.*

Cribellate Having a cribellum.

Cribellum A broad plate (or pair of plates when divided) that rests just anterior to the spinnerets and produces cribellate silk. The plural is *cribella.*

Cuspule A minute, thickened, spine-like structure found on the legs or mouthparts of some spiders. The plural is *cuspules.*

Cymbium Cuplike structure at the end of the adult male spider's palp that houses its copulatory organs. The plural is *cymbia.*

Embolus Organ of the male palp used to pull sperm into the sperm reservoir and deliver it into the female's spermathecae. The plural is *emboli.*

Endite The large platelike extension of the palpal coxa that rests one on either side of the labium. The plural is *endites.*

Entelegyne Relating to araneomorph spiders whose females have external copulatory openings, usually in the form of a sclerotized epigynum, and those males that have complex reproductive organs on their palps.

Epigastric furrow A transverse groove across the ventroanterior portion of the abdomen that houses the gonopore.

Epigynum A sclerotized structure resting between the book lungs and anterior to the epigastric furrow that covers the female's internal genitalia. It is found almost entirely on entelegyne spiders. The plural is *epigyna.*

Exuvia The exoskeleton that is shed after a molt. The plural is *exuviae.*

Femur The third segment from the base on a spider's leg or palp. The femur is between the trochanter and patella. The plural is *femora.*

Foliate Lobed and leaf-like. This pattern is found on the abdomen of many spider species.

Gonopore The genital opening within the epigastric furrow. The plural is *gonopores.*

Haplogyne Relating to those spiders whose females have internal copulatory openings and normally lack an epigynum. The males have simple bulb-like reproductive structures on their palps. This is the more primitive reproductive style and is found on mygalomorphs and numerous araneomorph families.

Holarctic Relating to the ecozone consisting of the northern continental landmasses around the world, including North America and Eurasia.

Instar A developmental stage prior to maturity. The plural is *instars.*

Labium The anterior extension of the sternum bordered on either side by the endites. The plural is *labia.*

Lamelliform Having a spreading or leaflike structure.

Laterigrade Relating to the orientation of the limbs in which they are sidewise from the body and rotated so that the ventral surface faces outward toward the front of the spider (e.g., in Thomisidae and Sparassidae).

Mesic Relating to a damp or humid environment.

Metatarsus The sixth segment from the base on a spider's leg. The metatarsus is located between the tibia and tarsus. The plural is *metatarsi.*

Mygalomorph A member of the infraorder Mygalomorphae, the more evolutionarily conservative of the two North American spider suborders. Mygalomorphae contains the tarantulas and trapdoor spiders.

Nearctic Relating to the ecozone covering the vast majority of North America south to southern Mexico.

Palp A modified portion of the mouthparts, and the most anterior leglike appendage, that on an adult male carries the copulatory organs. Also called *pedipalp.* The plural is *palps.*

Paracymbium A rigid (occasionally flexible) outgrowth off the cymbium. The plural is *paracymbia.*

Patella The fourth segment from the base on a spider's leg or palp. The patella is found between the femur and the tibia. The plural is *patellae.*

Pedicel The stalk that connects the cephalothorax to the abdomen. The plural is *pedicels.*

Pedipalp See *palp.* The plural is *pedipalps.*

Pleurite A lateral sclerite. The plural is *pleurites.*

Plumose Feather-like.

Preening brush A thick collection of setae found at the ventrodistal end of the posterior metatarsi on some spiders.

Preening comb A transverse row of stiff setae along the ventrodistal edge of the posterior metatarsi on some spiders.

Procurved Curved in a way that makes the ends anterior to the middle of the bend. When eyes are procurved, the lateral eyes lie in front of the median eyes.

Prograde Relating to the orientation of the limbs in which legs I and II point forward and legs III and IV point backward, which is the most common spider leg arrangement (e.g., in Gnaphosidae).

Promargin The front or anterior margin of the cheliceral fang furrow.

Rastellum A stiff rake-like feature located near the fang base in some mygalomorphs. The plural is *rastella.*

Recurved Curved in a way that makes the ends posterior to the middle of the bend. When eyes are recurved, the lateral eyes lie behind the median eyes.

Retromargin The rear or posterior margin of the cheliceral fang furrow.

Scape A sclerotized projection down the midline of the epigynum on some spiders. The plural is *scapes.*

Sclerite A sclerotized plate. The plural is *sclerites.*

Scopula A thick brush of hairs found on the ventral surface of the tarsi and metatarsi of some spiders. Also the dense patches of setae on the endites and labium that assist in the filtering of food. The plural is *scopulae.*

Scute A shiny, sclerotized plate found on the abdomen on some spiders. The plural is *scuta.*

Serrula A row of minute teeth found along the anterior edge of the endite on some spiders. The plural is *serrulae.*

Spermatheca A sac for sperm storage, found on any adult female spider. The plural is *spermathecae.*

Spination The number and arrangement of spines on a spider's leg.

Spinneret Tubular or conical silk-extruding organ at the posterior end of the abdomen. The plural is *spinnerets.*

Stabilimentum A thick band of silk (or debris) strung through the hub of the web by some orb-weaving spiders. The plural is *stabilimenta.*

Sternum The large ventral plate on the cephalothorax. The plural is *sterna.*

Stridulation The production of sound made by rubbing stiff setae or tubercles across a row of fine ridges.

Synanthropic Adapted to live among people.

Tarsus The seventh and most distal of a spider's leg segments. Also the sixth and final segment on a spider's palp. The plural is *tarsi.*

Thoracic furrow See *thoracic groove.*

Thoracic groove The depression or invagination on the carapace where the stomach muscles attach. Also called *thoracic furrow.*

Tibia The fifth segment from the base on a spider's leg or palp. The tibia is found between the patella and metatarsus on the leg and between the patella and tarsus on the palp. The plural is *tibiae.*

Tracheal spiracle The entrance to the tracheal respiratory system, found on the ventral abdominal surface.

Trichobothrium A long, slender sensory hair that emerges from a small pit and generally stands perpendicular to the leg segment from which it arises. The plural is *trichobothria*.

Trochanter The second segment from the base on the spider's leg or palp. The trochanter rests between the coxa and femur. The plural is *trochanters*.

Troglobitic Adapted to a cave-dwelling lifestyle.

Tubercle A rounded or nodule-like growth. The plural is *tubercles*.

Urticating hair A barbed, irritating defensive hair found on the abdomen of a tarantula (Theraphosidae).

Venter The underside of an organism. On spiders, this can include the labium, endites, sternum, coxae, and the ventral surface of the abdomen.

Xeric Relating to a dry or arid habitat.

ADDITIONAL RESOURCES

There is a great deal of information about spiders available both in print and on the World Wide Web. Unfortunately, much of this information is incorrect, and it can be difficult to determine which resources to trust. While not exhaustive, the following books and Internet sites are well regarded and can provide accurate information on spiders and other arachnids.

Books (see full citations in References)

Spiders of North America: An Identification Manual (2005) by Darrel Ubick, Pierre Paquin, Paula E. Cushing, and Vincent D. Roth (eds.). This incredibly detailed manual allows one to accurately identify North America's spider fauna to the family and genus levels. It is not a field guide but a scientific manual for preserved specimens and nearly always requires examining spiders under a dissecting microscope.

The Common Spiders of North America (2013) by Richard Bradley. This richly illustrated book describes nearly 500 of the most widespread and distinctive spiders in North America.

Spiders and Their Kin: A Golden Guide (2002) by Herb and Lorna Levi. This is a pocket-size field guide to many of North America's common spiders, with additional information on other arachnid orders from around the world.

Arachnids (2009) by Jan Beccaloni. This beautiful full-color book covers the diversity, biology, distribution, and ecology of the class Arachnida. While broad in its coverage, it does an exemplary job reviewing what is known about each of the arachnid orders and contains many stunning photos.

Biology of Spiders (1996) by Rainer F. Foelix. This book provides an in-depth review of many aspects of spider natural history, including their ecology, development, and physiology.

Internet Resources

World Spider Catalog
http://research.amnh.org/iz/spiders/catalog/INTRO1.html
This continuously updated website provides the most up-to-date taxonomic coverage of the world's spiders along with their accompanying bibliographic citations.

American Arachnological Society
www.americanarachnology.org/
This is the home page for the American Arachnological Society, with a link to the *Journal of Arachnology*. Most of the journal's articles are available for free download.

American Museum of Natural History Scientific Publications
http://library.amnh.org/scientific-publications
This site allows free access and downloads of the American Museum of Natural History publications. These include *American Museum Novitates* and the *Bulletin of the American Museum of Natural History*, both of which have published numerous articles of arachnological interest.

Biodiversity Heritage Library
www.biodiversitylibrary.org/Default.aspx
This free database allows access to thousands of publications, including many important out-of-print journal articles on spider biology and taxonomy.

Spider Research, University of California at Riverside
http://spiders.ucr.edu/
Dispelling myths and rumors, the information on this site reviews many of the stories and urban legends that have sprung up around spiders over the years, providing instead factual information regarding the perceived "threats" from many spider species.

BugGuide
http://bugguide.net/
Unlike the sites above, BugGuide is a photo collection publicly edited by both professional entomologists and amateur enthusiasts. Hosted by Iowa State University's Department of Entomology, this photo gallery and archive is organized by an active community of people interested in entomology and arachnology. While it is not a "professionally" curated photo collection, many of the

contributors are quite knowledgeable, and a great deal of good information is freely shared.

Arachnology Resources

BioQuip
www.bioquip.com/
While numerous companies sell equipment and materials for biological research, BioQuip is the largest company specializing in entomology and arachnology supplies and books.

REFERENCES

Agnarsson, I. 2004. Morphological phylogeny of cobweb spiders and their relatives (Araneae, Araneoidea, Theridiidae). *Zoological Journal of the Linnean Society* 141: 447–626.

Ayoub, N. A., S. E. Riechert, and R. L. Small. 2005. Speciation history of the North American funnel web spiders *Agelenopsis* (Araneae: Agelenidae): Phylogenetic inferences at the population-species interface. *Molecular Phylogenetics and Evolution* 36: 42–57.

Banks, N. 1894. Two families of spiders new to the United States. *Entomological News* 5: 298–300.

Barnes, R. D. 1958. North American jumping spiders of the subfamily Marpissinae (Araneae, Salticidae). *American Museum Novitates* 1867: 1–50.

Beatty, J. 1970. The spider genus *Ariadna* in the Americas (Araneae, Dysderidae). *Bulletin of the Museum of Comparative Zoology, Harvard* 139: 433–518.

Beccaloni, J. 2009. *Arachnids.* University of California Press, Berkeley and Los Angeles, California. 320 pp.

Bennett, R. G. 1988. The spider genus *Cybaeota* (Araneae, Agelenidae). *Journal of Arachnology* 16: 103–119.

Bennett, R. G. 1991. The systematics of North American Cybaeid Spiders (Araneae, Dictynoidea, Cybaeidae). Unpublished Ph.D. thesis, University of Guelph, Ontario, Canada. 308 pp.

Bennett, R. G. 2005a. Cybaeidae. p. 85 *in* D. Ubick, P. Paquin, P. E. Cushing, and V. Roth (eds.). *Spiders of North America: An Identification Manual.* American Arachnological Society, Keene, New Hampshire.

Bennett, R. G. 2005b. Dictynidae. p. 95 *in* D. Ubick, P. Paquin, P. E. Cushing, and V. Roth (eds.). *Spiders of North America:*

An Identification Manual. American Arachnological Society, Keene, New Hampshire.

Bennett, R. G. 2005c. Hahniidae. p. 112 *in* D. Ubick, P. Paquin, P. E. Cushing, and V. Roth (eds.). *Spiders of North America: An Identification Manual.* American Arachnological Society, Keene, New Hampshire.

Bennett, R. G. 2006. Ontogeny, variation, and synonymy in North American *Cybaeus* spiders (Araneae: Cybaeidae). *Canadian Entomologist* 138: 473–492.

Bennett, R. G. and L. J. Brumwell. 1996. *Zora hespera* in British Columbia: A new spider family record for Canada (Araneae: Zoridae). *Journal of the Entomological Society of British Columbia* 93: 105–109.

Bennett, R. G. and D. Ubick. 2005. Agelenidae. p. 56 *in* D. Ubick, P. Paquin, P. E. Cushing, and V. Roth (eds.). *Spiders of North America: An Identification Manual.* American Arachnological Society, Keene, New Hampshire.

Bentzien, M. M. 1976. Biosystematics of the Spider Genus *Brachythele* Ausserer (Araneida: Dipluridae). Unpublished Ph.D. thesis, University of California, Berkeley. 136 pp.

Berman, J. D. and H. W. Levi. 1971. The orb weaver genus *Neoscona* in North America (Araneae: Araneidae). *Bulletin of the Museum of Comparative Zoology, Harvard* 141(8): 465–500.

Binford, G. J. 2001. An analysis of geographic and intersexual variation in venoms of the spider *Tegenaria agrestis* (Agelenidae). *Toxicon* 39: 955–968.

Blackledge, T. A., J. A. Coddington, and R. G. Gillespie. 2003. Are three-dimensional spider webs defensive adaptations? *Ecology Letters* 6: 13–18.

Blanke, R. 1975. Untersuchungen zum Sexualverhalten von *Cyrtophora cicatrosa* (Stoliczka) (Araneae, Araneidae). *Zeitschrift für Tierpsychologie* 37: 62–74.

Blanke, R. 1986. Im Fortpflanzungsverhalten von Kreuzspinnen (Araneae, Araneidae), und deren Interpretation im Kontext von Systematik und der Existenz von Artbarrieren. pp. 69–94 *in* J.A. Barrientos (ed.). *Proceedings of the 10th International Arachnological Congress.* vol. I. Jaca, Spain.

Bonaldo, A. B. 2000. Taxonomia da subfamília Corinninae (Araneae, Corinnidae) nas regiões Neotropica e Neárctica. *Iheringia (Serie Zoologia)* 89: 3–148.

Bond, J. E. 1999. Systematics and evolution of the Californian trapdoor spider genus *Aptostichus* Simon (Araneae: Mygalomorphae, Euctenizinae). Ph.D. thesis, Virginia Polytechnic Institute and State University, Blacksburg, Virginia. 242 pp.

Bond, J. E. 2004. Systematics of the Californian euctenizine spider genus *Apomastus* (Araneae: Mygalomorphae: Cyrtaucheniidae): The relationship between molecular and morphological taxonomy. *Invertebrate Systematics* 18: 361–376.

Bond, J. E. 2012. Phylogenetic treatment and taxonomic revision of the trapdoor spider genus *Aptostichus* Simon (Araneae, Mygalomorphae, Euctenizidae). *Zookeys* 252: 1–209.

Bond, J. E., D.A. Beamer, T. Lamb, and M. Hedin. 2006. Combining genetic and geospatial analyses to infer population extinction in mygalomorph spiders endemic to the Los Angeles region. *Animal Conservation* 9: 145–157.

Bond, J. E., M. C. Hedin, M. J. Ramirez, and B. D. Opell. 2001. Deep molecular divergence in the absence of morphological and ecological change in the California coast dune endemic trapdoor spider *Aptostichus simus*. *Molecular Ecology* 10: 899–910.

Bond, J. E., B. E. Hendrixson, C. A. Hamilton, and M. Hedin. 2012. A Reconsideration of the classification of the spider infraorder Mygalomorphae (Arachnida: Araneae) based on three nuclear genes and morphology. PLOS ONE 7(6): e38753. doi:10.1371/journal.pone.0038753.

Bond, J. E. and B. D. Opell. 1997. Systematics of the spider genera *Mallos* and *Mexitlia* (Araneae, Dictynidae). *Zoological Journal of the Linnean Society* 119: 389–445.

Bond, J. E. and B. D. Opell. 2002. Phylogeny and taxonomy of the genera of south-western North American Euctenizinae trapdoor spiders and their relatives (Araneae: Mygalomorphae, Cyrtaucheniidae). *Zoological Journal of the Linnean Society* 136: 487–534.

Bond, J. E. and A. K. Stockman. 2008. An integrative method for delimiting cohesion species: Finding the population-species interface in a group of Californian trapdoor spiders with extreme genetic divergence and geographic structuring. *Systematic Biology* 57(4): 628–646.

Boulton, A. M. and G. A. Polis. 2002. Brood parasitism among spiders: Interactions between salticids and *Diguetia mojavea*. *Ecology* 83(1): 282–287.

Bowling, T. A. and R. J. Sauer. 1975. A taxonomic revision of the crab spider genus *Coriarachne* (Araneida, Thomisidae) for North America north of Mexico. *Journal of Arachnology* 2: 183–193.

Bradley, R. A. 2013. *The Common Spiders of North America*. University of California Press, Berkeley and Los Angeles, California. 271 pp.

Brady, A. R. 1964. The lynx spiders of North America, north of Mexico (Araneae: Oxyopidae). *Bulletin of the Museum of Comparative Zoology, Harvard* 131: 429–518.

Brady, A. R. 1980. Nearctic species of the wolf spider genus *Trochosa* (Araneae: Lycosidae). *Psyche, Cambridge* 86: 167–212.

Brady, A. R. 2007. *Sosippus* revisited: Review of a web-building wolf spider genus from the Americas (Araneae, Lycosidae). *Journal of Arachnology* 35: 54–83.

Brechbühl, R., J. Casas, and S. Bacher. 2010. Ineffective crypsis in a crab spider: A prey community perspective. *Proceedings of the Royal Society B* 277: 739–746. doi:10.1098/rspb.2009.1632.

Brignoli, P. M. 1979. Recherches en Afrique de l'Institut de Zoologie de l'Aquila (Italie) II. *Reo latro* nov. gen., nov. sp. du Kenya (Araneae: Mimetidae). *Revue de zoologie africaine* 93: 919–928.

Brignoli, P. M. 1981. Studies on the Pholcidae, I. Notes on the genera *Artema* and *Physocyclus* (Araneae). *Bulletin of the American Museum of Natural History* 170: 90–100.

Bryant, E. B. 1944. Three species of *Coleosoma* from Florida (Araneae; Theridiidae). *Psyche, Cambridge* 51: 51–58.

Buckle, D. J., D. Carroll, R. L. Crawford, and V. D. Roth. 2001. Linyphiidae and Pimoidae of America north of Mexico: Checklist, synonymy, and literature. Part 2. pp. 89–191 *in* P. Paquin and D.J. Buckle (eds.). *Contributions á la Connaissance des Araignées (Araneae) d'Amérique du Nord. Fabreries, Supplément* 10. Association des entomologistes amateurs du Québec.

Carico, J. E. 1973. The Nearctic species of the genus *Dolomedes* (Araneae: Pisauridae). *Bulletin of the Museum of Comparative Zoology, Harvard* 144: 435–488.

Carico, J. E. 1976. The spider genus *Tinus* (Pisauridae). *Psyche, Cambridge* 83: 63–78.

Catley, K. M. 1994. Descriptions of new *Hypochilus* species from New Mexico and California with a cladistic analysis of the Hypochilidae (Araneae). *American Museum Novitates* 3088: 1–27.

Chamberlin, R. V. 1921. A new genus and a new species of spiders in the group Phrurolitheae. *Canadian Entomologist* 53: 69–70.
Chamberlin, R. V. 1923. The North American species of *Mimetus*. *Journal of Entomology and Zoology (Claremont)* 15: 3–9.
Chamberlin, R. V. 1924. The spider fauna of the shores and islands of the Gulf of California. *Proceedings of the California Academy of Sciences* 12: 561–694.
Chamberlin, R. V. 1940. New American tarantulas of the family Aviculariidae. *Bulletin of the University of Utah* 30(13): 1–39.
Chamberlin, R. V. 1948. The genera of North American Dictynidae. *Bulletin of the University of Utah* 38(15): 1–31.
Chamberlin, R. V. and W. J. Gertsch. 1930. On fifteen new North American spiders. *Proceedings of the Biological Society of Washington* 43: 137–144.
Chamberlin, R. V. and W. J. Gertsch. 1958. The spider family Dictynidae in America north of Mexico. *Bulletin of the American Museum of Natural History* 116: 1–152.
Chamberlin, R. V. and W. Ivie. 1932. A review of the North American spiders of the genera *Cybaeus* and *Cybaeina*. *Bulletin of the University of Utah* 23(2): 1–43.
Chamberlin, R. V. and W. Ivie. 1935. Miscellaneous new American spiders. *Bulletin of the University of Utah* 26(4): 1–79.
Chamberlin, R. V. and W. Ivie. 1936. Nearctic spiders of the genus *Wubana*. *Annals of the Entomological Society of America* 29: 85–98.
Chamberlin, R. V. and W. Ivie. 1937. New spiders of the family Agelenidae from western North America. *Annals of the Entomological Society of America* 30: 211–230.
Chamberlin, R. V. and W. Ivie. 1939. New tarantulas from the southwestern states. *Bulletin of the University of Utah* 29(11): 1–17.
Chamberlin, R. V. and W. Ivie. 1940. Agelenid spiders of the genus *Cicurina*. *Bulletin of the University of Utah* 30(18): 1–108.
Chamberlin, R. V. and W. Ivie. 1941a. North American Agelenidae of the genera *Agelenopsis*, *Calilena*, *Ritalena* and *Tortolena*. *Annals of the Entomological Society of America* 34: 585–628.
Chamberlin, R. V. and W. Ivie. 1941b. Spiders collected by L. W. Saylor and others, mostly in California. *Bulletin of the University of Utah* 31(8): 1–49.

Chamberlin, R. V. and W. Ivie. 1942a. Agelenidae of the genera *Hololena, Novalena, Rualena* and *Melpomene. Annals of the Entomological Society of America* 35: 203–241.

Chamberlin, R. V. and W. Ivie. 1942b. A hundred new species of American spiders. *Bulletin of the University of Utah* 32(13): 1–117.

Chamberlin, R. V. and W. Ivie. 1943. New genera and species of North American linyphiid spiders. *Bulletin of the University of Utah* 33(10): 1–39.

Cokendolpher, J. C., R. W. Peck, and C. G. Niwa. 2005. Mygalomorph spiders from southwestern Oregon, USA, with descriptions of four new species. *Zootaxa* 1058: 1–34.

Copley, C. R., R. Bennett, and S. J. Perlman. 2009. Systematics of Nearctic *Cybaeus* (Araneae: Cybaeidae). *Invertebrate Systematics* 23: 367–401.

Corey, D. T. and D. J. Mott. 1991. A revision of the genus *Zora* (Araneae, Zoridae) in North America. *Journal of Arachnology* 19: 55–61.

Coyle, F. A. 1971. Systematics and natural history of the mygalomorph spider genus *Antrodiaetus* and related genera (Araneae: Antrodiaetidae). *Bulletin of the Museum of Comparative Zoology, Harvard* 141: 269–402.

Coyle, F. A. 1974. Systematics of the trapdoor spider genus *Aliatypus* (Araneae: Antrodiaetidae). *Psyche, Cambridge* 81: 431–500.

Coyle, F. A. 1981. The mygalomorph spider genus *Microhexura* (Araneae, Dipluridae). *Bulletin of the American Museum of Natural History* 170: 64–75.

Coyle, F. A. and W. Icenogle. 1994. Natural history of the Californian trapdoor spider genus *Aliatypus* (Araneae, Antrodiaetidae). *Journal of Arachnology* 22(3): 225–255.

Crawford, R. L. 1988. An annotated checklist of the spiders of Washington. *Burke Museum Contributions in Anthropology and Natural History* 5: 1–48.

Crews, S. C. 2011. A revision of the spider genus *Selenops* Latreille, 1819 (Arachnida, Araneae, Selenopidae) in North America, Central America, and the Caribbean. *Zookeys* 105: 1–182.

Crews, S. C. and M. Hedin. 2006. Studies of morphological and molecular phylogenetic divergence in spiders (Araneae: *Homalonychus*) from the American southwest, including divergence along the Baja Peninsula. *Molecular Phylogenetics and Evolution* 38: 470–487.

Crews, S. C., E. Wienskoski, and R. G. Gillespie. 2008. Life history of the spider *Selenops occultus* Mello-Leitão (Araneae, Selenopidae) from Brazil with notes on the natural history of the genus. *Journal of Natural History* 42(43–44): 2747–2761.

Cushing, P. E. 1997. Myrmecomorphy and myrmecophily in spiders: A review. *Florida Entomologist* 80(2): 165–193.

Cutler, B. 1988. A revision of the American species of the antlike jumping spider genus *Synageles* (Araneae, Salticidae). *Journal of Arachnology* 15: 321–348.

Cutler, B. 1990. A revision of the western hemisphere *Chalcoscirtus* (Araneae: Salticidae). *Bulletin of the British Arachnological Society* 8(4): 105–108.

Danielson-François, A., C. A. Fetterer, and P. D. Smallwood. 2002. Body condition and mate choice in *Tetragnatha elongata* (Araneae, Tetragnathidae). *Journal of Arachnology* 30: 20–30.

Dondale, C. D. 1961. Life histories of some common spiders from trees and shrubs in Nova Scotia. *Canadian Journal of Zoology* 39: 777–787.

Dondale, C. D. 2005a. Lycosidae. p 164 *in* D. Ubick, P. Paquin, P. E. Cushing, and V. Roth (eds.) *Spiders of North America: An Identification Manual.* American Arachnological Society, Keene, New Hampshire.

Dondale, C. D. 2005b. Thomisidae. p. 246 *in* D. Ubick, P. Paquin, P. E. Cushing, and V. Roth (eds.). *Spiders of North America: An Identification Manual.* American Arachnological Society, Keene, New Hampshire.

Dondale, C. D. 2007. Two new species of wolf spiders in the *Pardosa modica* group (Araneae, Lycosidae) from North America. *Journal of Arachnology* 34: 506–510.

Dondale, C. D. and J. H. Redner. 1968. The *imbecillus* and *rufus* groups of the spider genus *Philodromus* in North America (Araneida: Thomisidae). *Memoirs of the Entomological Society of Canada* 55: 1–78.

Dondale, C. D. and J. H. Redner. 1969. The *infuscatus* and *dispar* groups of the spider genus *Philodromus* in North and Central America and the West Indies (Araneida: Thomisidae). *Canadian Entomologist* 101: 921–954.

Dondale, C. D. and J. H. Redner. 1975a. The *fuscomarginatus* and *histrio* groups of the spider genus *Philodromus* in North America (Araneida: Thomisidae). *Canadian Entomologist* 107: 369–384.

Dondale, C. D. and J. H. Redner. 1975b. The genus *Ozyptila* in North America (Araneida, Thomisidae). *Journal of Arachnology* 2: 129–181

Dondale, C. D. and J. H. Redner. 1975c. Revision of the spider genus *Apollophanes* (Araneida: Thomisidae). *Canadian Entomologist* 107: 1175–1192.

Dondale, C. D. and J. H. Redner. 1976a. A rearrangement of the North American species of *Clubiona,* with descriptions of two new species (Araneida: Clubionidae). *Canadian Entomologist* 108: 1155–1165.

Dondale, C. D. and J. H. Redner. 1976b. A review of the spider genus *Philodromus* in the Americas (Araneida: Philodromidae). *Canadian Entomologist* 108: 127–157.

Dondale, C. D. and J. H. Redner. 1978a. The insects and arachnids of Canada, Part 5. The crab spiders of Canada and Alaska, Araneae: Philodromidae and Thomisidae. *Research Branch, Agriculture Canada, Publications* 1663: 1–255.

Dondale, C. D. and J. H. Redner. 1978b. Revision of the Nearctic wolf spider genus *Schizocosa* (Araneida: Lycosidae). *Canadian Entomologist* 110: 143–181.

Dondale, C. D. and J. H. Redner. 1979. Revision of the wolf spider genus *Alopecosa* Simon in North America (Araneae: Lycosidae). *Canadian Entomologist* 111: 1033–1055.

Dondale, C. D. and J. H. Redner. 1982. The insects and arachnids of Canada, Part 9. The sac spiders of Canada and Alaska, Araneae: Clubionidae and Anyphaenidae. *Research Branch, Agriculture Canada, Publications* 1724: 1–194.

Dondale, C. D. and J. H. Redner. 1983a. Revision of the wolf spiders of the genus *Arctosa* C. L. Koch in North and Central America (Araneae: Lycosidae). *Journal of Arachnology* 11: 1–30.

Dondale, C. D. and J. H. Redner. 1983b. The wolf spider genus *Allocosa* in North and Central America (Araneae: Lycosidae). *Canadian Entomologist* 115: 933–964.

Dondale, C. D. and J. H. Redner. 1990. The insects and arachnids of Canada, Part 17. The wolf spiders, nurseryweb spiders, and lynx spiders of Canada and Alaska, Araneae: Lycosidae, Pisauridae, and Oxyopidae. *Research Branch, Agriculture Canada, Publications* 1856: 1–383.

Dondale, C. D., A. L. Turnbull, and J. H. Redner. 1964. Revision of the Nearctic species of *Thanatus* C. L. Koch (Araneae: Thomisidae). *Canadian Entomologist* 96: 636–656.

Draney, M. L. and D. J. Buckle. 2005. Linyphiidae. p. 124 *in* D. Ubick, P. Paquin, P. E. Cushing, and V. Roth (eds.). *Spiders of North America: An Identification Manual.* American Arachnological Society, Keene, New Hampshire.

Edwards, G. B. 1979. Pantropical jumping spiders occurring in Florida. Florida Department of Agriculture and Consumer Services, Division of Plant Industry, Entomology Circular, No. 199: 1–2.

Edwards, G. B. 2000. Huntsman Spider, *Heteropoda venatoria* (Linnaeus) (Arachnida: Araneae: Sparassidae). EENY-160. University of Florida Institute of Food and Agricultural Sciences, Gainesville. http://edis.ifas.ufl.edu/in317.

Edwards, G. B. 2003. A review of the Nearctic jumping spiders (Araneae: Salticidae) of the subfamily Euophryinae north of Mexico. *Insecta Mundi* 16: 65–75.

Edwards, G. B. 2004. Revision of the jumping spiders of the genus *Phidippus* (Araneae: Salticidae). *Occasional Papers of the Florida State Collection of Arthropods* 11: i–vii +1–156.

Edwards, G. B. 2006. A review of described *Metacyrba,* the status of *Parkella,* and notes on *Platycryptus* and *Balmaceda,* with a comparison of the genera (Araneae: Salticidae: Marpissinae). *Insecta Mundi* 19: 193–226.

Edwards, R. J. 1958. The spider subfamily Clubioninae of the United States, Canada and Alaska (Araneae: Clubionidae). *Bulletin of the Museum of Comparative Zoology, Harvard* 118: 365–436.

Ennik, F. 1971. Laboratory observations on the biology of Loxosceles unicolor Keyserling (Araneae, Loxoscelidae). *Contributions in Science, Santa Barbara Museum of Natural History* 3: 1–16.

Exline, H. and H. W. Levi. 1962. American spiders of the genus *Argyrodes* (Araneae, Theridiidae). *Bulletin of the Museum of Comparative Zoology, Harvard* 127: 75–204.

Fergusson, I. C. 1972. Natural history of the spider *Hypochilus thorelli* Marx (Hypochilidae). *Psyche, Cambridge* 79: 179–199.

Fink, L. S. 1984. Venom spitting by the green lynx spider, *Peucetia viridans* (Araneae, Oxyopidae). *Journal of Arachnology* 12: 372–373.

Foelix, R. F. 1996. *Biology of Spiders.* 2nd ed. Oxford University Press, New York. 330 pp.

Forster, R. R. and N. I. Platnick. 1977. A review of the spider family Symphytognathidae (Arachnida, Araneae). *American Museum Novitates* 2619: 1–29.

Forster, R. R., N. I. Platnick, and M. R. Gray. 1987. A review of the spider superfamilies Hypochiloidea and Austrochiloidea (Araneae, Araneomorphae). *Bulletin of the American Museum of Natural History* 185: 1–116.

Fowler, H. G. 1984. Note on a clubionid spider associated with attine ants. *Journal of Arachnology* 12: 117–118.

Gertsch, W. J. 1933. Notes on American spiders of the family Thomisidae. *American Museum Novitates* 593: 1–22.

Gertsch, W. J. 1936. Further diagnoses of new American spiders. *American Museum Novitates* 852: 1–27.

Gertsch, W. J. 1937. New American spiders. *American Museum Novitates* 936: 1–7.

Gertsch, W. J. 1941. New American spiders of the family Clubionidae. I. *American Museum Novitates* 1147: 1–20.

Gertsch, W. J. 1942. New American spiders of the family Clubionidae. III. *American Museum Novitates* 1195: 1–18.

Gertsch, W. J. 1953. The spider genera *Xysticus, Coriarachne* and *Oxyptila* (Thomisidae, Misumeninae) in North America. *Bulletin of the American Museum of Natural History* 102: 415–482.

Gertsch, W. J. 1958a. The spider family Diguetidae. *American Museum Novitates* 1904: 1–24.

Gertsch, W. J. 1958b. The spider family Hypochilidae. *American Museum Novitates* 1912: 1–28.

Gertsch, W. J. 1958c. The spider family Plectreuridae. *American Museum Novitates* 1920: 1–53.

Gertsch, W. J. 1958d. The spider genus *Loxosceles* in North America, Central America, and the West Indies. *American Museum Novitates* 1907: 1–46.

Gertsch, W. J. 1960a. Descriptions of American spiders of the family Symphytognathidae. *American Museum Novitates* 1981: 1–40.

Gertsch, W. J. 1960b. The *fulva* group of the spider genus *Steatoda* (Araneae, Theridiidae). *American Museum Novitates* 1982: 1–48.

Gertsch, W. J. 1961. The spider genus *Lutica*. *Senckenbergiana Biologica* 42: 365–374.

Gertsch, W. J. 1964. The spider genus *Zygiella* in North America (Araneae, Argiopidae). *American Museum Novitates* 2188: 1–21.

Gertsch, W. J. 1982. The spider genera *Pholcophora* and *Anopsicus* (Araneae, Pholcidae) in North America, Central America and the West Indies. *Bulletin of the Association for Mexican Cave Studies* 8: 95–144.

Gertsch, W. J. 1984. The spider family Nesticidae (Araneae) in North America, Central America, and the West Indies. *Bulletin of the Texas Memorial Museum* 31: 1–91.

Gertsch, W. J. and F. Ennik. 1983. The spider genus *Loxosceles* in North America, Central America, and the West Indies (Araneae, Loxoscelidae). *Bulletin of the American Museum of Natural History* 175: 264–360.

Gertsch, W. J. and W. Ivie. 1955. The spider genus *Neon* in North America. *American Museum Novitates* 1743: 1–17.

Gertsch, W. J. and N. I. Platnick. 1979. A revision of the spider family Mecicobothriidae (Araneae, Mygalomorphae). *American Museum Novitates* 2687: 1–32.

Gertsch, W. J. and H. K. Wallace. 1935. Further notes on American Lycosidae. *American Museum Novitates* 794: 1–22.

Gilbert, C. and L. S. Rayor. 1985. Predatory behavior of spitting spiders (Araneae: Scytodidae) and the evolution of prey wrapping. *Journal of Arachnology* 13: 231–241.

Gillespie, R. G. 1987. The mechanism of habitat selection in the long-jawed orb-weaving spider *Tetragnatha elongata* (Araneae, Tetragnathidae). *Journal of Arachnology* 15: 81–90.

Glatz, L. 1967. Zur biologie und morphologie von *Oecobiis annulipes* Lucas. *Zeitschrift für Morphologie und Ökologie der Tiere* 61: 185–214.

Goddard, J., S. Upshaw, D. Held, and K. Johnson. 2008. Severe reaction from envenomation by the Brown Widow Spider, *Latrodectus geometricus* (Araneae: Theridiidae). *Southern Medical Journal* 101(12): 1269–1270.

Gregory, B. M., Jr. 1989. Field observations of *Gasteracantha cancriformis* (Araneae, Araneidae) in a Florida mangrove stand. *Journal of Arachnology* 17: 119–120.

Griswold, C. E. 1977. Araneae (spiders). pp. 136–143 *in* W. Barry and E. Schlinger (eds.). *Inglenook Fen: A Study and Plan*. State of California Department of Parks and Recreation, Sacramento.

Griswold, C. E. 1987. A revision of the jumping spider genus *Habronattus* F. O. P.-Cambridge (Araneae; Salticidae), with phenetic and cladistic analyses. *University of California Publications in Entomology* 107: 1–344.

Griswold, C. E., T. Audisio, and J. M. Ledford. 2012. An extraordinary new family of spiders from caves in the Pacific Northwest (Araneae, Trogloraptoridae, new family). *ZooKeys* 215: 77–102.

Griswold, C. E. and D. Ubick. 2001. Zoropsidae: A spider family newly introduced to the USA (Araneae, Entelegynae, Lycosoidea). *Journal of Arachnology* 29: 111–113.

Haynes, K. F., C. Gemeno, K. V. Yeargan, J. G. Millar, and K. M. Johnson. 2002. Aggressive chemical mimicry of moth pheromones by a bolas spider: How does this specialist predator attract more than one species of prey? *Chemoecology* 12: 99–105.

Hedin, M. 1997. Speciational history in a diverse clade of habitat-specialized spiders (Araneae: Nesticidae: *Nesticus*): Inferences from geographic-based sampling. *Evolution* 51(6): 1929–1945.

Hedin, M. 2001. Molecular insights into species phylogeny, biogeography, and morphological stasis in the relict spider genus *Hypochilus* (Araneae: Hypochilidae). *Molecular Phylogenetics and Evolution* 18: 238–251.

Heiss, J. S. and M. L. Draney. 2004. Revision of the Nearctic spider genus *Calymmaria* (Araneae, Hahniidae). *Journal of Arachnology* 32: 457–525.

Helsdingen, P. J., van. 1965. Sexual behaviour of *Lepthyphantes leprosus* (Ohlert) (Araneida, Linyphiidae), with notes on the function of the genital organs. *Zoologische Mededelingen* (Uitgegeven Door Het Rijksmuseum van Natuurlijke Historie, Leiden) 41: 15–42.

Helsdingen, P. J., van. 1969. A reclassification of the species of *Linyphia* Latreille based on the functioning of the genitalia (Araneida, Linyphiidae), I. *Zoologische Verhandelingen* (Uitgegeven Door Het Rijksmuseum van Natuurlijke Historie, Leiden) 105: 1–303.

Helsdingen, P. J., van. 1970. A reclassification of the species of *Linyphia* based on the functioning of the genitalia (Araneida, Linyphiidae), II. *Zoologische Verhandelingen* (Uitgegeven Door Het Rijksmuseum van Natuurlijke Historie, Leiden) 111: 1–86.

Hentz, N. M. 1847. Descriptions and figures of the araneides of the United States. *Boston Journal of Natural History* 5: 443–478.

Hippa, H. and I. Oksala. 1982. Definition and revision of the *Enoplognatha ovata* (Clerck) group (Araneae: Theridiidae). *Entomologica Scandinavica* 13(2): 213–222.

Hölldobler, B. 1970. *Steatoda fulva* (Theridiidae), a spider that feeds on harvester ants. *Psyche, Cambridge* 77: 202–208.

Hormiga, G. 1994. A revision and cladistic analysis of the spider family Pimoidae (Araneoidea: Araneae). *Smithsonian Contributions to Zoology* 549: 1–104.

Hormiga, G., D. J. Buckle, and N. Scharff. 2005. *Nanoa,* an enigmatic new genus of pimoid spiders from western North America (Pimoidae, Araneae). *Zoological Journal of the Linnean Society* 145: 249–262.

Horton, C. 1979. Apparent attraction of moths by the webs of araneid spiders. *Journal of Arachnology* 7: 88–89.

Huber, B. A., C. L. Deeleman-Reinhold, and A. Pérez G. 1999. The spider genus *Crossopriza* (Araneae, Pholcidae) in the New World. *American Museum Novitates* 3262: 1–10.

Jackson, R. R. 1977. Comparative studies of *Dictyna* and *Mallos* (Araneae, Dictynidae): III. Prey and predatory behavior. *Psyche, Cambridge* 84: 267–280.

Jackson, R. R. 1979. Comparative studies of *Dictyna* and *Mallos* (Araneae, Dictynidae): II. The relationship between courtship, mating, aggression, and cannibalism in species with differing types of social organization. *Revue Arachnologique* 2: 103–132.

Janowski-Bell, M. E. and N. V. Horner. 1999. Movement of the male brown tarantula, *Aphonopelma hentzi* (Araneae: Theraphosidae), using radio telemetry. *Journal of Arachnology* 27: 503–512.

Jenks, G. E. 1938. Marvels of metamorphosis. *National Geographic* 74(6): 807–828.

Kaston, B. J. 1938. North American spiders of the genus *Agroeca. American Midland Naturalist* 20: 562–570.

Kaston, B. J. 1946. North American spiders of the genus *Ctenium. American Museum Novitates* 1306: 1–19.

Kaston, B. J. 1965. Some little known aspects of spider behavior. *American Midland Naturalist* 73(2): 336–356.

Kasumovic, M. M. and M. C. B. Andrade. 2004. Discrimination of airborne pheromones by mate-searching male western black widow spiders (*Latrodectus hesperus*): Species and population-specific responses. *Canadian Journal of Zoology* 82(7): 1027–1034.

Kim, K. W., C. Roland, and A. Horel. 2000. Functional value of matriphagy in the spider *Amaurobius ferox. Ethology* 106: 729–742.

Klook, C. 2001. Diet and insectivory in the "araneophagic" spider, *Mimetus notius* (Araneae: Mimetidae). *American Midland Naturalist* 146: 424–428.

Knoflach, B. and S. P. Benjamin. 2003. Mating without sexual cannibalism in *Tidarren sisyphoides* (Araneae, Theridiidae). *Journal of Arachnology* 31: 445–448.

Lawler, N. 1972. Notes on the biology and behavior of *Mimetus eutypus* Chamberlin and Ivie (Araneae: Mimetidae). *Notes of the Arachnologists of the Southwest* 3: 7–10.

Ledford, J. M. 2004. A revision of the spider genus *Calileptoneta* Platnick (Araneae, Leptonetidae), with notes on morphology, natural history and biogeography. *Journal of Arachnology* 32: 231–269.

Ledford, J. M. and C. E. Griswold. 2010. A study of the subfamily Archoleptonetinae (Araneae, Leptonetidae) with a review of the morphology and relationships for the Leptonetidae. *Zootaxa* 2391: 1–32.

Leech, R. E. 1972. A revision of the Nearctic Amaurobiidae (Arachnida: Araneida). *Memoirs of the Entomological Society of Canada* 84: 1–182.

Lehtinen, P. T. 1967. Classification of the cribellate spiders and some allied families, with notes on the evolution of the suborder Araneomorpha. *Annales Zoologici Fennici* 4: 199–468.

Lehtinen, P. T. and Y. M. Marusik. 2008. A redefinition of *Misumenops* F. O. Pickard-Cambridge, 1900 (Araneae, Thomisidae) and review of the New World species. *Bulletin of the British Arachnological Society* 14: 173–198.

Levi, H. W. 1953. Spiders of the genus *Dipoena* from America north of Mexico (Araneae, Theridiidae). *American Museum Novitates* 1647: 1–39.

Levi, H. W. 1954. Spiders of the genus *Euryopis* from North and Central America (Araneae, Theridiidae). *American Museum Novitates* 1666: 1–48.

Levi, H. W. 1955. The spider genera *Coressa* and *Achaearanea* in America north of Mexico (Araneae, Theridiidae). *American Museum Novitates* 1718: 1–33.

Levi, H. W. 1956. The spider genera *Neottiura* and *Anelosimus* in America (Araneae: Theridiidae). *Transactions of the American Microscopical Society* 75: 407–422.

Levi, H. W. 1957a. The North American spider genera *Paratheridula, Tekellina, Pholcomma* and *Archerius* (Araneae: Theridiidae). *Transactions of the American Microscopical Society* 76: 105–115.

Levi, H. W. 1957b. The spider genera *Crustulina* and *Steatoda* in North America, Central America, and the West Indies (Araneae, Theridiidae). *Bulletin of the Museum of Comparative Zoology, Harvard* 117: 367–424.

Levi, H. W. 1957c. The spider genera *Chrysso* and *Tidarren* in America. *Journal of the New York Entomological Society* 63: 59–81.

Levi, H. W. 1957d. Spiders of the new genus *Arctachaea* (Araneae, Theridiidae). *Psyche, Cambridge* 64: 102–106.

Levi, H. W. 1957e. The spider genera *Enoplognatha, Theridion,* and *Paidisca* in America north of Mexico (Araneae, Theridiidae). *Bulletin of the Museum of Comparative Zoology, Harvard* 112: 1–124.

Levi, H. W. 1959. The spider genus *Latrodectus* (Araneae, Theridiidae). *Transactions of the American Microscopical Society* 78: 7–43.

Levi, H. W. 1963a. American spiders of the genera *Audifia, Euryopis* and *Dipoena* (Araneae: Theridiidae). *Bulletin of the Museum of Comparative Zoology, Harvard* 129: 121–185.

Levi, H. W. 1963b. American spiders of the genus *Achaearanea* and the new genus *Echinotheridion* (Araneae, Theridiidae). *Bulletin of the Museum of Comparative Zoology, Harvard* 129: 187–240.

Levi, H. W. 1963c. American spiders of the genus *Theridion* (Araneae, Theridiidae). *Bulletin of the Museum of Comparative Zoology, Harvard* 129: 481–589.

Levi, H. W. 1964. The spider genera *Stemmops, Chrosiothes,* and the new genus *Cabello* from America. *Psyche, Cambridge* 71: 73–92.

Levi, H. W. 1966. American spider genera *Theridula* and *Paratheridula* (Araneae: Theridiidae). *Psyche, Cambridge* 78: 123–130.

Levi, H. W. 1968. The spider genera *Gea* and *Argiope* in America (Araneae: Araneidae). *Bulletin of the Museum of Comparative Zoology, Harvard* 136(9): 319–352.

Levi, H. W. 1971a. The *diadematus* group of the orb-weaver genus *Araneus* north of Mexico (Araneae: Araneidae). *Bulletin of the Museum of Comparative Zoology, Harvard* 141: 131–179.

Levi, H. W. 1971b. The *ravilla* group of the orbweaver genus *Eriophora* in North America (Araneae: Araneidae). *Psyche, Cambridge* 77(3): 280–302.

Levi, H. W. 1972. The orb-weaver genera *Singa* and *Hypsosinga* in America (Araneae: Araneidae). *Psyche, Cambridge* 78(4): 229–256.

Levi, H. W. 1973. Small orb-weavers of the genus *Araneus* north of Mexico (Araneae: Araneidae). *Bulletin of the Museum of Comparative Zoology, Harvard* 145(9): 473–552.

Levi, H. W. 1974a. The orb-weaver genera *Araniella* and *Nuctenea* (Araneae: Araneidae). *Bulletin of the Museum of Comparative Zoology, Harvard* 146(6): 291–316.

Levi, H. W. 1974b. The orb-weaver genus *Zygiella* (Araneae: Araneidae). *Bulletin of the Museum of Comparative Zoology, Harvard* 146(5): 267–290.

Levi, H. W. 1975a. Additional notes on the orb-weaver genera *Araneus, Hypsosinga* and *Singa* north of Mexico (Araneae, Araneidae). *Psyche, Cambridge* 82(2): 265–274.

Levi, H. W. 1975b. The American orb-weaver genera *Larinia, Cercidia* and *Mangora* north of Mexico (Araneae, Araneidae). *Bulletin of the Museum of Comparative Zoology, Harvard* 147(3): 101–135.

Levi, H. W. 1977a. The American orb-weaver genera *Cyclosa, Metazygia* and *Eustala* north of Mexico (Araneae, Araneidae). *Bulletin of the Museum of Comparative Zoology, Harvard* 148(3): 61–127.

Levi, H. W. 1977b. The orb-weaver genera *Metepeira, Kaira* and *Aculepeira* in America north of Mexico (Araneae, Araneidae). *Bulletin of the Museum of Comparative Zoology, Harvard* 148(5): 185–238.

Levi, H. W. 1978. The American orb-weaver genera *Colphepeira, Micrathena* and *Gasteracantha* north of Mexico (Araneae, Araneidae). *Bulletin of the Museum of Comparative Zoology, Harvard* 148(9): 417–442.

Levi, H. W. 1980. The orb-weaver genus *Mecynogea,* the subfamily Metinae and the genera *Pachygnatha, Glenognatha* and *Azilia* of the subfamily Tetragnathinae north of Mexico (Araneae: Araneidae). *Bulletin of the Museum of Comparative Zoology, Harvard* 149: 1–74.

Levi, H. W. 1981. The American orb-weaver genera *Dolichognatha* and *Tetragnatha* north of Mexico (Araneae: Araneidae, Tetragnathinae). *Bulletin of the Museum of Comparative Zoology, Harvard* 149: 271–318.

Levi, H. W. 1999. The Neotropical and Mexican orb weavers of the genera *Cyclosa* and *Allocyclosa* (Araneae: Araneidae). *Bulletin of the Museum of Comparative Zoology, Harvard* 155: 299–379.

Levi, H. W. 2003. The bolas spiders of the genus *Mastophora* (Araneae: Araneidae). *Bulletin of the Museum of Comparative Zoology, Harvard* 157(5): 309–382.

Levi, H. W. 2005a. Araneidae. p. 68 *in* D. Ubick, P. Paquin, P. E. Cushing, and V. Roth (eds.). *Spiders of North America: An Identification Manual.* American Arachnological Society, Keene, New Hampshire.

Levi, H. W. 2005b. Tetragnathidae. p. 232 *in* D. Ubick, P. Paquin, P. E. Cushing, and V. Roth (eds.). *Spiders of North America: An Identification Manual.* American Arachnological Society, Keene, New Hampshire.

Levi, H. W. 2005c. Theridiidae. p. 235 *in* D. Ubick, P. Paquin, P. E. Cushing, and V. Roth (eds.). *Spiders of North America: An Identification Manual.* American Arachnological Society, Keene, New Hampshire.

Levi, H. W. and L. R. Levi. 2002. *Spiders and Their Kin: A Golden Guide.* St. Martin's Press, New York. 160 pp.

Lew, S. E. and D. J. Mott. 2005. Mimetidae. p. 171 *in* D. Ubick, P. Paquin, P. E. Cushing, and V. Roth (eds.). *Spiders of North America: An Identification Manual.* American Arachnological Society, Keene, New Hampshire.

Logunov, D. V. and B. Cutler. 1999. Revision of the genus *Paramarpissa* F. O. P.-Cambridge, 1901 (Araneae, Salticidae). *Journal of Natural History* 33: 1217–1236.

Logunov, D. V. and S. Koponen. 2002. Redescription and distribution of *Phlegra hentzi* (Marx, 1890) comb. n. (Araneae, Salticidae). *Bulletin of the British Arachnological Society* 12: 264–267.

Lowrie, D. C. and W. J. Gertsch. 1955. A list of the spiders of the Grand Teton Park area, with descriptions of some new North American spiders. *American Museum Novitates* 1736: 1–29.

MacKay, W. P. 1982. The effect of predation of Western Black Widows (Araneae: Theridiidae) on harvester ants (Hymenoptera: Formicidae). *Oecologia* 53: 406–411.

Maddison, W. 1986. Distinguishing the jumping spiders *Eris militaris* and *Eris flava* in North America (Araneae: Salticidae). *Psyche, Cambridge* 93: 141–149.

Maddison, W. 1987. *Marchena* and other jumping spiders with an apparent leg-carapace stridulatory mechanism (Araneae: Salticidae: Heliophaninae and Thiodinae). *Bulletin of the British Arachnological Society* 7: 101–106.

Maddison, W. P. 1996. *Pelegrina* Franganillo and other jumping spiders formerly placed in the genus *Metaphidippus* (Araneae: Salticidae). *Bulletin of the Museum of Comparative Zoology, Harvard* 154: 215–368.

Marusik, Y. M. and D. V. Logunov. 1998. Taxonomic notes on the *Evarcha falcata* species complex (Aranei Salticidae). *Arthropoda Selecta* 6(3/4): 95–104.

Marx, G. 1891. A contribution to the knowledge of North American spiders. *Proceedings of the Entomological Society of Washington* 2: 28–37.

Montgomery, T. H. 1904. Descriptions of North American Araneae of the families Lycosidae and Pisauridae. *Proceedings of the Academy of Natural Sciences of Philadelphia* 56: 261–325.

Morse, D. H. 1992. Dispersal of the spiderlings of *Xysticus emertoni* (Araneae, Thomisidae), a litter-dwelling crab spider. *Journal of Arachnology* 20: 217–221.

Müller, G. J. 1993. Black and brown widow spider bites in South Africa: A series of 45 cases. *South African Medical Journal* 83: 399–405.

Muma, M. H. 1953. A study of the spider family Selenopidae in North and Central America and the West Indies. *American Museum Novitates* 1619: 1–55.

Muma, M. H. 1971. Biological and behavioral notes on *Gasteracantha cancriformis* (Arachnida: Araneidae). *Florida Entomologist* 54: 345–351.

Muma, M. H. and W. J. Gertsch. 1964. The spider family Uloboridae in North America north of Mexico. *American Museum Novitates* 2196: 1–43.

Murphy, J. 2007. *Gnaphosid Genera of the World.* British Arachnological Society, Volume 1 (text): i–xii + 1–92; Volume 2 (plates): i–ii + 93–605. http://wiki.britishspiders.org.uk/.

Nieto-Castañeda, I. G. and M. L. Jiménez-Jiménez. 2009. Possible niche differentiation of two desert wandering spiders of the genus *Syspira* (Araneae: Miturgidae). Journal of Arachnology 37(3): 299–305.

Oi, R. 1960. Linyphiid spiders of Japan. *Journal of the Institutes of Polytechnics, Osaka City University* 11(Series D): 137–244.

Olmstead, J. V. 1975. A revision of the spider genus *Syspira* (Araneida: Clubionidae). Unpublished M.Sc. thesis, California State University, Long Beach. 89 pp.

Opell, B. D. and J. A. Beatty. 1976. The Nearctic Hahniidae (Arachnida: Araneae). *Bulletin of the Museum of Comparative Zoology, Harvard* 147: 393–433.

Paquin, P. and M. Hedin. 2005. Nesticidae. p. 178 *in* D. Ubick, P. Paquin, P. E. Cushing, and V. Roth (eds.). *Spiders of North America: An Identification Manual.* American Arachnological Society, Keene, New Hampshire.

Passmore, L. 1933. California Trapdoor Spider performs engineering marvels. *National Geographic* 64(2): 195–211.

Peckham, G. W. and E. G. Peckham. 1909. Revision of the Attidae of North America. *Transactions of the Wisconsin Academy of Sciences, Arts, and Letters* 16(1): 355–655.

Penniman, A. J. 1985. Revision of the *britcheri* and *pugnata* groups of *Scotinella* (Araneae, Corinnidae, Phrurolithinae) with a reclassification of Phrurolithine spiders. Unpublished Ph.D. dissertation, Ohio State University, ii–xx + 1–240 pp.

Piel, W. H. 2001. The systematics of Neotropical orb-weaving spiders in the genus *Metepeira* (Araneae: Araneidae). *Bulletin of the Museum of Comparative Zoology, Harvard* 157(1): 1–92.

Platnick, N. I. 1974. The Spider family Anyphaenidae in America north of Mexico. *Bulletin of the Museum of Comparative Zoology* 146(4): 205–266.

Platnick, N. I. 1975a. A revision of the Holarctic spider genus *Callilepis* (Araneae, Gnaphosidae). *American Museum Novitates* 2573: 1–32.

Platnick, N. I. 1975b. A revision of the spider genus *Eilica* (Araneae, Gnaphosidae). *American Museum Novitates* 2578: 1–19.

Platnick, N. I. 1984. On the Gnaphosidae (Arachnida, Araneae) of the California Channel Islands. *Journal of Arachnology* 11: 453–455.

Platnick, N. I. 1993. A new genus of the spider family Caponiidae (Araneae, Haplogynae) from California. *American Museum Novitates* 3063: 1–8.

Platnick, N. I. 1995. A revision of the spider genus *Orthonops* (Araneae, Caponiidae). *American Museum Novitates* 3150: 1–18.

Platnick, N. I. 2013. The World Spider Catalog, Version 13.5. American Museum of Natural History, New York. http://research.amnh.org/iz/spiders/catalog/INTRO1.html.

Platnick, N. I. and B. Baehr. 2006. A revision of the Australasian ground spiders of the family Prodidomidae (Araneae, Gnaphosoidea). *Bulletin of the American Museum of Natural History* 298: 1–287.

Platnick, N. I. and C. D. Dondale. 1992. The insects and arachnids of Canada, Part 19. The ground spiders of Canada and Alaska (Araneae: Gnaphosidae). *Research Branch, Agriculture Canada, Publications* 1875: 1–297.

Platnick, N. I. and N. Dupérré. 2009a. The American goblin spiders of the new genus *Escaphiella* (Araneae, Oonopidae). *Bulletin of the American Museum of Natural History* 328: 1–151.

Platnick, N. I. and N. Dupérré. 2009b. The goblin spider genera *Opopaea* and *Epectris* (Araneae, Oonopidae) in the New World. *American Museum Novitates* 3649: 1–43.

Platnick, N. I. and C. Ewing. 1995. A revision of the tracheline spiders (Araneae, Corinnidae) of southern South America. *American Museum Novitates* 3128: 1–41.

Platnick, N. I. and R. R. Forster. 1990. On the spider family Anapidae (Araneae, Araneoidea) in the United States. *Journal of the New York Entomological Society* 98: 108–112.

Platnick, N. I. and J. A. Murphy. 1984. A revision of the spider genera *Trachyzelotes* and *Urozelotes* (Araneae, Gnaphosidae). *American Museum Novitates* 2792: 1–30.

Platnick, N. I. and V. I. Ovtsharenko. 1991. On Eurasian and American *Talanites* (Araneae, Gnaphosidae). *Journal of Arachnology* 19: 115–121.

Platnick, N. I. and T. R. Prentice. 1999. A new species of the spider genus *Zelotes* (Araneae, Gnaphosidae) from California. *Journal of Arachnology* 27: 672–674.

Platnick, N. I. and M. U. Shadab. 1974a. A revision of the *bispinosus* and *bicolor* groups of the spider genus *Trachelas* (Araneae, Clubionidae) in North and Central America and the West Indies. *American Museum Novitates* 2560: 1–34.

Platnick, N. I. and M. U. Shadab. 1974b. A revision of the *tranquillus* and *speciosus* groups of the spider genus *Trachelas* (Araneae, Clubionidae) in North and Central America. *American Museum Novitates* 2553: 1–34.

Platnick, N. I. and M. U. Shadab. 1975a. A revision of the spider genus *Gnaphosa* (Araneae, Gnaphosidae) in America. *Bulletin of the American Museum of Natural History* 155: 1–66.

Platnick, N. I. and M. U. Shadab. 1975b. A revision of the spider genera *Haplodrassus* and *Orodrassus* (Araneae, Gnaphosidae) in North America. *American Museum Novitates* 2583: 1–40.

Platnick, N. I. and M. U. Shadab. 1976a. A revision of the spider genera *Drassodes* and *Tivodrassus* (Araneae, Gnaphosidae) in North America. *American Museum Novitates* 2593: 1–29.

Platnick, N. I. and M. U. Shadab. 1976b. A revision of the spider genera *Lygromma* and *Neozimiris* (Araneae, Gnaphosidae). *American Museum Novitates* 2598: 1–23.

Platnick, N. I. and M. U. Shadab. 1976c. A revision of the spider genera *Rachodrassus, Sosticus,* and *Scopodes* (Araneae, Gnaphosidae) in North America. *American Museum Novitates* 2594: 1–33.

Platnick, N. I. and M. U. Shadab. 1977. A revision of the spider genera *Herpyllus* and *Scotophaeus* (Araneae, Gnaphosidae) in North America. *Bulletin of the American Museum of Natural History* 159: 1–44.

Platnick, N. I. and M. U. Shadab. 1980a. A revision of the North American spider genera *Nodocion, Litopyllus,* and *Synaphosus* (Araneae, Gnaphosidae). *American Museum Novitates* 2691: 1–26.

Platnick, N. I. and M. U. Shadab. 1980b. A revision of the spider genus *Cesonia* (Araneae, Gnaphosidae). *Bulletin of the American Museum of Natural History* 165: 335–386.

Platnick, N. I. and M. U. Shadab. 1981. A revision of the spider genus *Sergiolus* (Araneae, Gnaphosidae). *American Museum Novitates* 2717: 1–41.

Platnick, N. I. and M. U. Shadab. 1982. A revision of the American spiders of the genus *Drassyllus* (Araneae, Gnaphosidae). *Bulletin of the American Museum of Natural History* 173: 1–97.

Platnick, N. I. and M. U. Shadab. 1983. A revision of the American spiders of the genus *Zelotes* (Araneae, Gnaphosidae). *Bulletin of the American Museum of Natural History* 174: 97–192.

Platnick, N. I. and M. U. Shadab. 1988. A revision of the American spiders of the genus *Micaria* (Araneae, Gnaphosidae). *American Museum Novitates* 2916: 1–64.

Platnick, N. I. and D. Ubick. 1989. A revision of the spider genus *Drassinella* (Araneae, Liocranidae). *American Museum Novitates* 2937: 1–12.

Platnick, N. I. and D. Ubick. 2001. A revision of the North American spiders of the new genus *Socalchemmis* (Araneae, Tengellidae). *American Museum Novitates* 3339: 1–25.

Platnick, N. I. and D. Ubick. 2005. A revision of the North American spider genus *Anachemmis* Chamberlin (Araneae, Tengellidae). *American Museum Novitates* 3477: 1–20.
Platnick, N. I. and D. Ubick. 2007. On a new species group in the spider genus *Socalchemmis* (Araneae, Tengellidae). *Journal of Arachnology* 35: 205–207.
Platnick, N. I. and D. Ubick. 2008. A revision of the endemic Californian spider genus *Titiotus* Simon (Araneae, Tengellidae). *American Museum Novitates* 3608: 1–33.
Porter, A. H. and E. M. Jacob. 1990. Allozyme variation in the introduced spider *Holocnemus pluchei* (Araneae, Pholcidae) in California. *Journal of Arachnology* 18: 313–319.
Porter, S. D. and D. A. Eastmond. 1982. *Euryopis coki* (Theridiidae), a spider that preys on *Pogonomyrmex* ants. *Journal of Arachnology* 10: 275–277.
Prentice, T. R. 1997. Theraphosidae of the Mojave Desert west and north of the Colorado River (Araneae, Mygalomorphae, Theraphosidae). *Journal of Arachnology* 25: 137–176.
Prentice, T.R., J.C. Burger, W.R. Icenogle, and R.A. Redak. 1998. Spiders from Diegan coastal sage scrub (Arachnida: Araneae). *Pan-Pacific Entomologist* 74(4): 181–202.
Prószyn'ski, J. 1971. Revision of the spider genus *Sitticus* Simon, 1901 (Aranei, Salticidae). II. *Sitticus saxicola* (C. L. Koch, 1848) and related forms. *Annales Zoologici, Warszawa* 28: 183–204.
Prószyn'ski, J. 1973. Revision of the spider genus *Sitticus* Simon, 1901 (Aranei, Salticidae). III. *Sitticus penicillatus* (Simon, 1875) and related forms. *Annales Zoologici, Warszawa* 30: 71–95.
Prószyn'ski, J. 1980. Revision of the spider genus *Sitticus* Simon, 1901 (Aranei, Salticidae). IV. *Sitticus floricola* (C. L. Koch) group. *Annales Zoologici, Warszawa* 36: 1–35.
Ramirez, M. G. 1995. Natural history of the spider genus *Lutica* (Araneae, Zodariidae). *Journal of Arachnology* 23: 111–117.
Ramirez, M. G. and R. D. Beckwitt. 1995. Phylogeny and historical biogeography of the spider genus *Lutica* (Araneae: Zodariidae). *Journal of Arachnology* 23: 177–193.
Ramirez, M. B. and B. Chi. 2004. Cryptic speciation, genetic diversity and gene flow in the California turret spider *Atypoides riversi* (Araneae: Antrodiaetidae). *Biological Journal of the Linnean Society* 82(1): 27–37.
Ramos, M., D. J. Irschick, and T. E. Christenson. 2004. Overcoming an evolutionary conflict: Removal of a reproductive organ

greatly increases locomotor performance. *Proceedings of the National Academy of Sciences of the United States of America* 101(14): 4883–4887.

Raven, R. J. 1985. The spider infraorder Mygalomorphae (Araneae): Cladistics and systematics. *Bulletin of the American Museum of Natural History* 182: 1–180.

Reiskind, J. 1969. The spider subfamily Castianeirinae of North and Central America (Araneae, Clubionidae). *Bulletin of the Museum of Comparative Zoology, Harvard* 138: 163–325.

Rezác, M., J. Král, and S. Pekár. 2007. The spider genus *Dysdera* (Araneae, Dysderidae) in central Europe: Revision and natural history. *Journal of Arachnology* 35: 432–462.

Rheims, C. A. 2010a. Notes on the neotropical genus *Macrinus* (Araneae: Sparassidae). *Zoologia* 27: 440–444.

Rheims, C. A. 2010b. On the native Nearctic species of the huntsman spider family Sparassidae Bertkau (Araneae). *Journal of Arachnology* 38: 530–537.

Rheims, C. A., A. D. Brescovit, and C. G. Durán-Barrón. 2007. Mexican species of the genus *Scytodes* Latreille (Araneae, Scytodidae). *Revista Ibérica de Aracnologia* 13: 93–119.

Richman, D. B. 1965. Jumping spiders (Salticidae) from Yuma County, Arizona, with a description of a new species and distributional records. *Southwestern Naturalist* 10: 132–135.

Richman, D. B. 1989. A revision of the genus *Hentzia* (Araneae, Salticidae). *Journal of Arachnology* 17: 285–344.

Richman, D. B. 2008. Revision of the jumping spider genus *Sassacus* (Araneae, Salticidae, Dendryphantinae) in North America. *Journal of Arachnology* 36: 26–48.

Richman D. B., G. B. Edwards, and B. Cutler. 2005. Salticidae. p. 205 *in* D. Ubick, P. Paquin, P. E. Cushing, and V. Roth (eds.). *Spiders of North America: An Identification Manual.* American Arachnological Society, Keene, New Hampshire.

Richman, D. B. and R. S. Vetter. 2004. A review of the spider genus *Thiodina* (Araneae, Salticidae) in the United States. *Journal of Arachnology* 32: 418–431.

Roddy, L. R. 1966. New species, records, of clubionid spiders. *Transactions of the American Microscopical Society* 85: 399–407.

Roddy, L. R. 1973. American spiders of the *Clubiona canadensis* group (Araneae; Clubionidae). *Transactions of the American Microscopical Society* 92: 143–147.

Ross, K. and R. L. Smith. 1979. Aspects of the courtship behavior of the Black Widow spider, *Latrodectus hesperus* (Araneae: Theridiidae), with evidence for the existence of a contact sex pheromone. *Journal of Arachnology* 7: 69–77.

Roth, V. D. 1952. Notes and a new species in *Cybaeina*. *Pan-Pacific Entomologist* 28: 195–201.

Roth, V. D. 1956. Revision of the genus *Yorima* Chamberlin and Ivie (Arachnida, Agelenidae). *American Museum Novitates* 1773: 1–10.

Roth, V. D. 1968. The spider genus *Tegenaria* in the Western Hemisphere (Agelenidae). *American Museum Novitates* 2323: 1–33.

Roth, V. D. 1981. A new genus of spider (Agelenidae) from California exhibiting a third type of leg autospasy. *Bulletin of the American Museum of Natural History* 170: 101–105.

Roth, V. D. 1984. The spider family Homalonychidae (Arachnida, Araneae). *American Museum Novitates* 2790: 1–11.

Roth, V. D. and P. L. Brame. 1972. Nearctic genera of the spider family Agelenidae (Arachnida, Araneida). *American Museum Novitates* 2505: 1–52.

Roth, V. D. and W. L. Brown. 1975. Comments on the spider *Saltonia incerta* Banks (Agelenidae?). *Journal of Arachnology* 3: 53–56.

Sabath, L. E. 1969. Color change and life history observations of the spider *Gea heptagon* (Araneae: Araneidae). *Psyche, Cambridge* 76: 367–374.

Sauer, R. J. and N. I. Platnick. 1972. The crab spider genus *Ebo* (Araneida: Thomisidae) in the United States and Canada. *Canadian Entomologist* 104(1): 35–60.

Schenkel, E. 1950. Spinnentiere aus dem westlichen Nordamerika, gesammelt von Dr. Hans Schenkel-Rudin. *Verhandlungen der Naturforschenden Gesellsschaft, Basel* 61: 28–92.

Schick, R. X. 1965. The crab spiders of California (Araneae, Thomisidae). *Bulletin of the American Museum of Natural History* 129: 1–180.

Shear, W. A. 1970. The spider family Oecobiidae in North America, Mexico, and the West Indies. *Bulletin of the Museum of Comparative Zoology* 140(4): 129–164.

Slowik, J. 2009. A review of the cellar spider genus *Psilochorus* Simon 1893 in America north of Mexico (Araneae: Pholcidae). *Zootaxa* 2144: 1–53.

Slowik, J. and P. E. Cushing. 2007. Redescription of *Hogna coloradensis* (Banks 1894) from the southwestern United States (Araneae, Lycosidae). *Journal of Arachnology* 35: 46–53.

Smith, A. M. 1994. *Tarantula Spiders: Tarantulas of the U.S.A. and Mexico.* Fitzgerald, London. 196 pp.

Spagna, J. C., S. C. Crews, and R. G. Gillespie. 2010. Patterns of habitat affinity and Austral/Holarctic parallelism in dictynoid spiders (Araneae: Entelegynae). *Invertebrate Systematics* 24: 238–257. http://dx.doi.org/10.1071/IS10001.

Starr, C. K. 1988. Sexual behavior in *Dictyna volucripes* (Araneae, Dictynidae). *Journal of Arachnology* 16: 321–330.

Starrett, J. and M. Hedin. 2007. Multilocus genealogies reveal multiple cryptic species and biogeographic complexity in the California turret spider *Antrodiaetus riversi* (Mygalomorphae, Antrodieatidae). *Molecular Ecology* 16: 583–604.

Stockman, A. K. and J. E. Bond. 2007. Delimiting cohesion species: Extreme population structuring and the role of ecological interchangeability. *Molecular Ecology* 16(16): 3374–3392.

Stockman, A. K. and J. E. Bond. 2008. A taxonomic review of the trapdoor spider genus *Promyrmekiaphila* Schenkel (Araneae, Mygalomorphae, Cyrtaucheniidae, Euctenizinae). *Zootaxa* 1823: 25–41.

Suter, R. B., G. Doyle, and C. M. Shane. 1987. Oviposition site selection by *Frontinella pyramitela* (Araneae, Linyphiidae). *Journal of Arachnology* 15: 349–354.

Suter, R. B. and G. Renkes. 1984. The courtship of *Frontinella pyramitela* (Araneae, Linyphiidae): Patterns, vibrations and function. *Journal of Arachnology* 12: 37–54.

Ubick, D. 2005a. Amaurobiidae. p 60 *in* D. Ubick, P. Paquin, P. E. Cushing, and V. Roth (eds.). *Spiders of North America: An Identification Manual.* American Arachnological Society, Keene, New Hampshire.

Ubick, D. 2005b. Gnaphosidae. p. 106 *in* D. Ubick, P. Paquin, P. E. Cushing, and V. Roth (eds.). *Spiders of North America: An Identification Manual.* American Arachnological Society, Keene, New Hampshire.

Ubick, D. 2005c. Key to the spider families of North America north of Mexico. p. 25 *in* D. Ubick, P. Paquin, P. E. Cushing, and V. Roth (eds.). *Spiders of North America: An Identification Manual.* American Arachnological Society, Keene, New Hampshire.

Ubick, D. 2005d. New genera and species of cribellate coelotine spiders from California (Araneae: Amaurobiidae). *Proceedings of the California Academy of Sciences* 56: 305–336.

Ubick, D. 2005e. Oonopidae. p. 185 *in* D. Ubick, P. Paquin, P. E. Cushing, and V. Roth (eds.). *Spiders of North America: An Identification Manual.* American Arachnological Society, Keene, New Hampshire.

Ubick, D. and P. R. Craig. 2005. Zodariidae. p. 254 *in* D. Ubick, P. Paquin, P. E. Cushing, and V. Roth (eds.). *Spiders of North America: An Identification Manual.* American Arachnological Society, Keene, New Hampshire.

Ubick, D. and M. J. Moody. 1995. On males of Californian *Talanites* (Araneae, Gnaphosidae). *Journal of Arachnology* 23: 209–211.

Ubick, D., P. Paquin, P. E. Cushing, and V. Roth (eds.). 2005. *Spiders of North America: An Identification Manual.* American Arachnological Society, Keene, New Hampshire. 377 pp.

Ubick, D. and N. I. Platnick. 1991. On *Hesperocranum,* a new spider genus from western North America (Araneae, Liocranidae). *American Museum Novitates* 3019: 1–12.

Ubick, D. and D. B. Richman. 2005a. Corinnidae. p. 79 *in* D. Ubick, P. Paquin, P. E. Cushing, and V. Roth (eds.). *Spiders of North America: An Identification Manual.* American Arachnological Society, Keene, New Hampshire.

Ubick, D. and D. B. Richman. 2005b. Liocranidae. p. 162 *in* D. Ubick, P. Paquin, P. E. Cushing, and V. Roth (eds.). *Spiders of North America: An Identification Manual.* American Arachnological Society, Keene, New Hampshire.

Ubick, D. and R. S. Vetter. 2005. A new species of *Apostenus* from California, with notes on the genus (Araneae, Liocranidae). *Journal of Arachnology* 33: 63–75.

Uetz, G. W. and K. R. Cangialosi. 1986. Genetic differences in social behavior and spacing in populations of *Metepeira spinipes* F.O. Pickard-Cambridge (Araneae: Araneidae), a communal-territorial orb weaver. *Journal of Arachnology* 14: 159–173.

Uetz, G. W. and S. P. Hartstock. 1987. Prey selection in an orb-weaving spider: *Micrathena gracilis* (Araneae: Araneidae). *Psyche, Cambridge* 94: 103–116.

Uetz, G. W., A. D. Johnson, and D. W. Schemske. 1978. Web placement, web structure and prey capture in orb-weaving spiders. *Bulletin of the British Arachnological Society* 4: 141–148.

Uhl, G. and J. Maelfait. 2008. Male head secretion triggers copulation in the dwarf spider *Diplocephalus permixtus. Ethology* 114(8): 760–767.

Vanacker, D., J. V. Borre, A. Jonckheere, L. Maes, S. Pardo, F. Hendrickx, and J. Maelfait. 2003. Dwarf spiders (Erigoninae, Linyphiidae, Araneae): Good candidates for evolutionary research. *Belgian Journal of Zoology* 133(2): 143–149.

Vetter, R. S. 1996. The extremely rare *Prodidomus rufus* Hentz (Araneae, Prodidomidae) in California. *Journal of Arachnology* 24: 72–73.

Vetter, R. S. 1998. Envenomation by a spider, *Agelenopsis aptera* (family: Agelenidae) previously considered harmless. *Annals of Emergency Medicine* 32: 739–741.

Vetter, R. S. 2001. Revision of the spider genus *Neoanagraphis* (Araneae, Liocranidae). *Journal of Arachnology* 29: 1–10.

Vetter, R. S. 2008. Spiders of the genus *Loxosceles* (Araneae, Sicariidae): A review of biological, medical and psychological aspects regarding envenomations. *Journal of Arachnology* 36: 150–163.

Vetter, R. S. 2012. A large European combfoot spider, *Steatoda nobilis* (Thorell 1875) (Araneae: Theridiidae), newly established in Ventura County, California. *Pan-Pacific Entomologist* 88(1): 92–97.

Vetter, R. and A. Antonelli. 2002. *How to Identify (or Misidentify) the Hobo Spider.* Washington State University, Puyallup. http://pep.wsu.edu/pdf/pls116_1.pdf.

Vetter, R. S. and D. K. Barger. 2002. An infestation of 2,055 brown recluse spiders (Araneae: Sicariidae) and no envenomations in a Kansas home: Implications for bite diagnosis in nonendemic areas. *Journal of Medical Entomology* 39: 948–951.

Vetter, R. S. and J. C. Cokendolpher. 2000. *Homalonychus theologus* (Araneae, Homalonychidae): Description of eggsacs and a possible defensive posture. *Journal of Arachnology* 28: 361–363.

Vetter, R. S. and G. K. Isbister. 2006. Verified bites by the woodlouse spider, *Dysdera crocata. Toxicon* 47(7): 826–829.

Vetter, R. S., G. K. Isbister, S. P. Bush, and L. J. Boutin. 2006. Verified bites by yellow sac spiders (*Cheiracanthium*) in the United States and Australia: Where is the necrosis? *American Journal of Tropical Medicine and Hygiene* 74(6): 1043–1048.

Vetter, R. S., L. S. Vincent, J. E. Berrian, and J. K. Kempf. 2008. *Metaltella simoni* (Araneae, Amphinectidae): Widespread in coastal southern California. *Pan-Pacific Entomologist* 84(2): 146–149.

Vetter, R. S. and P. K. Visscher. 1994. A non-native spider, *Metaltella simoni,* found in southern California (Araneae, Amaurobiidae). *Journal of Arachnology* 22: 256.

Vincent, L. S. 1993. The natural history of the California Turret Spider *Atypoides riversi* (Araneae: Antrodiaetidae): Demographics, growth rates, survivorship, and longevity. *Journal of Arachnology* 21(1): 29–39.

Vincent, L. S., R. S. Vetter, W. J. Wrenn, J. K. Kempf, and J. E. Berrian. 2008. The brown widow spider *Latrodectus geometricus* C. L. Koch, 1841, in southern California. *Pan-Pacific Entomologist* 84(4): 344–349.

Vogel, B. R. 2004. A review of the spider genera *Pardosa* and *Acantholycosa* (Araneae, Lycosidae) of the 48 contiguous United States. *Journal of Arachnology* 32: 55–108.

Waldron, W. G., M. B. Madon, and T. Suddarth. 1975. Observations on the occurrence and ecology of *Loxosceles laeta* (Araneae: Scytodidae) in Los Angeles County, California. *Vector News* 22(4): 29–36.

Wallace, H. K. 1942. A revision of the burrowing spiders of the genus *Geolycosa*. *American Midland Naturalist* 27: 1–62.

Wallace, H. K. and H. Exline. 1978. Spiders of the genus *Pirata* in North America, Central America and the West Indies (Araneae: Lycosidae). *Journal of Arachnology* 5: 1–112.

Weng, J. L., G. Barrantes, and W. G. Eberhard. 2006. Feeding by *Philoponella vicina* (Araneae, Uloboridae) and how uloborid spiders lost their venom glands. *Canadian Journal of Zoology* 84(12): 1752–1762.

Wing, K. 1983. *Tutelina similis* (Araneae: Salticidae): An ant mimic that feeds on ants. *Journal of the Kansas Entomological Society* 56(1): 55–58.

Wise, D. H. 1984. Phenology and life history of the Filmy Dome Spider (Araneae: Linyphiidae) in two local Maryland populations. *Psyche, Cambridge* 91: 267–288.

Wunderlich, J. 2004. The fossil spiders (Araneae) of the families Tetragnathidae and Zygiellidae n. stat. in Baltic and Dominican amber, with notes on higher extant and fossil taxa. pp. 899–955 *in* J. Wunderlich. *Vol. 3. Fossil Spiders in Amber and Copal.* Beiträge zur Araneologie. Publishing House Joerg Wunderlich, Hirschberg, Germany.

Yeargan, K. V., and L. W. Quate. 1996. Juvenile bolas spiders attract psychodid flies. *Oecologia* 106: 266–271.

Yeargan, K. V. and L. W. Quate. 1997. Adult male bolas spiders retain juvenile hunting tactics. *Oecologia* 112: 572–576.

Yoshida, H. 2008. A revision of the genus *Achaearanea* (Araneae: Theridiidae). *Acta Arachnologica Tokyo* 57: 37–40.

INDEX

Page numbers in **boldface type** refer to main discussion of spider species.

ABOUT THE AUTHORS

Photo by Michelle C. Torres-Grant

R. J. Adams is a special education teacher, naturalist, and wildlife tour guide who currently lives along the central California coast. He received his B.S. in biology from Humboldt State University and his M.S. in biology from the University of Utah, where he studied host-parasite coevolution and insect systematics. He has authored or coauthored numerous papers on the natural history and diversity of bird lice and has described nearly a dozen new species.

Tim D. Manolis is an artist, illustrator, and biological consultant who received his Ph.D. in biology from the University of Colorado. From 1986 to 1990, he was the editor and art director of the magazine *Mainstream*. His papers on birds and his bird illustrations have appeared in many journals and magazines. He is the author of *Dragonflies and Damselflies of California* and illustrator of *Butterflies of the San Francisco Bay and Sacramento Valley Regions.*

California Natural History Guides

From deserts and grasslands to glaciers and the spectacular Pacific coast, California is a naturalist's paradise. The authors of the California Natural History Guides have walked into wildfires, plunged into shark-infested waters, scaled the Sierra Nevada, peeked under rocks, and gazed into the sky to present the most extensive California environmental education series in existence. Packed with photographs and illustrations and compact enough to take on the trail, they are essential reading for any California adventurer. For a complete list of guides, please visit http://www.ucpress.edu/go/cnhg.